RHYTHMIC EEG ACTIVITIES AND CORTICAL FUNCTIONING

Developments in Neuroscience

Volume 1 - Control of Gaze by Brain Stem Neurons, edited by R. Baker and A. Berthoz, 1977

Volume 2 - Mechanisms, Regulation and Special Functions of Protein Synthesis in the Brain, edited by S. Roberts, A. Lajtha and W.H. Gispen, 1977

Volume 3 - Senile Dementia: A Biomedical Approach, edited by Kalidas Nandy, 1978

Volume 4 - Characteristics and Function of Opioids, edited by J.M. Van Ree and L. Terenius, 1978

Volume 5 - Transmethylation, edited by Earl Usdin, Ronald T. Borchardt and Cyrus R. Creveling, 1979

Volume 6 - Cerebro-Cerebellar Interactions, edited by J. Massion and K. Sasaki, 1979

Volume 7 - Phenothiazines and Structurally Related Drugs: Basic and Clinical Studies, edited by Earl Usdin, Helmut Eckert and Irene S. Forrest, 1980

Volume 8 - Catecholamines and Stress Recent Advances, edited by Earl Usdin, Richard Kvetnansky and Irwin J. Kopin, 1980

Volume 9 - Multidisciplinary Approach to Brain Development, edited by C. Di Benedetta, R. Balázs, G. Gombos and G. Porcellati, 1980

Volume 10 - Rhythmic EEG Activities and Cortical Functioning, edited by Gert Pfurtscheller, Pierre Buser, Fernando H. Lopes da Silva and Hellmuth Petsche, 1980

RHYTHMIC EEG ACTIVITIES AND CORTICAL FUNCTIONING

Proceedings of the International Symposium on Event-Related Changes in Cortical Rhythmic Activities. Behavioural Correlates, held in Schladming, Austria, September 19-22, 1979.

Edited by

Gert Pfurtscheller
Pierre Buser
Fernando H. Lopes da Silva

and

Hellmuth Petsche

1980

ELSEVIER/NORTH-HOLLAND BIOMEDICAL PRESS
AMSTERDAM • NEW YORK • OXFORD

ISBN for this volume: 0-444-80247-9
ISBN for the series : 0-444-80028-X

Published by:
Elsevier/North-Holland Biomedical Press
335 Jan van Galenstraat, P.O. Box 211
Amsterdam, The Netherlands

Sole distributors for the USA and Canada
Elsevier North-Holland Inc.
52 Vanderbilt Avenue
New York. N.Y. 10017

Library of Congress Cataloging in Publication Data

International Symposium on Event-Related Changes in
Cortical Rhythmic Activities--Behavioural Correlates,
Schladming, Austria, 1979.
Rhythmic EEG activities and cortical functioning.

(Developements in neuroscience ; v. 10)
Bibliography: p.
Includes indexes.
1. Cerebral cortex. 2. Electroencephalography.
3. Evoked potentials (Electrophysiology) I. Pfurtscheller, Gert. II. Title. III. Series.
QP376.I58 1979 612'.825 80-15945
ISBN 0-444-80247-9

Printed in The Netherlands

PREFACE

The functioning of the brain can be studied through recording and analyzing bioelectrical phenomena belonging to two main categories: brain rhythms and evoked potentials. Whereas a large amount of data has accumulated on the latter category, only little information is available yet on "event-related" changes of brain rhythms. It was Berger who proposed that the rhythmic electrical activity of the cortex must reflect changes related to external stimulation and even psychic processes. This original idea was not further elaborated until recently, taking advantage of modern computer evaluation techniques. The emphasis of this book is precisely placed on the functional significance and the behavioural correlates of some of these event-related modifications of brain rhythms, as obtained from animal experiments and human studies. Neurophysiologically these changes can be classified in two main groups: those related to movement and those which occur during sensory stimulation, in parallel to the evoked responses. These findings may well modify the classical assumption that the background EEG remains stationary during event-related responses. On the other hand, movement related changes in rhythmic activities have proved to be a general phenomenon of the same category as the classical alpha blocking response to light, but with different topography. Grey Walter's idea of several types of alpha rhythms related to different physiological conditions and anatomical locations has thereby been confirmed.
The book is based on contributions by participants in an international symposium held at Schladming, Austria (Sept 19-22, 1979). This symposium would not have taken place without the generous support of the "Fonds zur Foederung der wissenschaftlichen Forschung in Oesterreich" concerning part of experimental work which is now material of this book and the stimulating interest of Dr. A. Aranibar. We also thank Mrs. M. Kainz for helping in the organization and for typing most of the manuscripts.

September 1979

The Editors

CONTENTS

ORIGIN OF EEG RHYTHMS

Rhythmic EEG Activities and Cortical Functioning
G. Pfurtscheller et al., editors

NEUROANATOMICAL CORRELATES OF SPONTANEOUS AND EVOKED EEG ACTIVITIES IN THE VISUAL AREA OF THE RABBIT

HELLMUTH PETSCHE, PETER RAPPELSBERGER, HELMUT POCKBERGER, OTTO PROHASKA and IRENE-BEATE MÜLLER-PASCHINGER
Institute of Neurophysiology, University of Vienna, and Institute of Brain Research, Austrian Academy of Sciences, Währinger Str. 17, A-1090 Wien(Austria)

INTRODUCTION

In previous studies on the genesis of electrical activities the event occurring at the cortical surface have proved to be best understood when studied by means of as many small electrodes as possible, closely spaced in the form of rectangular grids. The difficulties arising from the complicated geometry of a gyrated cortex have been overcome by choosing the rabbit's striate area,which is nearly flat. In this area, the spatio-temporal characteristics of both spontaneous and seizure activities have been studied by plotting potential fields and observing their behaviour[1,2]. The embarrassing variety of behaviour of these potential fields also prompted an investigation of the intracortical events which proved to be even more complex than surface recordings.

The results of these studies suggested that the histological structure of the cortex must play a leading role in the formation of local electrical characteristics. Thus, the question arose whether it might be possible to find electroarchitectonic pecularities in order to achieve, finally, the construction of cortico-electroarchitectonic maps. This novel concept of electroarchitectonics was coined in analogy to such old-established concepts as cyto-, myelo- and angioarchitestonics and so forth, which stands for the description of pecularities of the cortical structure, as manifested by the study of its cellular organization, its myelination and the arrangement of its vessels.

For the study of electroarchitectonics, several technical problems had to be solved, concerning micro-electroencephalographic technology. Under "Micro-EEG" we understand the methods of recording field potentials in an order of magnitude of electrode sizes and distances somewhere between the ones used in the recording of membrane potentials and those used in the clinical routine EEG. By both the increase of the number of recording electrodes and the reduction of their distances, a great deal of data may be obtained from relatively small cortical volumes, and these data may be brought in relation to such structural properties as cortical lamination and distribution of different populations of nerve cells.

MATERIALS AND METHODS

This paper deals mainly with results from intracortical recordings. For this purpose, multi-electrodes have been used; they consist of 8 Ag-AgCl contacts of 50 x 50 μm^2 at 300 μm distances or of 16 contacts of 10 x 10 μm^2 at 150 μm distances, which had been applied to lancet-shaped glass carriers of 0.1 mm thickness by thin-film technology[3,4]. The dimensions of the multi-electrodes were designed to just cover the entire cortical diameter. They were inserted by a micro-drive into the striate cortex of rabbits which are artificially respirated and kept under Nembutal and Alloferin and additional local anaesthesia by 2 % Procaine. All traces are common-reference recordings with respect to the nose bone via impedance transformers, the outputs of which are fed into a multichannel EEG machine. Simultaneously, the data are stored on tape for off-line processing in a HP 9603.

After each experiment, silver is deposited electrolytically from each of the 8 contacts and made visible by a photochemical method[5]. Since the rate of shrinking by the histological procedure is not uniform, the position of each individual electrode of each experiment has to be identified. It is entered into a schematic drawing of the cortex with the lamination drawn in, to enable the study of the contribution of the single laminae to electrogenesis.

The data have been subjected to Fourier analysis. Moreover, potential-chronotopograms have been calculated[6]: in these graphs, the electrical parameter is plotted as a system of contour lines in a coordinate system of time (abscissa) with regard to cortical diameter (ordinate). The intracortical potential profiles required for this method are calculated every 4 msec and are used to compute the distribution of current sources and sinks by calculating the second derivative in space[7].

RESULTS

Spontaneous ECoG and "photic driving". The spontaneous activity varies considerably. In contrast to humans, an alpha rhythm is absent. The ECoG of the striate area consists of groups of irregular waves in the delta-range with a few hundred μV amplitude, which only rarely take more than two seconds and alternate with a lower, irregular activity in the alpha- and theta-range. In the deepest layers of the striate area the activity is similar to the surface activity and often shows phase-reversal. These two activities are separated from each other by a zone of minimum activity[8].

Fig. 1a shows a typical sample of the spontaneous activity and its analysis. The zone of minimum activity is at contact 3. The upper right part of Fig. 1a shows the conventional ECoG pattern. To the right is a schematic drawing of the electrode track with the positions of the 7 intracortical contacts with respect to the cortical layers. The different distances of the contacts are caused by the different degrees of shrinking of the layers by the histological procedure. Beneath is the power spectrum of one minute of spontaneous activity. There is a peak between 0.5 and 2 Hz in all recordings, with one maximum near the surface and another near the white matter.

Fig. 1a (to the left below) shows the power profile at 0.5 Hz throughout the cortex, plotted on a linear scale. The largely extended power minimum around contact 3 concerns layer IV and III. The squared coherence (Fig. 1a, middle above) between supra- and infragranular activities (i.e. above and below the granular layer IV) is statistically not different from zero. The phase-plot (Fig. 1a, upper left) shows a phase-reversal between supra- and infragranular layers; this phenomenon, however, cannot be interpreted unambiguously because the coherence is statistically not different from zero.

Next, the changes of spontaneous activity by flicker-light are discussed. Continuous flicker stimuli were applied to the contralateral eye by a fibre-glass rod. The changes of the corticogram produced by two different flickering frequencies are shown in Fig. 1b and c. The intracortical activities are now more regular than in spontaneous activities. The power spectra again show maxima at 0.5 Hz (Fig. 1b and c), the intracortical power distribution of which is not different from spontaneous activity. Additionally, however, a peak of 3.5 and 7.5 Hz, respectively, caused by the flickering frequency, becomes prominent. Moreover, in the higher frequency range, numerous harmonics become prominent, particularly in contact 4 and 5, but also in contact 1, which have more power.

In the power profiles at the flicker-frequencies (Fig. 1b and c, lower left) the deep power maximum shows a tendency to shift towards the surface (from contact 7 in the spontaneous record to contact 4 at a flickering rate of 7.5 Hz). Simultaneously, the width of the zero-zone decreases and the minimum power also shifts from contact 3 to contact 2.

The squared coherence also undergoes considerable changes: with increasing flickering rate, it becomes larger and larger, thus indicating an increasing synchronization of the electrical events in the cortical diameter.

The phase-plot (Fig. 1b and c) shows a phase-shift of about 180 degree between surface and deep activities.

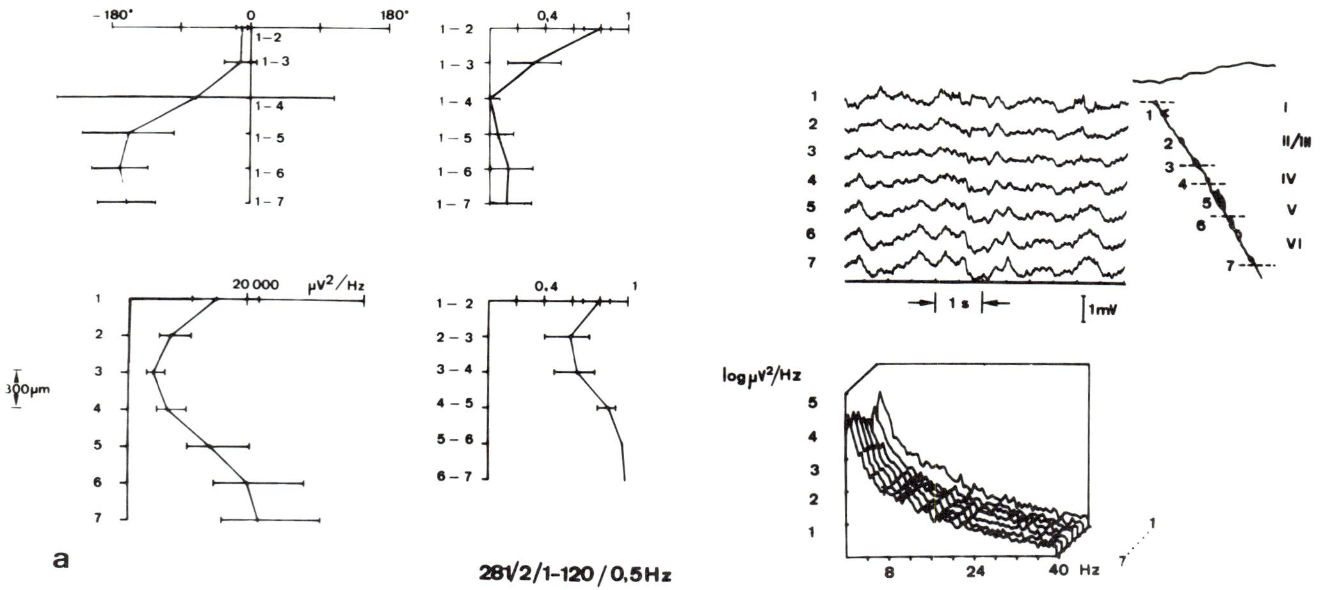

Fig. 1a. Spontaneous activity recorded from the rabbit's striate area (see Fig. 2, inset, 281). For each of these activities the ECoGs are found in the right upper part, the power spectra in the right lower part. The inset to the right shows the position of the recording contacts with respect to the cortical laminae. The left-sided graphs show the phase relationships with respect to the uppermost contact (upper left) and the power profiles through the cortex at 0.5 Hz for the spontaneous acitivty. To the right of these graphs are the squared coherences: with respect to the uppermost contact (upper part of each partial figure) and with respect to the adjacent contact (lower part).

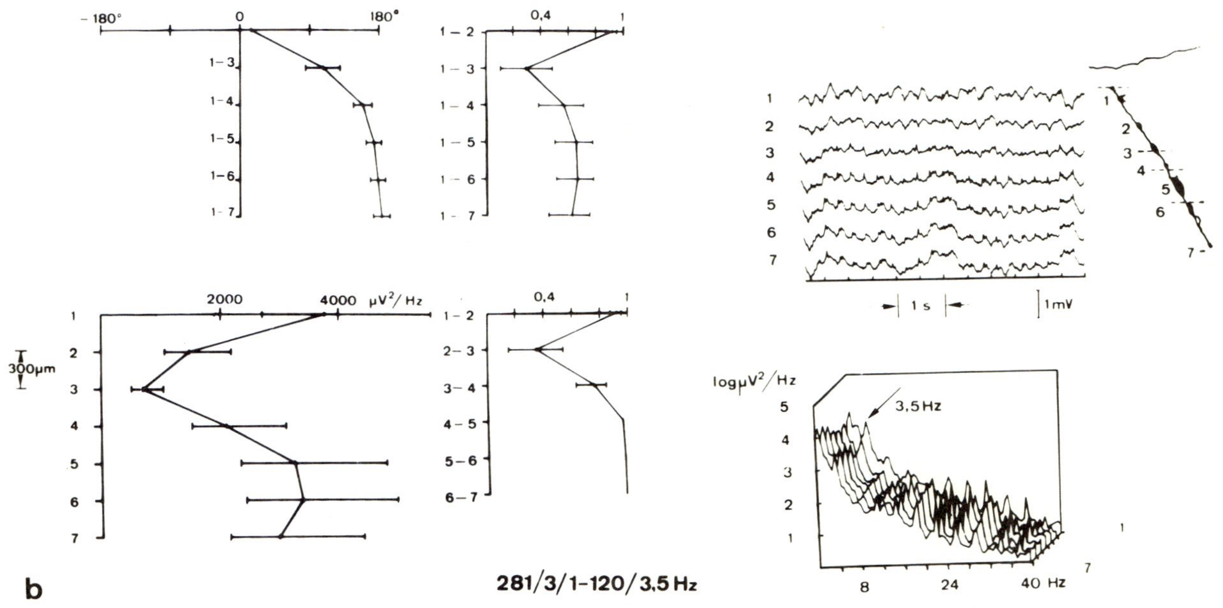

Fig. 1b. As in Fig. 1a, but for photic driving at 3.5 Hz, phase-, power- and coherence profiles through the cortex at 3.5 Hz.

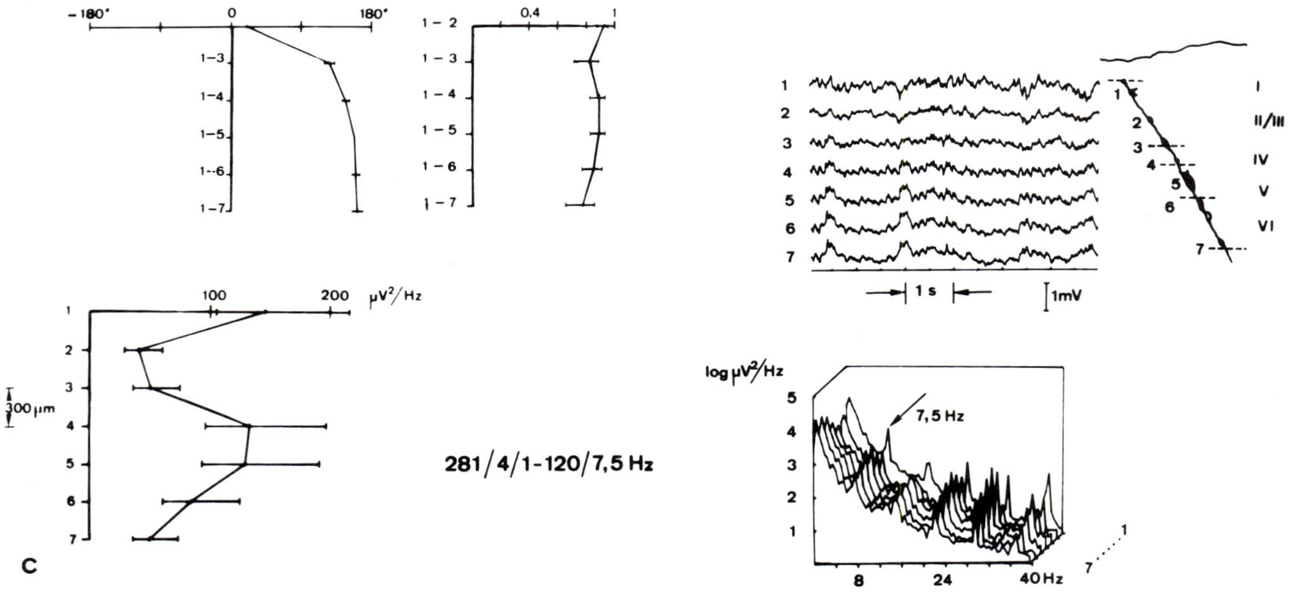

Fig. 1c. As in Fig. 1b, but for photic driving at 7.5 Hz and profiles at 7.5 Hz. With increased flicker frequency, the power maximum shifts upwards and the coherence approaches 1 (compare Fig. 1a, b and c).

Let us now consider the histological structures possibly involved in these electrical pecularities.

For the spontaneous activity, two generators seem to be essential: one in the upper part of layer II and probably in I, the deeper one in layer IV. Layers III to V hardly seem to contribute to the generation of the spontaneous activity. These two generators are only loosely coupled with one another. This results also from the observation that the two activities in the superficial and the deepest layers of the cortex do not always occur simultaneously but each one can also appear on its own.

By FL, the intracortical spectral distribution is significantly changed: with increasing flicker-frequency, the layer VI power maximum shifts further and further upwards to reach the uppermost region of layer V, whereas the superficial power maximum remains unaltered. Additionally, the degree of intracortical synchronization, i.e. the interconnectivity between the uppermost and the deep cortical layers largely increases.

Visually evoked potentials (VEP). After these global effects by light stimulation, the effects of single flashes on the striate area are discussed. In studies of the topography of VEP by multiple surface electrodes, it was demonstrated[9] that the different components of the response have a different spatiotemporal behaviour, even when averaged. One of the main results of this paper was that the first component P 1 reaches maximum medially and caudally of the zone of optimum visual acuity which extends in rabbits[10] as shown in the inset of Fig. 2, whereas the components following P 1 reach maximum beyond this zone. Since the cytoarchitectonics of the rabbit's cortex have not yet been explored sufficiently - the latest reference relating only to Nissl stainings being Rose (1931)[11] - it must be left unsolved whether these two components may be related to the striate and the peristriate region. Be that as it may, the different location of P 1 and P 2 induced us to study these two locations by intracortical multi-electrodes.

Fig. 2 shows two averaged VEPs (293 and 281), formed by the superimposition of 100 individual signals and recorded from the two regions indicated in the inset.

First, the VEP recorded from the caudal part of the striate is discussed (Fig. 2a, b and c). P 1 at the surface corresponds to an initial negativity in the deep layers with shortest latencies and largest amplitudes in contact 4 and 5, which corresponds to layer V. This steeply rising and slowly falling transient (Fig. 2b) is followed by several oscillations, most distinct in about the same depth.

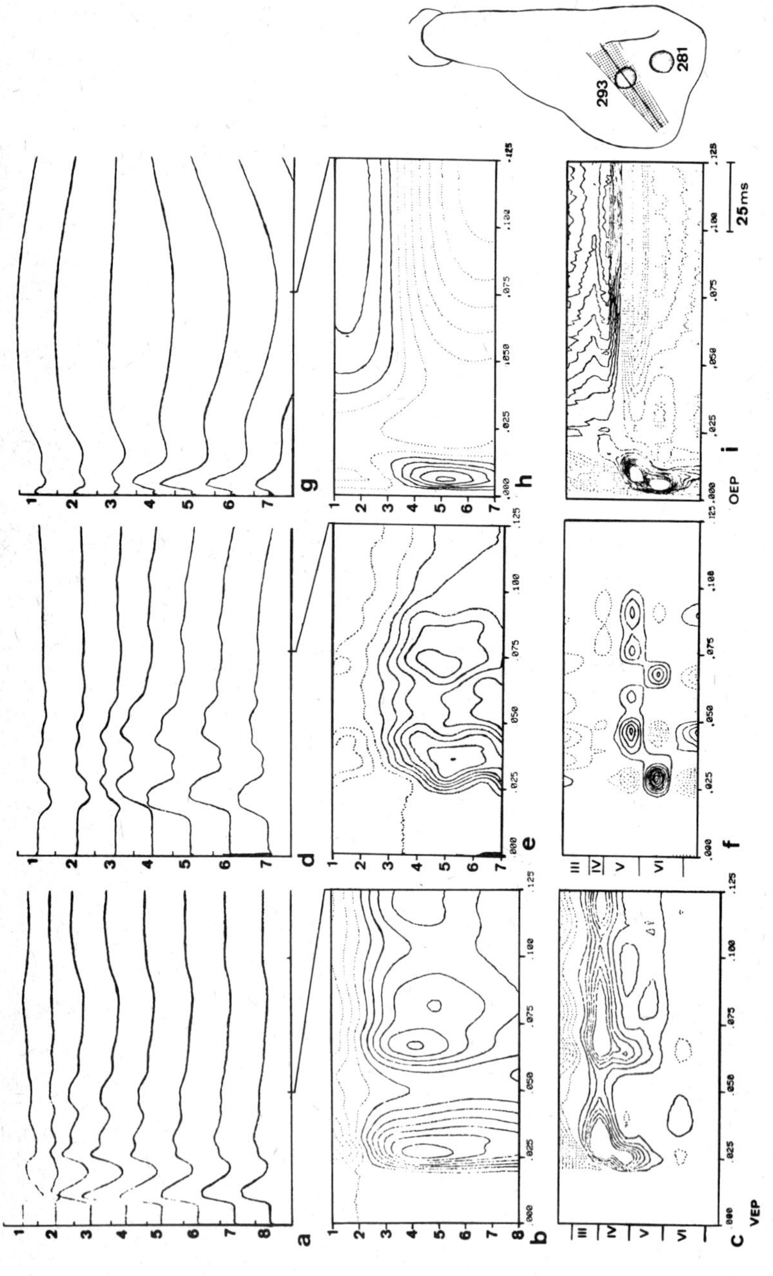
281
293
a
b
c
d
e
f
g
h
i
VEP
OEP
III
IV
V
VI
25ms
.000
.025
.050
.075
.100
.125

In contrast to this, the VEP recorded from the anterior electrode position (Fig. 2d, e and f) is quite different. The latencies now seem to be shortest in contacts 5 and 7,and the potential is more susceptible to oscillation. The intracortical phase-relationships are more complex as demonstrated by the chrono-topogram.

The current-source-density plots (CSD, Fig. 2c and f) demonstrate, however, that in both cases the primary event is a sink which appears most distinctly, in 281, in layers IV - V, but in 293 somewhat deeper, namely in layer VI, where it is accompanied by adjacent and simultaneous sources above and below. The initial sinks are in both experiments immediately followed by additional sinks above which show some tendency to oscillation. Since these considerable topographical differences in the configuration of the VEP turned out to be a consistent finding, it is likely that differences in the cortical structure underly them.

Since even the retinal circuit probably contributes to some reverberation of the signal with the effect that, instead of one short train of pulses, a fairly long-lasting pattern of action potentials arrives at the geniculate and is finally carried on to the cortex,contralateral electrical stimulation of the optic nerve (OEP, Fig. 2g, h and i) was also applied. The OEP differs from the VEP insofar as the primary component is by far the most prominent event; after a second, lower and slower potential, a long-lasting positivity appears in the depth, accompanied by a comparable negativity at the surface (Fig. 2g and h). The sink causing N 1 has less temporal dispersion in the OEP than in the VEP, as the CSD analysis shows. It is also accompanied by sources above and below and immediately followed by another strong sink above, of shorter latency than in the VEP. A further, smaller sink follows in the upper part of layer V. The long-lasting after-swing has its isoelectric zone at the upper limit of layer IV.

This analysis demonstrates that the trains elicited by optic nerve stimulation and those produced by flicker-light are not elaborated in the cortex in

---

Fig. 2. Visual evoked potentials (VEP, a-f) and potentials evoked by electrical stimulation of the contralateral optic nerve (OEP, g-i), recorded by the 8-fold intracortical electrode (in VEP and OEP 293 only the 7 uppermost contacts are in the cortex). 100 superimposed responses at two seconds intervals. The VEP of experiment 293 were recorded from within Thompson's zone of maximum visual acuity, the VEP and OEP of experiment 281 were recorded medio-caudally from this zone. a, d, g: the raw data: Contact 2 was just below cortical surface, 8 and 7, respectively, next to the white matter. Cal.: 600 μV; time marks: 125 msec. b, e, h: potential-chronotopograms of these evoked potentials. Time 125 msec. Potential contour lines drawn at 100 μV distances. The first dotted line next to a full line is zero potential. c, f, i: current source density contour lines (c: 0.5 $\mu A/mm^2$, f: 2.0 $\mu A/mm^2$; i: 0.25 $\mu A/mm^2$).

the same way. The only common denominator is the primary excitation zone occurring in layer V.

DISCUSSION

These findings are to be discussed in the context of cortical circuitry which is based on Golgi-technique on the rabbit.

Fig. 3 shows a drawing of the Golgi configuration of the rabbit's visual cortex in a frontal section. The description starts with the deepest layer (VI). Three types of cells are found in this layer: (a) interneurons with long horizontal dendrites and axons. These interneurons are supplied with information probably by the axon collaterals of the pyramidal cells of layer V, perhaps also layer III, and evidently serve the horizontal exchange of information: (b) is a Martinotti-cell, a pyramidal cell the axon of which is reversed and supplies the upper cortical layers with information. Finally, pyramidal cells of average size (c) populate layer VI.
Between layers V and VI there is a dense fibre plexus which consists predominantly of axon collaterals of pyramidal cells, and axons and dendrites of the above mentioned interneurons of layer VI. These pyramidal axon collaterals reach as far as 3 mm and terminate on the spines of the dendrites of other pyramidal cells and interneurons.

The horizontal collaterals of the layer VI pyramidal fibres also have ascending fibres.

The next layer V is characterized by pyramidal cells of different sizes (d). Their axon collaterals are twofold: either, they ascend vertically and terminate in the immediate neighbourhood of the ascending dendrite by so-called cartridges[12]. To these cartridges, several additional elements contribute, such as Golgi-type-II-cell dendrites, horse-tail-axon terminals of the cells with double dendritic trees and other terminal axons. These cartridges seem to be one of the main input processing structures of the pyramidal cells[12].

The other form of pyramidal collaterals extends horizontally, with vertical branches extending up to the uppermost layer. The abundance of these axon collaterals and their range is shown by the Scheibels[13]: in their Golgi drawings, the total extension of all axon collaterals of a group of three adjacent pyramidal cells is 3 mm. This corresponds to a cortical volume of 14 mm^3, whereas the volume covered by their dendritic trees (the feeding area of information) is only 0.39 mm^3, which is the 36th part of the volume where information is distributed.

Another characteristic element of layer V are basket cells (e) the dendrites

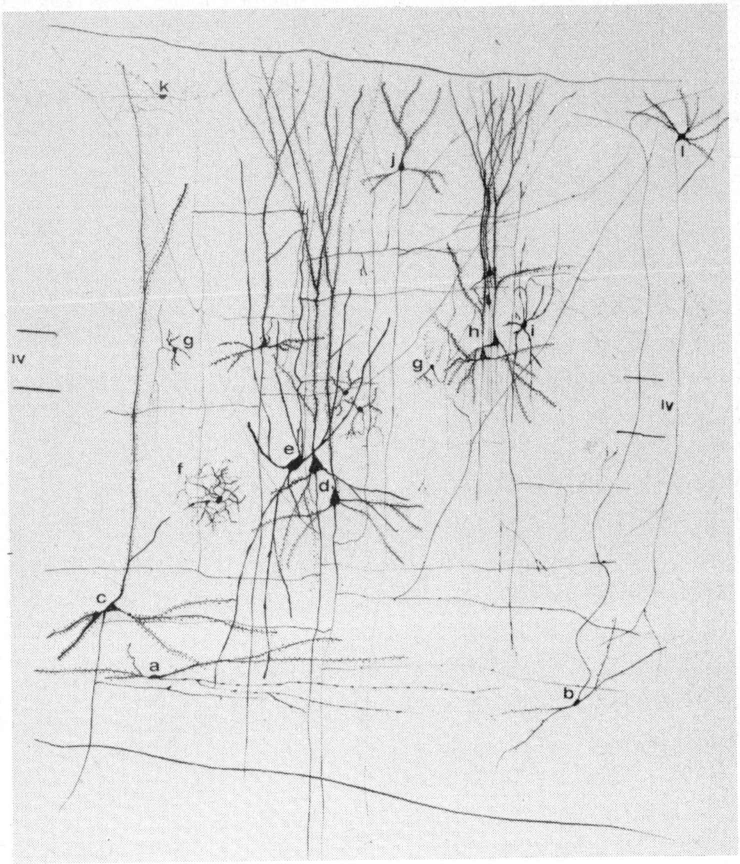

Fig. 3. Circuitry of the rabbit's striate area, according to Golgi studies (Tömböl 1979). Description see text.

of which have no spines, but fusiform bulges, and reach up to layer III and deeply down into layer VI. Their axons ascend and send off horizontal collaterals as fas as 500 µm and, additionally, vertical collaterals towards above and below. The terminals of these basket axons contribute to the formation of basket-like terminations around the somata of pyramidal cells[12,14]. Such baskets are usually formed by the axon terminals of several basket cells. There is convincing evidence that these baskets produce an inhibition of pyramidal cells. Such basket cells are met with relatively frequently.

Finally, layer V contains a great many neurogliform neurons, Golgi-type-II-cells (f) with a spheric dendritic tree and an abundantly branching axon termination within this tree. These are interneurons for local functions.

The following layer IV consists almost exclusively of star cells (g), Golgi-type-II-interneurons with only a few dendrites which spread out in all directions from the small cell body with only little branching; their axons sometimes run into layers III and V.

Next comes layer III which is characterized by pyramidal cells of different sizes (h). Their axons, axon collaterals and dendrites show about the same distribution as in layer V. Here again basket cells (i) are found: their axons ascend to layer II and descend down to layer V, giving off several horizontal collaterals as far as 500 μm.

The following layer II has fewer pyramidal cells (j) than III; they are smaller and have descending axons with only a few ascending collaterals.

Characteristic for the uppermost layer I is its abundance of horizontal fibres and its lack of cells: there are only Cajal's horizontal cells (k) which seem to guarantee horizontal exchange of information, and a few multiangular cells (l) with axons descending down to layer VI.

A closer view of the probable termination of the specific afferent fibres arising from the lateral geniculate body is given by Fig. 4: the specific afferent fibres ascend to branch abundantly and preponderantly in layer IV, where they have axo-dendritic contacts with the spines of the star-cells; (c) is a medium-size pyramidal cell of layer V which is contacted by star-cell-axons. The specific fibres also ascend to layer V where they have contacts with ascending pyramidal dendrites. The specific afferents also contact basket cell dendrites (d) and dendrites of fusiform cells with double dendritic trees and axons branching in the form of vertical horse-tails and terminating in the above-mentioned cartridges around the ascending dendrites of pyramidal cells[12].

This is the present available torso of our knowledge of the neuronal circuitry of the visual area of the rabbit's cortex.

Before speculating about the possible neuroanatomical basis of spontaneous and evoked activities in the striate area, Fig. 5 is presented to summarize the power profiles of the spontaneous activities of 5 experiments. All show a flat trough with a minimum in and near layer IV and two maxima, one in the uppermost layers and a deep one in layer VI. It is conspicuous that maximum power is found in layers the salient morphological feature of which is a great number of horizontal fibres rather than a dense packing of pyramidal cells: supragranularly, the maximum power coincides with the band of tangential fibres in layer I, and infragranularly with the tight fibre plexus of Baillarger's internal line. The histological characteristics thus suggest that the generation of spontaneous activities, which are likely to be composed of graded processes rather than of pyramidal cell discharges[15], is favoured by the presence of a great number of horizontally running fibres with mutual connections via interneurons branching at the same level. This fibre feltwork consists of dendrites of pyramidal cells and a large number of axon terminals from cells of layers V and VI with a tangential span of 6-8 mm, but also from association and from callosal afferents, moreover, 35 % of noradrenergic fibres from the locus coeruleus terminate in layer I[16]. In this fibre system, a continuous and abundant synaptic activity is present. This system may be compared with an accumulation of myriads of circuit elements, producing graded rather than all-or-none-discharges and having a high degree of interaction. In such systems, oscillations start as soon as a certain degree of interconnectivity is reached, as observations of sets of artificial neurons demonstrate[17], and spread through the medium. This is exactly the case with the potential fields on the cortical surface which propagate along the cortex in different directions[1,2]. Such "synergetic processes"[18] do always occur in systems of uniformly behaving and interacting elements.

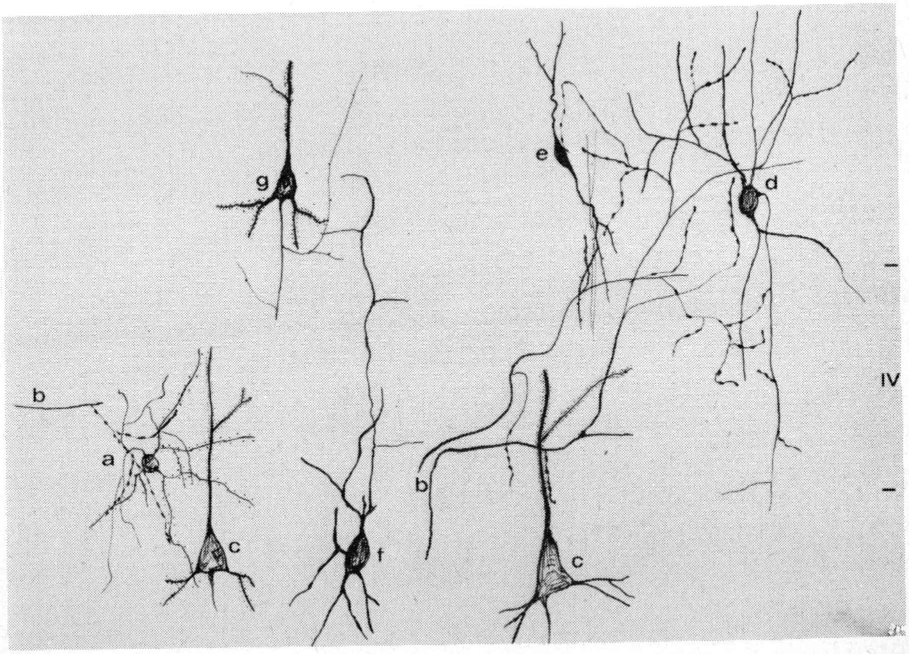

Fig. 4. Terminal arborization of the specific input into the rabbit's striate area, according to Golgi studies (Tömböl 1979). Details see text.

Both the size and duration of these oscillations depend - apart from the interconnectivity - on several other conditions, primarily on the patterns of information from other regions of the cortex and from extracortically. Particularly this latter seems to be of importance for spontaneous activity to arise, since, in undercut cortex, there is usually no,or almost no,spontaneous activity[19].

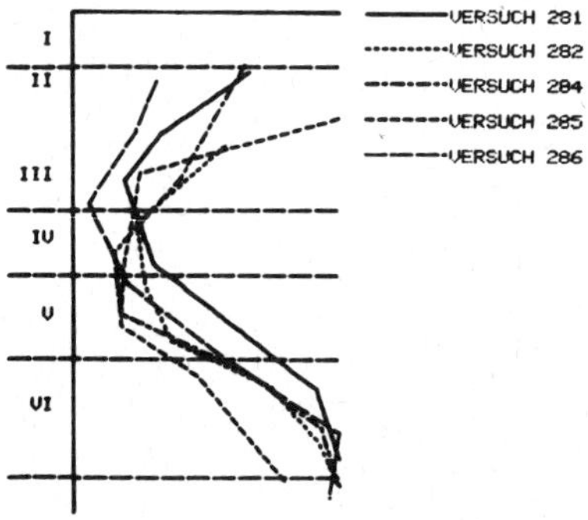

Fig.5. A superimposition of the power profiles through the cortex of 5 experiments. Maximum activity is found in the two external layers of the cortex.

This hypothesis finds support from the concept of cortical modules. This concept[20,21,22], comprises the observation that cortical neurons in sensory areas are arranged in columns of roughly 0.5 mm diameter with uniform function. These columns, however, do not extend throughout the entire cortex but mainly concern layers II to V. Therefore, spontaneous activity, which is largely uniform over extended cortical regions, is much less probable in these layers which serve special functions,than in the two limiting layers with their horizontal feltwork where the fibres extend over several millimeters. In these layers, therefore, the cortical functions can no longer be so clearly parcelled as in the middle cortical layers.

Additional evidence in favour of this view is provided by previous studies[23]. In this paper,an attempt was made at measuring the minimum cortical volume in which two field-potentials of a seizure appear at exactly the same time. For this purpose two closely-spaced microelectrodes of equal length were used. The cortical volume in which two signals appear exactly synchronous was found to be smallest in layers II-V and much larger in the layers above and below. In other words: the potential gradients, measured horizontally, are steepest in the middle layers of the cortex where the horizontal interconnectivity is smallest.

A still unsolved question concerns the relationships between these two layers of maximum spontaneous activity. There are some cells which connect the two limiting layers of the cortex: these are the Martinotti-cells in layer VI (Fig. 3b) and the multiangular cells in layer I (Fig. 3l), the axons of which traverse the entire cortex. Moreover, there are also reverse pyramidal cells (not drawn in Fig. 3) the axons of which turn upwards. It is not unlikely that these cells guarantee a certain interdependence between the two generating layers. Nevertheless, the question as to the reason of the phase shift of nearly 180^{0} must remain unanswered.

The genesis of evoked potentials, on the other hand, seems to be basically different. In contrast to spontaneous activities, cellular discharges seem to play an additional role in evoked activities. Common to the different kinds of stimulation of the optic system is an initial current sink in the deep layers of the cortex. The location of this sink, however, depends on the position of the recording area with respect to Thompson's zone[10]. This sink may be interpreted as a depolarization caused either by soma discharges or dendritic EPSPs or both. An interpretation as to the population of cells which may give rise to this phenomenon cannot yet be given. The star-cells of layer IV, which may be expected to be activated first by light stimuli, theoretically do not contribute to a vertical current flow since their axon terminals are mainly within their globular dendritic trees and, therefore, the fields they produce are closed.

With a more synchronized input, as in the case of OEP (Fig. 2i), the recruitment of the neuronal elements of layers V and VI is faster.

The next event is a depolarization in the next higher layer, which, in the VEP, tends to oscillate. In the OEP, on the other hand, oscillations do not occur but, instead, a long-lasting sink develops in the upper part of layer V, accompanied by a source immediately above. The considerable differences of the VEP whether recorded within or off Thompson's region[10] are most likely due to structural differences within the striate area and require further histological studies.

The data offered by this study still leave many questions open and additional work, particularly with still small electrode distances, will be required to fully understand the significance of these phenomena and to enable their interpretation in histological terms. In this paper, a first step was attempted to construct a bridge to cross the large gap between cortical neuroanatomy and electrophysiology.

ACKNOWLEDGEMENTS

This work was supported by the Fonds zur Foerderung der wissenschaftlichen Forschung in Oesterreich, Project No. S 25/01.

We gratefully acknowledge the valuable assistance of Prof.Dr.Teres Tömböl, II. Anatomical Institute of the University, Budapest, who kindly provided us with drawings of her Golgi studies of the rabbit's striate area.

REFERENCES

1. Petsche, H., Nagypal, T.,Prohaska, O., Rappelsberger, P. and Vollmer, R. (1975) Approaches to the spatio-temporal analysis of seizure patterns. In: CEAN-Computerized EEG-Analysis (Eds.: G. Dolce and H. Künkel), Fischer, Stuttgart, pp. 111-127.
2. Petsche, H., Rappelsberger, P., Lapins, R., Vollmer, R. (1979) Rhythmicity in seizure patterns: topographical aspects. In: Origin of Cerebral Field Potentials (Eds.: E.J. Speckmann and H.Caspers),Thieme, Stuttgart, pp.80-97.
3. Prohaska, O., Olcyatug, F., Womastek, K., Petsche, H. (1977) A multielectrode for intracortical recordings produced by thin-film technology. Electroenceph.clin.Neurophysiol. 42, 421-422.
4. Prohaska, O., Pacha, F., Pfundner, P., Petsche, H. A 16-fold semi-microelectrode for intracortical recording of field-potentials.Electroenceph. clin.Neurophysiol. (in print).
5. Müller-Paschinger, I.B. (1979) Korrelationen zwischen Morphologie und elektrischer Tätigkeit einzelner Rindenbereiche am Kaninchengroßhirn, (Thesis) Salzburg, 109 pp.
6. Rêmond, A. et Conte, C. (1964) Evolution temporelle et valeur moyenne de l'intensitê efficace d'un EEG diffêrentiel, Rev. Neurol. 111, 283.
7. Rappelsberger, P., Pockberger, H., Nechvile, Ch.,and Petsche, H. Current source density analysis of simultaneously recorded intracortical field potentials. Pflügers Arch. (in preparation).
8. Petsche, H., Müller-Paschinger, I.B., Pockberger, H., Prohaska, O., Rappelsberger, P., and Vollmer, R. (1978) Depth profiles of electrocortical activities and cortical architectonics. In: Architectonics of the Neocortex (Eds: Brazier, M.A.B., and Petsche, H.), Raven Press, New York, pp. 257-280.
9. Pockberger, H., Petsche, H., Rappelsberger, P., Müller-Paschinger,I.B., and Prohaska, O. Epi- und intrakortikale Aspekte visuell evozierter Potentiale, EEG-EMG (in print).
10. Thompson, J.M., Woolsey, C.N., and Talbot, S.A. (1950) Visual areas I and II of the cerebral cortex of rabbit, J. Neurophysiol. 13, 277-288.
11. Rose, M. (1931) Cytoarchitektonischer Atlas der Großhirnrinde des Kaninchens, J.Psychol.Neurol. (Leipzig) 43, 353-440.
12. Szentágothai, J.(1973) Synaptology of the visual cortec. In: Handbook of Sensory Physiology, VII/3, Central Processing of Visual Information. Part B (Ed.:R.Jung), Springer, Berlin-Heidelberg-New York, pp. 269-325.
13. Scheibel,M.E., and Scheibel, A.B.(1970) Elementary processes in selected thalamic and cortical subsystems - the structural substrates. In: The Neurosciences (Ed.:F.O.Schmitt), Rockefeller, New York, pp. 443-457.

14. Cajal, R.Y. (1911) Histologie du systême nerveux de l'homme et des vertébrês, Maloine, Paris.

15. Elul, R.(1972) The genesis of the EEG, Internat. Rev.Neurobiol.15,227-273.

16. Lapierre, Y., Beaudet, A., Demianczuk, N., Descarries,L. (1973) Noradrenergic axon terminals in the cerebral cortex of rat,Brain Res.63,165-182.

17. MacGregor,R.J., Palasek,R.L.(1974) Computer simulation of rhythmic oscillations in neuron pools,Kybernetik 16, 70-86.

18. Haken, H.(1977) Synergetics. An Introduction, Springer, Berlin-Heidelberg-New York, 325 pp.

19. Sherwin, I. (1965) Differential effects of hyperventilation on the excitability of intact and isolated cortex. Electroenceph.clin.Neurophysiol. 8, 599-607.

20. Mountcastle, V.B.(1957) Modality and topographic properties of single neurons of cat's somatic sensory cortex, J.Neurophysiol. 20, 408-434.

21. Kuffler, S.W. (1953) Discharge patterns and functional organization of the mammalian retina, J. Neurophysiol. 16, 37-68.

22. Hubel, D.H., Wiesel, T.N. (1974) Sequence, regularity and geometry of orientation columns in the monkey striate cortex, J.Comp.Neurol.158,267-294.

23. Petsche, H., Prohaska, O., Rappelsberger, P., Vollmer, R.(1975) The possible role of dendrites in EEG synchronization. In:Physiology and Pathology of Dendrites (Ed.: G.W.Kreutzberg), Raven Press, New York, pp.53-70.

Rhythmic EEG Activities and Cortical Functioning
G. Pfurtscheller et al., editors

ELECTROGENESIS OF EVENT-RELATED SLOW POTENTIALS IN THE CEREBRAL CORTEX OF CONSCIOUS ANIMALS

JAMES E. SKINNER and G. L. KING
Section of Neurophysiology, Department of Neurology, Baylor College of Medicine
Houston, Tx 77030 (U.S.A.)

INTRODUCTION

When a conscious animal is presented with a meaningful stimulus a cortical extracellular slow potential (SP) response occurs. This cerebral SP does not respond to the physical attributes of the stimulus, like the sensory evoked potential does, but occurs only in response to features of the stimulus that have meaning because of the context or event of the stimulus presentation. That is, stimuli which signal something about to occur, or are novel, or are psychologically stressful at that moment will all evoke the cortical SP response, but if these same stimuli are presented in a context in which they have no immediate significance for the animal, because they are not signals or are no longer novel or fearful, then no SP response occurs.

The unfamiliar surroundings of the laboratory produce a type of psychologic stress in conscious pigs that appears to constitute a necessary factor for the initiation of ventricular fibrillation following coronary artery occlusion.[1] If the animal is adapted to the laboratory for 6-8 days, so that the objects and personnel in it no longer are novel stimuli, then coronary artery occlusion does not result in ventricular fibrillation and death. Thus, environmental contexts in which stimuli are psychologically stressful increase the vulnerability of the heart to arrhythmias. This finding suggests why stresses such as bereavement, job insecurity, new surroundings, etc. can be significant risk factors in humans for cardiac sudden death.[2]

By mapping the structures of the brain that manifest SPs in response to a psychologic stress (i.e. a tone that forewarns electric shock) and by stimulating and blocking the interconnecting pathways between these responsive parts of the brain, we have established that the frontal granular cortex orchestrates the occurrence of the stress-evoked SP, as well as other types of event-related responses, such as EEG desynchronization and sensory evoked potential amplitude increase [3]. The frontal cortex has also been shown by our laboratory 1) to regulate the sensory channels by controlling the tonic inhibition exerted on the thalamic relay nuclei an 2) to be necessary for the initiation of a cardiac response to a conditioned signal. Thus, this brain mechanism centered around the frontal lobes fulfills the neurophysiologic requirements for Cannon's early conceptualization [4] of a cerebral defense system, that is, the joint regulation of the sensory and autonomic channels. Cannon thought that such a neural mechanism not only existed, but was intimately involved in the rapid evolution of the higher vertebrate brain. This brain mechanism, he claimed, enabled the animal to prepare itself better to respond to attended objects in advance of the requirement to do so in either flight or fight behavior, and thus had important survival value for the species. In this regard, it is important to note that the frontal granular cortex is the most recently encephalized part of the brain and the most distinctive feature of the human cerebrum.

Recently, we have shown that cryogenic blockade of a frontocortical-brainstem pathway in the conscious pig will prevent the occurrence of ventricular fibrillation following coronary artery occlusion, even though the animal is psychologically stressed by being unadapted to the laboratory [5]. Thus, if we could develop a pharmacologic intervention that would equal our physical cryogenic blockade, then we could prevent the lethal consequences of myocardial ischemia in animals and humans who cannot avoid an environment in which many of the stimuli evoke stress. We have now turned to the cellular analysis of the event-related SP in our search for such a salutary pharmacologic intervention. We believe that if we can understand the neurochemical mechanisms operating within the cerebral defense system, then we may discover how to control its functioning by chemical therapeutic means. The search for the electrogenic basis of the event-related SP is of great importance in guiding our research toward this goal.

Our current hypothesis is that the electrogenic mechanism for the cerebral SP is the same as that known for the neurons of the isolated sympathetic ganglion. This latter structure is regulated by both the cerebral defense system and the autonomic reflex mechanisms. Because of its structural simplicity, the sympathetic ganglion has served as a useful biological model of the vertebrate neuron and has been studied extensively for many years (for a recent review see Kuba and Koketsu, 1978[6]). In the isolated ganglion, stimulation of one of the two synaptic input channels results in cholinergic-muscarinic receptor activation, an increase in postsynaptic membrane resistance, the occurrence of a slow EPSP, and an increase in intracellular cyclic GMP; stimulation of the other synaptic input results in dopaminergic receptor activation, an increase in postsynaptic membrane resistance associated with, in some cases, a slow IPSP, an increase in intracellular cyclic AMP, and heterosynaptic potentiation of the cholinergic-muscarinic responses. That similar types of monaminergic mechanisms operate in the association cerebral cortex is suggested in a study by Stone, Taylor and Bloom [7]. They showed that the local microiontophoretic applications of the putative neurotransmitters and cyclic nucleotides used in the ganglion had analogous excitatory and inhibitory effects in the cortex, as evidenced by changes in the spike firing rates of the neurons close to the site of injection. A second line of support for analogous central and peripheral mechanisms was demonstrated by our laboratory[8] in which it was found that cortical cyclic AMP levels covary with the amplitude of an event-related SP recorded in the same region. What is currently missing in this comparison between sympathetic ganglion cell mechanisms and those present in the cortical neurons is some intracellular data that would show whether or not slow membrane potential and input resistance responses occur in the cortical cells in response to stimuli that would evoke the extracellular event-related SP. We will now present data that will show that the majority of the neurons in the frontal granular cortex of the conscious cat do show slow membrane potential shifts and input resistance increases in response to a tone that forewarns mild cutaneous shock. We had to develop a new technology for maintaining long term intracellular recordings in order to collect this data; this method will also be described briefly.

METHODS

We developed a method of "floating" the micropipette tip in the brain in order to achieve long term intracellular recordings in the conscious animal. The method essentially entails mechanical coupling of a 15 mm micropipette tip (broken off from its larger shaft) to a hydraulic microdrive with ice being the coupling medium. A cryogenic device, through which -60°C methanol can be circulated, made of 0.5 mm diameter stainless steel tubing is attached to the microdrive. A sterotaxically aligned pipette tip is then frozen to the cryogenic device with a bead of water. The electrical contact between the filled pipette tip and the amplifier is via a very fine 25 micron diameter silver-chlorided wire. After impaling a neuron, the ice is rapidly melted thus enabling the almost massless pipette to "float" in the brain tissue. The fine amplifier lead helps stabilize the upper end of the pipette during shallow cortical penetrations. The use of water as the coupling medium enables the change of state from a solid to a liquid without producing movement of the pipette tip and the consequent loss of the intracellular recording.

The micropipettes were pulled on a David Kopf instrument and were made from

Frederick Haer 2mm diameter pipette tubes that contained a single capillary filament. The 3-molar potassium-chloride-filled tips typically had a resistance of 50-60 megohms, a finding which indicated that the tip diameters were all less than 0.5 microns. Both potassium acetate and potassium citrate were also used as electrolytes to control for chloride ion effects, but since no differences in membrane properties were ever noted, we recorded the majority of the cells with potassium-chloride-filled micropipettes.

Five cats were first prepared under general anesthesia. A permanent dental cement crown was attached to each animal's skull with stainless steel screws. Trephine holes were drilled in the cranium and filled with bone wax. After recovery from surgery, the animal's crown was temporarily attached to a David Kopf Instruments heavy duty stereotaxic frame. This was performed on a daily basis until the animal adapted to the head restraint and would sit quietly for a few hours without struggling. The animals were rewarded with sardines for sitting quietly while they were attached to the stereotaxic frame.

After adaptation the animals were subjected to tone-shock conditioning until they showed, to the tone onset, both sustained potential shifts on the cortex surface and changes in cardiac rate. Following acquisition of conditioning the bone wax was removed from the trephine holes, the dura slit, a pressor foot placed atop the cortex and a floating micropipette lowered into the tissue. No behavioral struggling reactions occurred to any of these procedures.

After impaling a neuron the membrane potential was continuously recorded on a Precision Instruments FM tape recorder (bandwidth: DC to 10kHz,at 37.5 inches per sec tape speed). The extracellular zero potential was determined later when the intracellular record was spontaneously lost or the electrode just barely withdrawn from the cell. Constant current pulses were passed once per sec using a WPI balanced-bridge microprobe electrometer. These current pulses were used to measure the membrane resistance and were 50 msec in duration and ranged from 0.5 to 3,0 nA in both positive and negative directions. After the animal and cell were determined to meet control standards, as explained below, the tone was then presented and 4 secs later a very mild cutaneous shock was delivered through electrodes permanently attached to the skin of the back. During acquisition of conditioning a stronger shock was used, one that just elicited back-arching movements. During intracellular recording the shock was adjusted to a lower level to help prevent loss of the cell due to this particular movement.

Immediately after achieving impalement of a cell a hyperpolarizing current was passed for a few minutes until the membrane potential no longer drifted. The hyperpolarizing current was then removed and 0,5 nA, 50 msec current pulses were passed once every second to measure the membrane input resistance. The bridge was balanced so that the voltage drop across the electrode was cancelled.

The control standards required of the cell and animal prior to delivery of a tone-shock trial were: 1) a stable baseline membrane potential with no large voltage excursions , 2) a stable baseline input resistance with no large resistance changes, 3) symmetrical turn-on and turn-off transients produced by the bridge unbalance (i.e. no rectification by the electrode), 4) linear plots of the steady-state current vs. voltage relationship between plus and minus 1,0 nA current pulses (i.e. ohmic resistance measurements devoid of any voltage-dependent conductance changes), and 5) animal awake, with eyes open and sitting quietly. It usually took 10 min of recording before the first trial was delivered. A second or third trial was administered if the intracellular recording was still maintained, but never with an intertrial interval of less than 3 minutes. We were able to maintain intracellular recordings of the cells from 10 to 40 min.

At the end of the intracellular recording some cells were injected with 3% Lucifer Yellow CH dissolved in the pipette electrolyte (obtained from Dr. Walter Stewart of the National Institute of Health, Bethesda). This dye was iontophoretically injected with 10 to 15 nA hyperpolarizing currents for 15 to 30 secs. The entire tissue block beneath the pressor foot was then rapidly frozen by circulating -60°C methanol through the 7 mm diameter pressor foot that was constructed from 0.7 mm diameter stainless steel tubing. This small block of tissue was then removed in the frozen state from the brain without any untoward behavioral reaction by the animal. The frozen tissue was then fixed by freeze substitution in 4% formaldehyde in 50% methanol at -15°C. It was then rehydrated and cut on a freezing microtome into 50 micron sections and immediately inspected under ultraviolet light with a phase-optics microscope. The brilliant yellow glow of this UV-stimulated procion dye was quickly detected and the tissue was then immediately photographed.

The data was printed out on a 100 Hz bandwidth Beckman chart recorder at a tape speed that was 1/100 of the recording speed, thus conserving the DC to 10 kHz bandwidth of the data. The data was also played back at real time and photographed from a Tektronics 502 oscilloscope with a Grass C4 camera.

The tones and shocks were delivered by BRS programming equipment.

RESULTS

A total of 46 cells have been successfully recorded to date with our "floating" micropipette method. The details of these results will be reported elsewhere [9], and only a preliminary report will be made here. In each case the cell was held long enough to measure membrane potential (Em) and input resistance (Rm) responses to the tone and the mild cutaneous shock. The cells could easily be divided into two groups based on their resting Em level and the size of their action potentials. The average resting Em for a group of 39 cells that showed action potentials smaller than their resting level (undershoot cells) was 35 mV. The average resting Em of the other 7 cells that all had action potentials as large as or larger than teir resting level (overshoot cells) was 6o mV. Our use of the term, undershoot, is not to be confused with that used to describe the potassium current that follows an action potential and causes the membrane potential to hyperpolarize and undershoot the resting membrane potential. The undershoot phenomenon discussed here is thought to arise from dendritic impalements in which a part of the amplitude of the somatically generated action potential is electrotonically reduced at the site of recording by the membrane resistance of the dendrite as described by Woody and Gruen [10].

Of the seven overshoot cells three showed no Rm changes with their Em responses and two showed no responses to the tone or shock whatsoever. Most of the undershoot cells showed large and striking shifts in both Rm and Em, as will be described below. Large 10 mV EPSP's and IPSP's were apparent in the overshoot cells, but none of the undershoot records ever showed any such postsynaptic potentials. The average resting Rm of the undershoot cells was 50 megohms and that of the overshoot cells was 5 megohms.

Figure 1 shows a typical undershoot cell response to the tone onset. Examples of the raw data from the collection of cells have been reported elsewhere [11] in preliminary form. The slow hyperpolarizing shift of Em in the cell shown in Figure 1 lasted for approximately 60 sec before returning to baseline. This shift was accompanied by a slow increase in the input resistance (Rm), but the two responses were not perfectly correlated. Figure 2 shows the lack of a significant response of this same cell during a second trial. Note that during the second trial, the input resistance is generally lower, the amplitude of the action potentials (AP) is smaller and more variable, and the spontaneous firing frequency of the cell (FREQ) is much greater. The resting Em, however, has not changed. Also during the second trial there seems to be a reversal of polarity in the extracellular slow potential recorded immediately above the cell on the cortical surface (EEG).

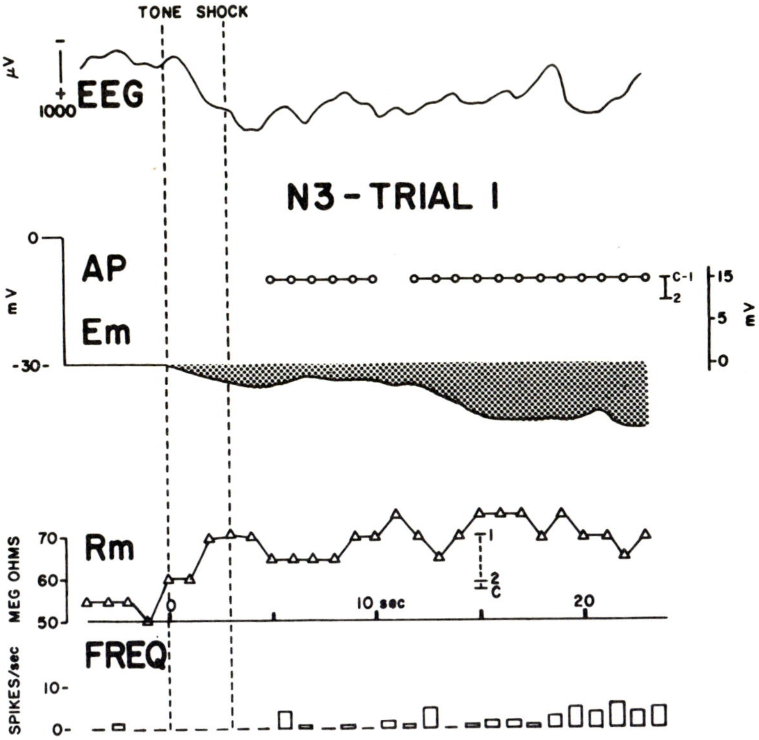

FIG. 1. Membrane responses of frontal cortex cell N3 during the first tone-shock conditioning trial recorded during its impalement. EEG shows the positive extracellular event related potential. AP shows a constant action potential amplitude of 15 mV (scale at right). Em shows the hyperpolarizing membrane response from the -30mV resting level. Rm shows the input resistance increase. FREQ shows the low spontaneous rate of action potentials. The timebase is shown just below the Rm curve and extends to 25 sec following the tone onset.

Of the 39 undershoot cells recorded, only 11 failed to show an Em or Rm response. The results of the cell shown in Figures 1 and 2 suggest that had repeated trials been possible in these 11 unresponsive cells, they also might have shown the slow Em and Rm responses. Ten of the undershoot cells had repeated trials; of these, 3 showed Em and Rm responses on one trial and no responses on the other, while 5 others showed the same response in both cases. Two cells showed different polarity Em responses for the two trials.

Seventeen undershoot cells showed Rm increases of 28% or more. Of these, 4 were associated with Em depolarizations and 13 with Em hyperpolarizations. Ten cells showed Rm decreases, all of which were associated with Em depolarizations. Four undershoot cells showed both Rm increases and decreases during

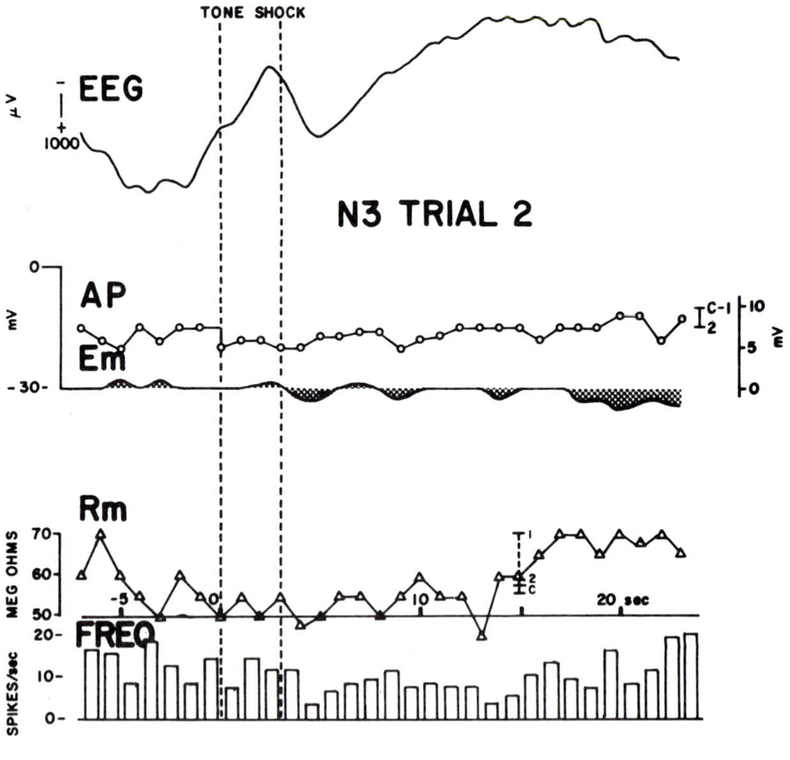

FIG. 2. Membrane responses of the same frontal cortex cell shown in Figure 1, but during the second tone-shock conditioning trial recorded during its impalement. Note that compared to the first trial the EEG response has reversed in polarity, the AP amplitude is reduced and is variable, and Em response is not significant, the Rm response is not significant and the action potential frequency has increased.

the same trial, with one response to the tone and the other to the shock. All of the Em responses of these 17 cells were 5 mV or more and were statistically significantly different ($p < .01$, t-test) from the 10 sec baseline period that preceded the tone.

In our study to date, only three undershoot cells have been injected with Lucifer Yellow CH. In all of these cases the dye has been located only in cell processes and not in cell bodies. Woody and Gruen [10] have reported dye injection into the dendrites of cells through micropipettes that showed undershoot, a finding which, like ours, suggests that the undershoot phenomenon is attributable to dendritic impalements.

DISCUSSION

Our data clearly show that slow membrane potential shifts are evoked in cortical neurons in response to meaningful stimuli. These results suggest that the electrogenic mechanism of the extracellular event-related slow potentials may be neuronal in origin in contrast to the glial cell mechanism proposed by other theorists, Castellucci and Goldring [12], Ranson and Goldring [13, 14, 15] and Somjen [16, 17] have all suggested that extracellular sustained potential shifts may be produced by increases in extracellular potassium from active neurons that causes slow depolarizations of electrotonically coupled glial cells. These glial cell depolarizations, according to Somjen [17], then result in closed-field dipoles forming within the glial syncytium that generate the extracellular currents of the sustained potential shifts. Our data suggests that slow membrane potential changes within certain regions of a cortical neuron may also contribute to such extracellular currents. Preliminary evidence from our laboratory, in confirmation of Woody and Gruen [10], suggest that it is the dendritic region of the neuron that manifests the slow membrane potential response to a meaningful stimulus.

The reasons that such slow membrane potentials have not been observed in the past by other experimenters could be, 1) they are not observed in anesthetized animals, because of the effect of the drug on the neural membrane, 2) they are not observed in highly stressed animals immobilized by neuromuscular junction blockade, 3) they were recorded, but were disregarded because they were associated with low resting membrane potentials and small action potentials, both of which were thought to be signs of an injured cell, or 4) natural meaningful stimuli were not used to activate the recorded cell, only the electric activation of a limited number of presynaptic fibers was employed. The reasons we believe that our recorded Em and Rm responses of the undershoot cells are legitimate measures of in vivo neuronal responses are, 1) the Em and Rm shifts returned to baseline within the same time domain as the event-related slow potentials that were recorded from the cortical surface, 2) the values of the 10 to 15 mV action potentials and the 50 megohm input resistances are what one would expect from dentritic impalements in which the length constant of the dendrites is 1.5 to 2.0 (see Jack, Nobel and Tsien [18]) and the impalement was midway between the soma and distal end of the dendrite branch (see Rall and Rinzel [19]), 3) the Em and Rm responses were doubly dissociated from any movements of the animal that could have caused injury to the cell (i.e., responses did not occur between trials, though the animals did move, and responses occurred during trials in which cases the animals did not move), 4) the same baseline control values of Em and Rm were maintained over extended 20 to 40 min periods of intracellular recording, and 5) many replications of the observed responses were made in different cells in the same and in different animals. We have no explanation of why the average resting membrane potential of the undershoot cells was 35 mV, but this exact same value was reported by Takato and Goldring [20] as the average membrane potential of neurons they impaled and injected with either Lucifer Yellow CH or horseradish peroxidase for later anatomical visualization of that same single cell. Woody and Gruen [10] also reported similar low resting potentials for many of their undershoot cell records. Our conclusion is that our "floating" micropipette method has enabled us to observe a particular set of membrane responses to meaningful stimuli that occurs only in neurons in the conscious animal brain.

The independent changes in Em and Rm, both within and between cells, may be of special significance for theoretical models of dendritic integration. Barratt [21] has reviewed the literature bearing on how the independent modulation of the length constant of the dendrites could affect the efficacy of dendritic potentials on the voltage across the membrane of the soma where the axon trigger zone is located. If the membrane resistance of the dendrites (which is related to the dendritic input resistance) is increased, then the net dendritic potential there will be more effective in its influence on the somatic potential. Although our data presented in Figures 1 and 2 do not show a relationship between Rm and the action potential firing frequency, we did observe such an effect in some of our cells. Presumably the particular dendritic branch that we may have impaled in neuron N3 in Figures 1 and 2 did not have a major influence on the somatic membrane potential.

Whether or not the slow Em and Rm changes that we observed are related to cholinergic-muscarinic and catecholaminergic receptor activation remains to be demonstrated. What our data does show is a third line of support linking the cerebral slow potential electrogenic mechanism to that of the sympathetic ganglion cells. It has now been demonstrated that the cerebral event-related slow potentials are correlated with dynamic shifts in the cyclic nucleotide levels and membrane potentials of the underlying cells in the recorded tissue. Since small amounts of acetylcholine and norepinepherine will produce 15 to 60 sec effects on the firing frequency of nearby cortical neurons, as Stone, Taylor and Bloom [7] demonstrated, it seems quite likely that these substances may be the synaptic chemicals that produce the slow 15 to 60 sec membrane potential shifts that occur in neurons in response to meaningful stimuli. If this is the case, and if we can eventually understand their postsynaptic mechanisms of action, then we may have an important clue in our search for a biochemical intervention in the cerebral defense mechanism that will prevent the lethal consequences of myocardial ischemia and severe psychological stress.

ACKNOWLEDGEMENTS

This research was supported by grant HL17907 from the National Institute of Health.

REFERENCES

1. Skinner, J. E. , Lie, J. T., and Entmann, M. L. (1975) Modification of Ventricular Fibrillation Latency Following Coronary Artery Occlusion in the Conscious Pig: The Effects of Psychologic Stress and Beta-adrenergic Blockade. Circulation, 51: 656-667.

2. Rahe, R. H., Bennett, L., Romo, M., Seltanen, P., and Arthur, R. S. (1973) Subject's Recent Life Changes and Coronary Heart Disease in Finland, American J. Psychiat. 130: 1222-1226.

3. Skinner, J. E., and Yingling, C. D. (1977) Central Gating Mechanisms That Regulate Event-Related Potentials and Behavior: A Neural Model for Attention. In J. E. Desmedt (Ed.) Progress in Clinical Neurophysiology, 1:30-69, Karger, Basel.

4. Cannon, W. B., (1931) Again the James-Lange and the Thalamic Theories of Emotion, Psychol. Rev., 38: 281-295.

5. Skinner, J. E., (1980) The Role of the Frontal Cortex in the Regulation of Cardiac Vulnerability to Ventricular Fibrillation: A Neurophysiological Model of Cannon's Cerebral Defense System. In, E. Donchin, G. Galbraith and M. Kietzman (Eds.) Neurophysiology and Psychology: Basic Mechanisms and Clinical Applications, Academic Press. (In Press)

6. Kuba, K., and Koketsu, K., (1978) Synaptic Events in Sympathetic Ganglia, Progress in Neurobiology, II: 77-169

7. Stone, T. W. , Taylor, D. A., and Bloom, T. E. (1975), Cyclic AMP and Cyclic GMP May Mediate Opposite Neuronal Responses in the Rat Cerebral Cortex. Science, 187: 845-847

8. Skinner, J. E., Reed, J. C., Welch, K. M. A., and Nell, J. (1978) Cutaneous Shock Produces Correlated Shifts in Slow Potential Cortex of the Conscious Rat. J. Neurochem. 30: 699-704.

9. King, G. L., and Skinner, J. E., (1980) Neurons in the Cortex of the Conscious Cat Show Long Lasting Shifts in Membrane Potential and Input Resistance in Response to Meaningful Stimuli. J. Neurophysiol., (submitted for publication)

10. Woody, C. D. and Gruen, E., (1978) Characterization of Electrophysiological Properties of Intracellularly Recorded Neurons in the Neocortex of Awake Cats: A Comparison of the Response to Injected Current in Spike Overshoot and Undershoot Neurons. Brain Res., 158: 343-357.

11. Skinner, J. E., and King, G. L., (1979) Contribution of Neuron Dendrites to Extracellular Sustained Potential Shifts, 5th International Symposion on Electrical Potentials Related to Motivation, Motor and Sensory Processes of the Brain, Ulm.

12. Castellucci, V. F., and Goldring, S., (1970) Contribution to Steady Potential Shifts of Slow Depolarization in Cells Presumed to be Glia. Electro-enceph. Clin. Neurophysiol., 28: 109-118.

13. Ranson, B. R., and Goldring, S. (1973a) Ionic Determinants of Membrane Potential of Cells Presumed to be Glia in Cerebral Cortex of Cat. J. Neurophysiol., 36: 855-868.

14. Ranson, B. R., and Goldring, (1973b) Slow Depolarization in Cells Presumed to be Glia in Cerebral Cortex of Cat. J. Neurophysiol., 36: 869-878.

15. Ranson, B. R., and Goldring, (1973c) Slow Hyperpolarization in Cells Presumed to be Glia in Cerebral Cortex of Cat. J. Neurophysiol., 36: 879-892.

16. Somjen, G. G., (1973) Electrogenesis of Sustained Potentials. Prog. Neurobiol., 1: 199-237.

17. Somjen, G. G. (1978) Contribution of Neuroglia to Extracellular Sustained Potential Shifts. In D. A. Otto (Ed.) Multidisciplinary Perspectives in Event-related Brain Potential Research. U. S. Environmental Protection Agency, Washington, 19-24.

18. Jack, T. T. B., Nobel, D., and Tsien, R. W., (1975) Electric Current Flow In Excitable Cells. Clarendon Press, Oxford.

19. Rall, W., and Rinzel, J., (1973) Branch Input Resistance and Steady Attenuation for Input to One Branch of a Dentritic Neuron Model. Biophys. J., 13: 648-667.

20. Takato, M., and Goldring, S., (1979) Intracellular Marking with Lucifer Yellow CH and Horseradish Peroxidase of Cells Electrophysiologically Characterized as Glia in the Cerebral Cortex of the Cat, Neurol., 186: 173-188.

21. Barrett, J. N., (1975) Motoneuron Dendrites: Role in Synaptic Integration. Red. Proc., 34: 1398-1407.

Rhythmic EEG Activities and Cortical Functioning
G. Pfurtscheller et al., editors

PARTIAL COHERENCE ANALYSIS OF THALAMIC AND CORTICAL ALPHA RHYTHMS IN DOG - A CONTRIBUTION TOWARDS A GENERAL MODEL OF THE CORTICAL ORGANIZATION OF RHYTHMIC ACTIVITY

F.H. LOPES DA SILVA[1)], J.E. VOS[2)], J. MOOIBROEK[2)] and A. VAN ROTTERDAM[1)]
1) Brain Research Department, Institute of Medical Physics TNO, Da Costakade 45, Utrecht (The Netherlands)
2) Department of Developmental Neurology, University of Groningen (The Netherlands)

INTRODUCTION

A good deal of electroencephalographic research in man pertains to the alpha rhythms as recorded from the scalp. In order to understand the neurophysiology of alpha rhythms one has, however, to resort to studies in other species which may offer an experimental model of the human alpha rhythm, for example the cat[1,2,3] and dog[4,5].

Our previous investigations in dog provided new findings which led to a number of basic ideas regarding the origin of alpha rhythms, namely the following: a) alpha rhythms of the same peak frequency, bandwidth and reactivity can be recorded as well from the visual cortex as from the lateral thalamus (namely from the lateral geniculate nucleus (LGN) and the pulvinar (PUL)[5]); b) alpha rhythms of the visual cortex are generated by an equivalent dipole layer centered at the level of the bodies and basal dendrites of neurons of layers IV and V[6]; c) cortical alpha rhythms appear to be generated in relatively small cortical areas which act as epicenters; from these, alpha activity spreads around in different directions up to distances of 4 mm with an "apparent velocity of propagation" over the cortical surface of about 0.33 $m.sec^{-1}$ [6]; d) coherences between alpha rhythms simultaneously recorded in thalamus and cortex are in most cases significantly different from zero (largest value for the squared coherence was found to be 0.56 at 12.1 Hz, 95% confidence limits 0.75-0.32)[5]; e) coherences between alpha rhythms recorded between electrodes placed in the cortex at relatively small distances ($\sim$2 mm) are usually large (>0.81 squared coherence) and decrease with distance.

The above mentioned studies have revealed that intracortical coherences are in general larger than any thalamo-cortical coherence measured in the same animal. This entails the following question: are those large intracortical coherences exclusively due to an intracortical process independently of thalamic sources, or not? In other words, the question is to measure the amount of influence of thalamic sources on the large intracortical coherences. One possible way to solve this problem would be by way of thalamic lesions: one could measure the amount of intracortical coherence before and after such a lesion. This method, however, would certainly lead to a radical change of the whole physiological state responsible for the generation of the alpha rhythm and therefore it would not be suited to the purpose. An alternative approach to this problem is by way of a signal-analytical technique by means of which a sort of "theoretical thalamic deafferentation" can be achieved. This technique resorts to the computation of *partial coherence functions*. These type of functions have been applied already in a few cases to neurophysiological problems such as that of spread of epileptiform activity[7], of the organization of EEG during development[8] and of the analysis of seizure patterns within the cortex[9].

Models

In these introductory remarks it is useful to consider two possible models which may be proposed to describe thalamo-cortical and cortico-cortical relationships which underly the process of alpha rhythm generation. According to what we may call model I (point-to-point correspondence) each defined *cortical area* receives an input from one specific *thalamic neuronal population*; besides, there are cortico-cortical processes which are responsible for the spread of alpha activity between different cortical areas. These cortico-cortical processes are independent of the activity of specific thalamic populations. According to model II each thalamic neuronal population projects to several cortical areas by way either of ramifications of thalamo-cortical axons and/or of projections within the cortex arising from the neurons which receive the thalamic afferents; besides, the cortical areas communicate also among themselves due to the same type of intracortical processes postulated above.

Both models could explain the fact that the coherences found between alpha rhythms recorded from different, albeit neighbouring, cortical areas are larger than those found between thalamic sites and cortical areas. These models are schematically given in Fig. 1. It should be stressed that in the sense of this presentation cortical areas or thalamic populations are considered as those neural masses the electrical activity of which is recorded by one of our electrodes.

The fundamental question we addressed ourselves to in the present investigation was which of the above mentioned models should be preferred. In our previous

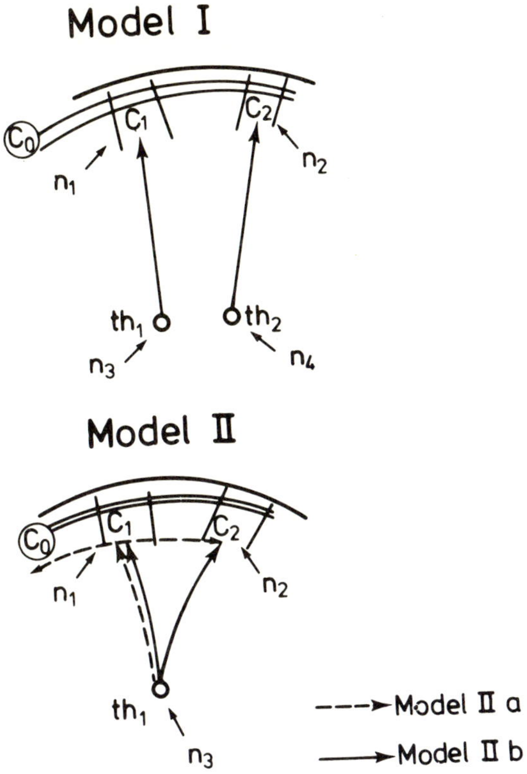

Fig. 1. Two models of thalamo-cortical relationships; in model I a point-to-point correspondence is assumed while th_1 and th_2 are supposed to be unrelated; in model II each thalamic population projects to several cortical areas. Symbols explained in text and Appendix.

investigations of alpha rhythms it was not possible to answer this question because the classic method of coherence analysis is not powerful enough to discriminate between the two models. The computation of partial coherences by way of a new computer program[8] appeared to offer the possibility of solving the question formulated above.

Signal analysis: theoretical considerations

Computing partial coherences implies first eliminating from each of the two signals of a pair (e.g. two cortical signals) that part that can be considered as being determined by, or predictable on the basis of, a third signal (e.g. a thalamic signal). Let us consider the following cases: we record alpha rhythms from two cortical areas and one thalamic area; the former are the signals C_1 and C_2; the latter is the signal th_1. According to model II both C_1 and C_2 may contain contributions which are determined by th_1, and for which linear least-square predictions can be computed based on the knowledge of th_1 and on the cross power spectra between both th_1 and C_1, and th_1 and C_2. When we correct C_1 and C_2 for these "predictable" contributions we can compute the (partial) coherence between the residual signals as follows:

$$\hat{Coh}_{th_1 \cdot C_1C_2}(f) = \frac{\left|\hat{Cr}_{th_1 \cdot C_1C_2}(f)\right|^2}{\hat{A}_{th_1 \cdot C_1C_1}(f)\ \hat{A}_{th_1 \cdot C_2C_2}(f)} \tag{1}$$

where the residual spectra are defined as follows:

$$\hat{Cr}_{th_1 \cdot C_1C_2}(f) = \hat{Cr}_{C_1C_2}(f)\left[1 - \frac{\hat{Cr}_{th_1C_1}(f)\ \hat{Cr}_{th_1C_2}(f)}{\hat{A}_{th_1th_1}(f)\ \hat{Cr}_{C_1C_2}(f)}\right] \tag{2}$$

and

$$\hat{A}_{th_1 \cdot C_1C_1}(f) = \hat{A}_{C_1C_1}(f)\left[1 - \hat{Coh}_{th_1C_1}(f)\right] \tag{3}$$

and

$$\hat{A}_{th_1 \cdot C_2C_2}(f) = \hat{A}_{C_2C_2}(f)\left[1 - \hat{Coh}_{th_1C_2}(f)\right] \tag{4}$$

where $\hat{Cr}$ is the estimated cross power spectra, $\hat{A}$ is the estimated autospectra and $\hat{Coh}_{th_1C_1}$, $\hat{Coh}_{th_1C_2}$ are the estimated magnitude-squared coherences, just called coherences in the remainder of this paper. The estimated coherences between C_1 and C_2 partialized on th_1 is given by $\hat{Coh}_{th_1 \cdot C_1C_2}(f)$.
In case $\hat{Coh}_{th_1 \cdot C_1C_2}(f)$ will be significantly *smaller* than the straightforward coherence between C_1 and C_2 ($\hat{Coh}_{C_1C_2}(f)$) we may decide that there are good arguments to prefer model II. However, it is possible in case model II holds, to find partial coherences which are larger than the ordinary coherence as explained further on. In order to understand the meaning of partial coherences for the type of problem discussed here we give in an Appendix the results

obtained with a computer simulation corresponding to model II descibed above.

MATERIAL AND METHODS

The investigation was done in the same dogs as used before[5]. The relevant epochs of the EEGs of thalamic and cortical sites, recorded for the previous study, had been stored on digital tape and therefore could be used again for computing partial coherence functions. In summary we used recordings of alpha rhythms obtained by way of two types of electrodes: (a) bundles of 3-7 wires (100 μm diameter) insulated to 0.5 (thalamic sites) or 1 mm (cortical sites) from the tip; (b) sheets of wires with uninsulated surfaces of 0.15-0.20 mm which were embedded in a polyethylene sheet less than 0.5 mm thick. The bundles by way of which alpha rhythms were recorded, were inserted in the marginal gyrus (dog C) and in the lateral geniculate nucleus (LGN) and pulvinar nucleus. The sheets were placed under the dura either over the occipital marginal gyrus or over the mesial cortex above the calcarine region. The positions of the electrodes were confirmed post-mortem by histological examination (sections of the brain 15 μm thick stained by Klüver or Nissl methods). Derivations were made in general against one common electrode, a stainless steel pin placed in the frontal bone at the level of the frontal sinus. As some electrodes were placed in areas free of alpha rhythms the degree of "indifference" of the common electrode as regards the phenomenon being studied could be determined. Recording was by means of a 16-channel radio-telemetry system[9]. The signals were further amplified by a 16-channel EEG Offner-Beckman Dynograph and stored on magnetic tape (EMI) simultaneously with a time code. The epochs of EEG used for analysis were selected on basis of the presence of alpha rhythms. They were digitized by way of an 8-channel analogue-to-digital converter. The signals were passed by anti-aliasing filters with a band between 0.5 and 30 c/sec. The sampling frequency was 80/sec. Analysed epochs consisted either of 1024 (duration 12.8 sec) or of 256 sample points (duration 3.2 sec). The digitized signals were stored on DEC tape (PDP 9 computer). The same tapes as used in the previous study were analysed again by way of the PDP 15 computer at the University of Groningen using the partial coherence program developed there[8].

The statistical evaluation of the results was done as follows: in order to improve the reliability of the estimates of spectra and coherences we performed ensemble averaging. Coherence functions should be treated statistically by way of the z-transformation in a similar way as usually employed for the correlation coefficient. In this way the square root of the coherence function (i.e. of the coherence as defined by eq. 1) was first taken and then the z-transform was computed. The corresponding standard deviation was calculated using the formula: $1/\sqrt{N-3}$ (Dixon and Massey, 1957, p. 201[10]) where N was 80 for the cases of dogs D and R, and 40 for dog C. In order to quantify in an explicit way the results obtained by using partial coherences we decided to formulate these results as the difference between the ordinary coherence for a particular frequency and the corresponding partial coherence; these differences were then expressed in terms of the standard deviation (sd) of the ordinary coherence. Moreover we tested whether the z-transformed partial coherence values differed significantly from the corresponding values of the ordinary coherence using the t-test. $t = \left[(\overline{x}_1-\overline{x}_2) / (\sqrt{2/N}\cdot sd)\right]$ should be > 2.358 in order to accept a difference between 2 means as significant at $p < 0.01$ ($df = 2\cdot 80-2$ for the cases of dogs D and R, and $df = 2\cdot 40-2$ for the case of dog C). In our case $\overline{x}_1-\overline{x}_2 = \Delta z$; for the cases where $N = 80$ a value $\Delta z > 0.372\cdot sd$ should be reached in order for the difference to be accepted as significant at $p < 0.01$; for the case where $N = 40$ it should be $\Delta z > 0.527\cdot sd$.

RESULTS

The operation of partialisation means the removal of the contribution to the ordinary coherence between a pair of signals of a third signal recorded from another site. On the one hand, the coherence which remains after partialisation, or *partial coherence* indicates the strength of the common components in a signal pair which are independent of the contributions from a third signal. On the other hand, the difference here called Δz, between the ordinary coherence and that computed after partialisation is a measure of the strength of the contribution of the third signal to a signal pair. Examples of ordinary and partial coherences for a number of cortico-cortical signal pairs representative of 3 dogs are shown in Figs. 2, 3 and 4. Figures 2, 3 and 4 illustrate the ordinary coherence between a pair of cortical signals and the effect of eliminating the influence on such a pair of other cortical and also thalamic (pulvinar and lateral geniculate nucleus) signals. It can be seen in Fig. 2, that in the alpha frequency band (the alpha peak was as usually found in dog C between 12 and 13 Hz) the coherence between two cortical signals (recorded from C13 and C10) decreased markedly when the influence of the signal from C9 (another cortical signal) was eliminated. In Fig. 3 a pronounced decrease in the alpha band coherence was found when the signal removed was D15 (cortex) (for placement of electrodes see Fig. 5) and to a lesser extent D27 (pulvinar); in Fig. 4 the most pronounced decrease in the coherence between the cortical pair (R14 & R15)

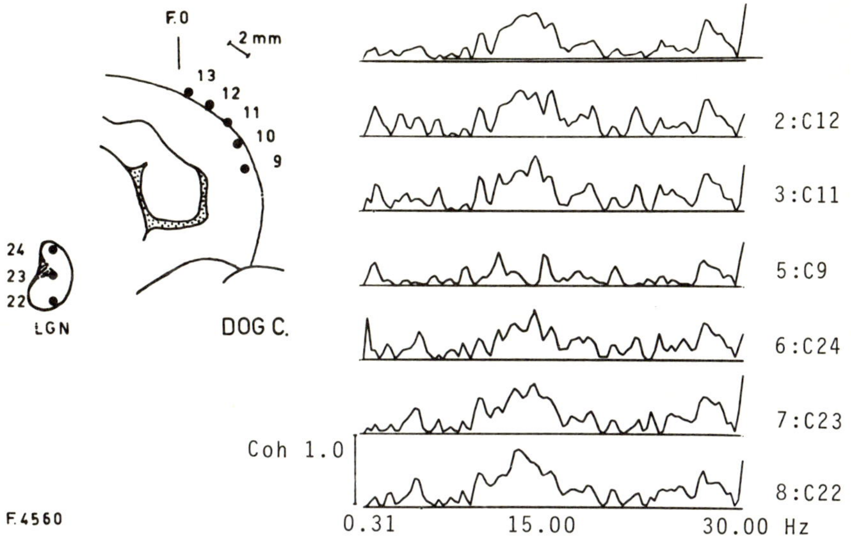

Fig. 2.Ordinary and partial coherences; the EEG signals are indicated by numbers 1... 8. First row:ordinary coherence 1:C13 x 4:C10; other rows: partial coherences where the 3rd signal is as indicated on the right. 1: intracortical EEG signal recorded at C13; 4: intracortical signal at C10; partialisation has been carried out using other intracortical signals (C12,C11, C9,C10) and LGN signals (C24, C23, C22). Note that the coherence between C13 and C10 decreased markedly when the influence of the signal at C9 was eliminated.

(see also Fig. 7) can be seen to occur when the influence of two other cortical signals was eliminated (R13 & R12). From these figures it can also be concluded that the effects of partialisation were not evenly distributed over the whole frequency range investigated (up to 30 Hz). For example Fig. 2 shows that around the alpha peak some frequency components were affected in different ways depending on the signal used for partialisation; in this case the elimination of the influence of the cortical signal C9 affected much more the frequencies around the alpha peak (12-13 Hz) than the neighbouring components at around 10.5 or 14 Hz. In order to evaluate the effects of partialisation the data are summarized in Tables I and II. These results will be described under two headings: (1) thalamic influence on cortico-cortical coherences within the alpha frequency band; (2) influence of different cortical alpha signals upon

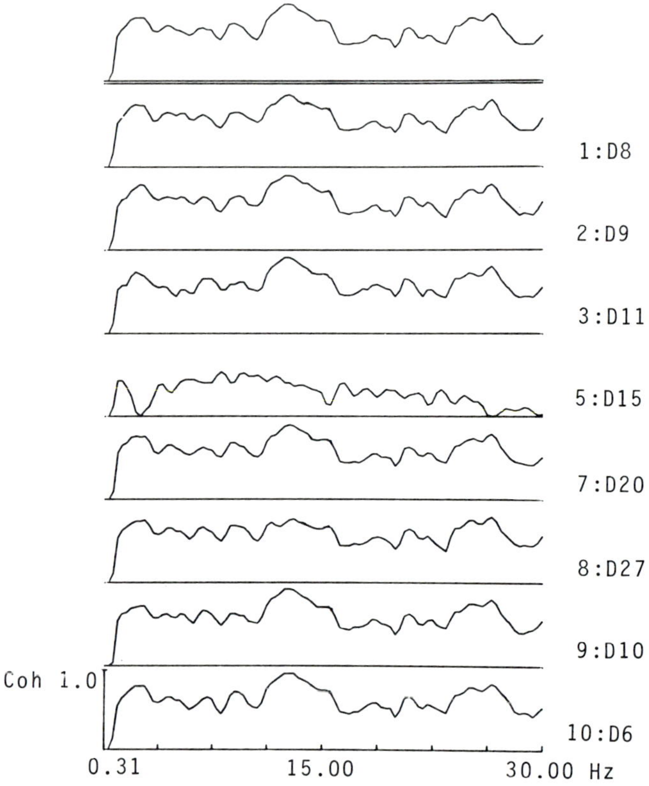

Fig. 3. Ordinary and partial coherences as in Fig.2 but for another animal. First row: ordinary coherence 4:D13 x 6:D17; other rows: partial coherences where the 3rd signal is as indiated on the right. The position of the cortical electrodes (D13,D17,D8,D9,D11,D10) as well as that of the thalamic ones (D20-LGN and D27-pulvinar) is indicated in Fig. 5; D6 is an electrode site on the contralateral occipital cortex about symmetric to D8. Note the change in coherence due to partialisation with the cortical signal D15 and also with the pulvinar D27.

cortico-cortical coherences. Finally we will describe (3) the results obtained by way of analysis of phase spectra because these data are important in order to interpret the partial coherence analysis.

1) Effects of eliminating thalamic components on cortico-cortical coherences in the alpha band

Examination of Table I shows that for most pairs of cortical signals investigated there was a significant ($p < 0.01$) *decrease* of the coherence after

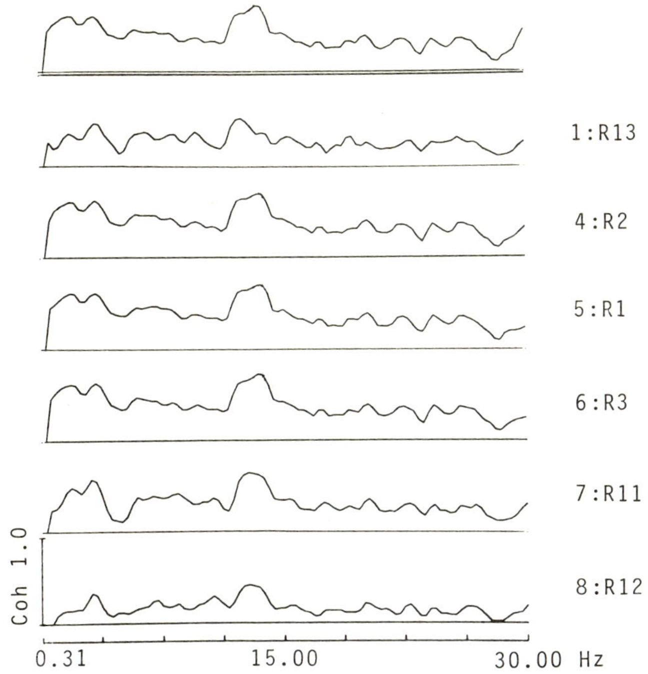

Fig. 4.Ordinary and partial coherences as in Fig. 2 but for another animal. First row: ordinary coherence 2:R14 x 3:R15; other rows: partial coherences where the 3rd signal is as indiated on the right. The position of the cortical electrodes (R14, R15, R13, R11, R12) as well as that of the thalamic ones (R2, R1, R3 all at the LGN) is indicated in Figs. 7 and 8. Note that the largest decrease in coherence is seen on partialisation with R13 and R12.

partialisation indicating that the thalamic signal contributed to the alpha rhythms recorded from both cortical sites; there were, however, two exceptions:

- in the case of the pair D9 & D11 (for placement of the electrodes see Fig. 5) the influence of the LGN (D20) was not found to be significant; it should be noted that of all cases illustrated in Table I this was the case where the largest distance between cortical electrodes was present ($\simeq$ 5 mm); however eliminating the influence of the signal recorded from pulvinar (D27) led to a small, although significant decrease of ordinary coherence between D9 & D11. This indicates that the influence of the LGN site here investigated, upon cortical alpha rhythms spreads over a more restricted area than that of the pulvinar;
- in some cases we found that partialisation led to an *increase* in coherence

TABLE I

SUMMARY OF THE EFFECT ON CORTICO-CORTICAL COHERENCES OF THALAMIC SIGNALS IN THE ALPHA BAND (For electrode placements see Figs. 5, 7)

SIGNALS FROM ELECTRODES	ORDINARY COHERENCES (Coh)			DIFFERENCE (Δz) BETWEEN THE Z-TRANSFORMED $(Coh)^{\frac{1}{2}}$ AFTER PARTIALISATION ON THE SIGNAL RECORDED FROM THALAMIC SITES AND THE Z_{Coh}: $\Delta z = q \cdot sd$	
	FREQUENCY HZ	Coh	Z-TRANSFORMED $(Coh)^{\frac{1}{2}}$ $= Z_{Coh}$	D27 (PULVINAR) q	D20 (LGN) q
D17, D15	12.2	0.94	2.09	-5.5	-1.8
D15, D13	12.2	0.96	2.30	-5.6	-1.2
D8, D9	12.2	0.70	1.22	-2.4	-0.6
D9, D11	12.2	0.57	0.97	-1.2	n.s.
D8, D15	12.2	0.46	0.83	-8.3	-1.8
D8, D17	12.2	0.42	0.78	-7.8	-1.8
D9, D13	12.2	0.29	0.60	-4.6	-1.0
D11, D13	12.5	0.30	0.62	-2.9	+0.8
D8, D6	11.9	0.17	0.42	-2.7	-0.4
				R2 (LGN) q	
R12, R11	12.5	0.64	1.10	n.s.	
R11, R13	12.5	0.46	0.83	n.s.	
R14, R13	12.5	0.86	1.66	-0.7	
R14, R12	12.7	0.80	1.42	-0.4	
R15, R14	12.9	0.72	1.26	-0.4	
R16, R17	12.5	0.30	0.62	-2.0	
R17, R18	12.5	0.38	0.73	-1.0	
R18, R19	12.6	0.40	0.74	+0.4	
R16, R12	12.5	0.50	0.89	-0.4	
R17, R11	12.5	0.48	0.85	-0.4	
				C23 (LGN) q	
C11, C10	12.9	0.28	0.59	+1.0	
C10, C9	13.2	0.81	1.47	-0.9	

NOTE: For cases D and R is the standard deviation of the z-transformed $(Coh)^{\frac{1}{2}}$ sd=0.10; a significant ($p<0.01$) effect of partialisation is considered when $\Delta z>0.372 \cdot sd$; for case C is the standard deviation of $z(Coh)^{\frac{1}{2}}$ sd=0.16; in this case is $\Delta z>0.527 \cdot sd$ at $p<0.01$; n.s.= nonsignificant or $p>0.01$.

instead of the usual *decrease*; this was the case for the pairs D11 & D13 (partialisation on D20), R18 & R19 (partialisation on R2) and C11 & C10 (partialisation on C23). In all these cases the ordinary coherence between one of the cortical signals and the thalamic signal was much larger than that between the latter and the other signal of the cortical pair. Indeed the (ordinary) coherence in the alpha band between D13 and the LGN signal (D20) was 0.4 whereas that between D11 & D20 was only 0.02; that between C10 and the LGN signal (C23) was 0.28 whereas that between C11 & C23 was 0.04; and that between R19 and the LGN signal (R2) was 0.14 whereas that between R18 & R2 was 0.07. This explains why the coherence increased after partialisation; the reason is that the alpha rhythms component common to the thalamic signal and *only one* of the cortical signals (the contribution to the other one being of no practical importance) was removed by the operation of partialisation. This is equivalent to eliminating a noise term as regards the relationship between both cortical signals within the alpha frequency band. Removing of one noise term from one of the signals of the pair leads to a relative increase of the frequency components shared by the signal pair (see also Appendix).

A remarkable feature of the results obtained by partialisation was that the effect of removing the influence of the LGN signals on cortico-cortical coherences was usually significant, although small, in all cases investigated. In contrast the effect of eliminating the influence of the signal from the pulvinar was, in all cases, much larger than that of removing the LGN signal. These findings are summarized in the topographic schemes of Figs. 5, 6, 7 and 8. From here it can be concluded that the extent of the influence of the pulvinar over the cortical alpha rhythms was much more considerable than that of the LGN. The consequence of this for the interpretation of thalamo-cortical relationships in terms of the models of Fig. 1 is discussed below.

The effect of thalamic signals on interhemispheric coherences was investigated only in one experiment (D8, D6 in Table I); D8 is a marginal gyrus site ipsilateral to the pulvinar (D27) and the LGN (D20) sites while D6 is placed on the contra-lateral site, as symmetric as possible. In this case it was possible to demonstrate a significant influence, although small, of both the pulvinar and the LGN on the interhemispheric alpha rhythm coherence.

A last point which should be examined is what was the value of the *remaining* cortico-cortical coherence after partialisation on a thalamic signal. In this case two groups of results may be distinguished: (a) those where the partial coherence remained appreciable and (b) those where it decreased to negligible values. Most cases studied were of the former type; the latter cases

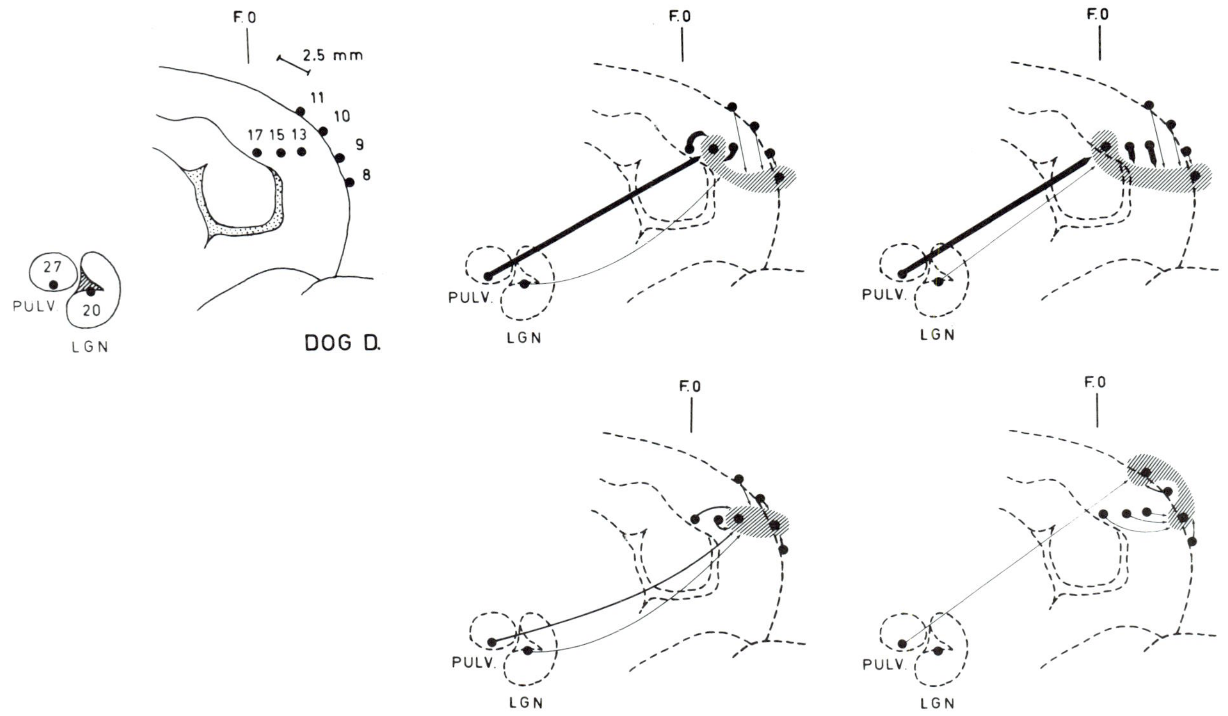

Fig. 5. Position of electrodes in dog D. Each scheme represents the influence of partialisation on the coherence between signals recorded from a pair of electrode sites which are indicated by the shaded area. The width of the lines is directly related to the difference Δz between ordinary and partial coherences; the thinnest lines represent a $\Delta z = 1 \cdot sd$; the thickest represent a $\Delta z = 8 \cdot sd$. Only differences which are significantly different from zero at $p < 0.01$ are shown. Note that the largest decreases in differences are caused by the signals recorded at the mesial cortex and the pulvinar (D27) when the combination investigated includes one mesial cortex site (D8 & D15, D8 & D17 and D9 & D13). The influence of the LGN (D20) is much smaller.

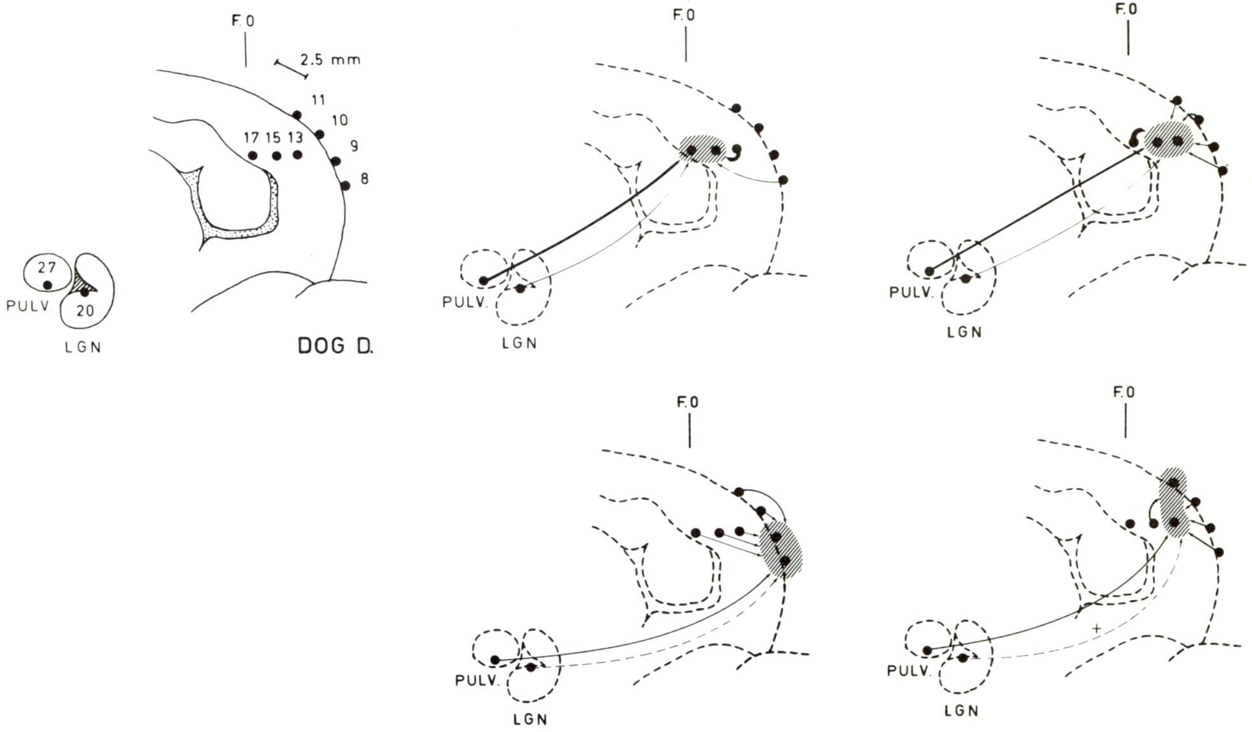

Fig. 6. As in Fig. 5 but for different electrode combinations (same conventions, with 2 additions: a broken line indicates a $\Delta z < 1 \cdot sd$; the sign + indicates that partialisation resulted in an increase in coherence). Note that larger decreases in coherence after partialisation are found for the mesial cortical combinations.

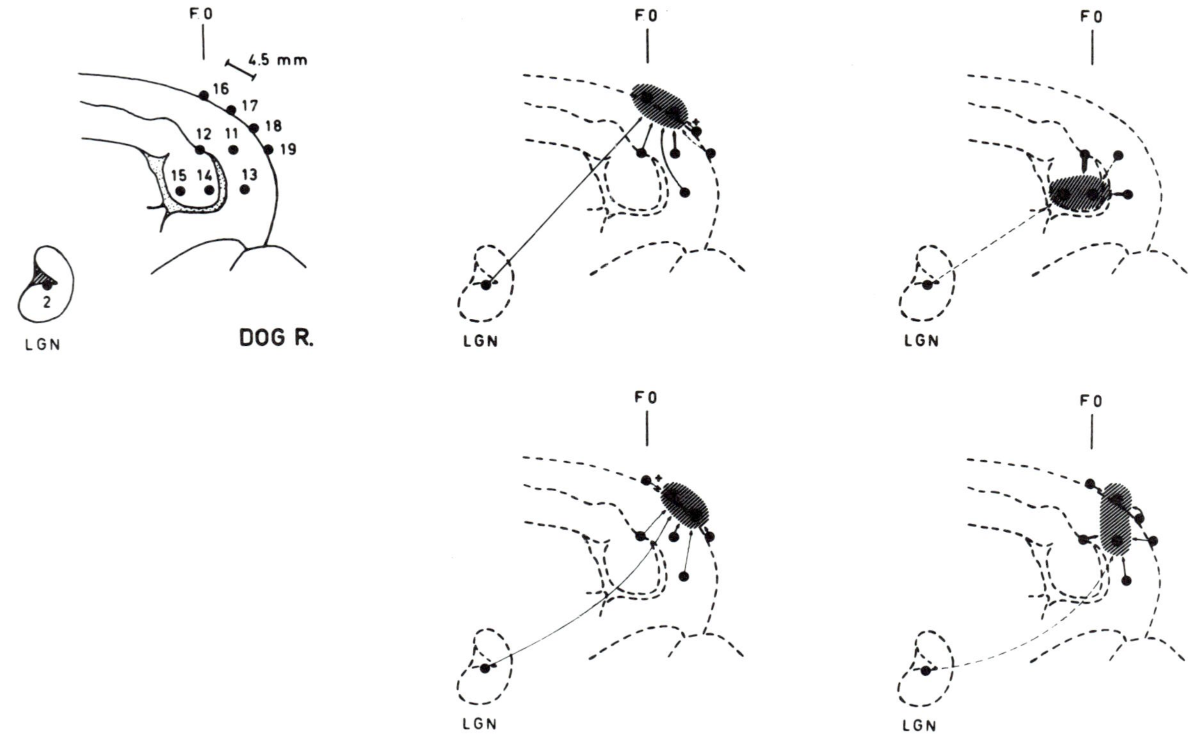

Fig. 7. As in Fig. 5 but for another dog (R). Same conventions. Note that the larger decreases in coherence after partialisation are found for combinations within the mesial cortex; the influence of the LGN (R2) is larger for the combination of electrodes lying more anteriorly (R16 & R17) than for R17 & R18, and slight for the combination R15 & R14.

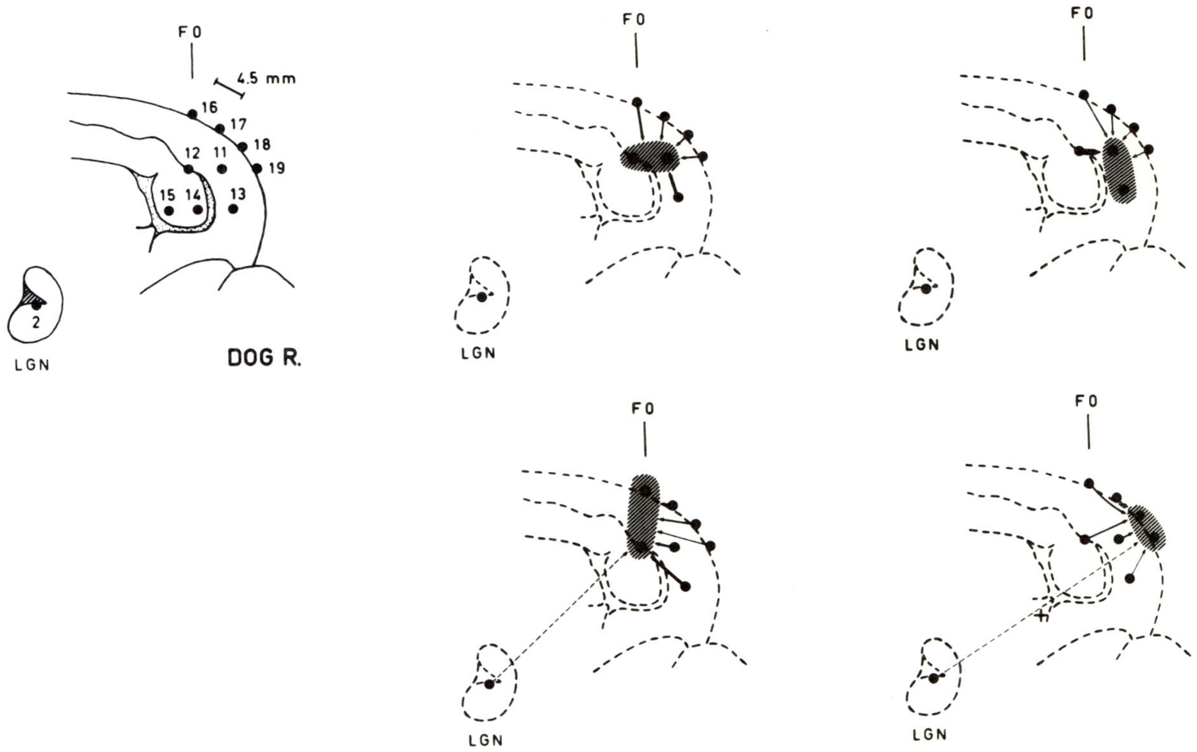

Fig. 8. As in Fig. 7 but for other electrode combinations. Note the same pattern as in Fig. 7; note also that the influence of the LGN (R2) does not reach a significant level for the combinations R12 & R11 and R11 & R13.

were all the combinations including the marginal gyrus and one mesial cortical signal in dog D when the influence of the pulvinar (D27) was eliminated. These findings lead to the following conclusions:

- although a remarkable effect of thalamic signals on the degree of coherence between cortical alpha rhythms could be demonstrated, there must exist other factors which are responsible for the remaining large values of intracortical partial coherences;
- in some cases, however, such as between the alpha rhythms of marginal gyrus and mesial cortex sites in dog D the coherence in the alpha band was determined almost exclusively by the influence of a thalamic signal, that of the pulvinar (D27). This experimental result is shown in Fig. 9; in this figure it can be seen that by eliminating the influence of D27 the coherence between D8 (marginal gyrus) and D15 (mesial cortex) in the alpha band (12-13 Hz) decreased down to negligible values; approximately the same happened by eliminating the influence of the signal recorded from neighbouring sites at the mesial cortex (D13 or D17). However, there was a difference between the partial coherences components at a frequency double that of the alpha band: the effect of partialisation by way of the thalamic signal (D27) was negligible at this frequency band whereas that of partialisation by way of the mesial cortex signals (D13 or D17) was appreciable, mainly that of D17. This dissociation of frequency components by way of partialisation is an indication that different spectral frequency components have different sources and spatial distributions.

2) Effects of eliminating cortical influences on cortico-cortical coherences in the alpha band

We will distinguish 3 particular cortical areas: i) within the mesial cortex; ii) within the marginal gyrus; iii) between mesial cortex and marginal gyrus. The corresponsing data are summarized in Table II and in Figs. 5, 6, 7 and 8.

i) The most salient finding was that within the *mesial cortex* the coherence between any pair of cortical signals was strongly reduced by partialisation using other signals recorded from the same mesial cortex. This indicates that the mesial cortex forms a more or less homogeneous area as regards the degree of coherence between signals in the alpha band. Moreover, the effect of the pulvinar signals on this cortical area is also conspicuous as described above.

ii) The effect of signals recorded from neighbouring sites on coherences of alpha rhythms recorded along the *marginal gyrus* was always significant ($p < 0.01$) although of a lesser degree than within the mesial cortex. In two

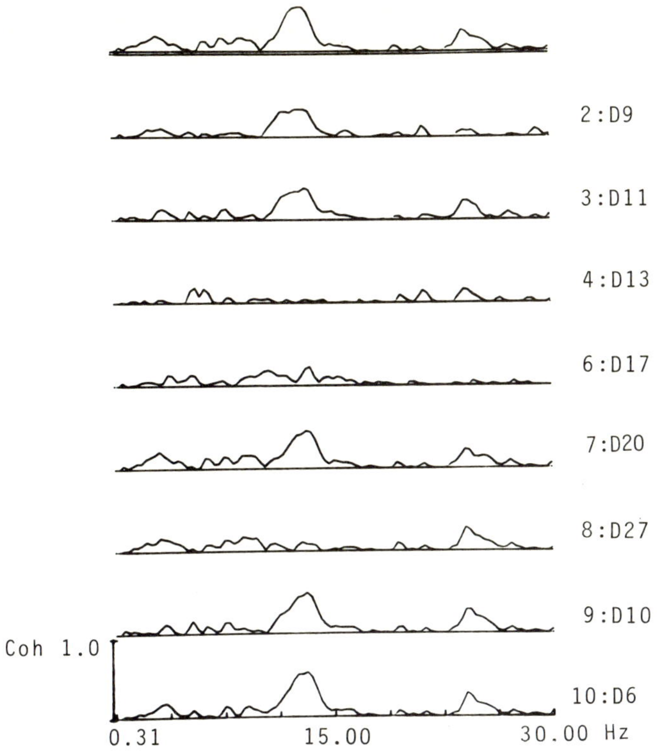

Fig. 9. Ordinary and partial coherences - the position of the electrodes is indicated in Fig. 5. First row: ordinary coherence 1:D8 x 5:D15; other rows: partial coherences where the 3rd signal is as indicated on the right. Note the large decrease in coherence between D8 & D15, resulting from partialisation using the other cortical signals D13, D17 and also D27 (pulvinar). Note that in case of D27 the coherence at the frequency components around 24 Hz remains unchanged.

cases an *increase* in coherence was encountered: R16 & R17, R17 & R18; this was due to the fact that, as in the case of the thalamus presented above, some frequency components in the alpha band were common only to two sites and not to all three.

iii) The analysis of the influence of other cortical signals on the coherences between marginal gyrus and mesial cortex signals within the alpha band evidences the relatively stronger influence of mesial cortex signals than of marginal gyrus signals on those coherences (Table II).

TABLE II

SUMMARY OF CORTICO-CORTICAL COHERENCES WITHIN THE ALPHA FREQUENCY BAND AND THE INFLUENCE OF OTHER CORTICAL SIGNALS UPON THEM (For the placement of electrodes see Figs. 5, 7)

	SIGNALS FROM ELECTRODES	FREQUENCY HZ For Coh and Z_{Coh} see TABLE I	DIFFERENCE (Δz) BETWEEN THE Z-TRANSFORMED $(Coh)^{\frac{1}{2}}$ AFTER PARTIALISATION ON THE SIGNAL RECORDED FROM THE FOLLOWING CORTICAL SITES AND THE Z_{Coh}: $\Delta z = q \cdot sd$; values of q:					
WITHIN MESIAL CORTEX	D17, D15	12.2	D13 -6.2	D8 -1.4	D9 n.s.	D10 n.s.	D11 n.s.	D6 n.s.
	D15, D13	12.2	D17 -7.7	D8 -3.2	D9 -1.0	D10 -2.0	D11 -1.9	D6 -2.0
	R12, R11	12.5	R13 -4.5	R16 -3.1	R17 -2.5	R18 -2.7	R19 -2.3	
	R11, R13	12.5	R12 -7.3	R16 -2.8	R17 -2.3	R18 -1.8	R19 -0.5	
	R14, R13	12.5	R11 -3.7	R12 -5.4	R15 -5.6			
	R14, R12	12.9	R11 -3.2	R13 -5.1	R15 -4.9			
	R15, R14	12.9	R11 -0.7	R12 -6.0	R13 -4.5			
WITHIN MARGINAL GYRUS	D8, D9	12.2	D10 -1.8	D11 -2.5	D17 -1.5	D15 -1.8	D13 -2.0	D6 n.s.
	D9, D11	12.2	D8 -3.2	D17 -1.0	D15 -1.2	D13 -1.4	D6 n.s.	D10 -3.0
	R16, R17	12.5	R18 +1.6	R19 -0.7	R11 -3.1	R12 -2.4	R13 -2.0	
	R17, R18	12.5	R16 +1.2	R19 -2.2	R11 -2.3	R12 -1.4	R13 -1.7	
	R18, R19	12.6	R16 -2.7	R17 -1.3	R11 -2.1	R12 -2.1	R13 -1.3	
BETWEEN MARGINAL GYRUS AND MESIAL CORTEX	D8, D15	12.2	D9 -2.3	D10 -1.2	D11 -1.8	D13 -7.3	D17 -8.3	D6 -0.4
	D8, D17	12.2	D9 -1.9	D10 -1.2	D11 -1.3	D13 -7.8	D15 -7.8	D16 -0.4
	D9, D13	12.2	D8 -4.6	D10 -1.8	D11 -2.3	D15 -4.0	D17 -3.6	D6 -0.6
	D11, D13	12.5	D8 -3.3	D9 -2.5	D10 -1.0	D15 -3.4	D6 n.s.	
	R16, R12	12.5	R17 -1.6	R18 -2.3	R19 -1.3	R11 -3.8	R13 -4.1	
	R17, R11	12.5	R16 -2.0	R18 -2.0	R19 -1.1	R12 -3.1	R13 -2.5	

NOTE: For cases D and R is the standard deviation of the z-transformed $(Coh)^{\frac{1}{2}}$ sd=0.10; a significant ($p<0.01$) effect of partialisation is considered when $\Delta z > 0.372 \cdot sd$.

3) Complementary observations to the analysis of coherences patterns: phase spectra

In case a large coherence between two alpha rhythm signals is present or an appreciable effect of partialisation is found, we may conclude that there are alpha rhythm components which are common to signals recorded from different sites. The question which in these cases must be answered is whether those common frequency components result from a passive process - volume conduction of the electrical potential - or whether they are determined by processes of active propagation of neural activity. In the former case the phase between the different signals should be zero; in the latter a phase which is significantly different from zero should be encountered. It is therefore of importance to examine which was the phase angle for those combinations of signals which presented large coherences (> 0.75) and the effect of partialisation was considerable (i.e. partial coherence decreased to a negligible value, < 0.10). In this respect two aspects should be distinguished: (a) the phase shifts over the mesial cortex where over relatively large areas (over ∿ 6 mm) common frequency components within the alpha band are found, showed the same pattern as previously described[6] for the marginal gyrus: thus the phase shift of the alpha activity (12.9 Hz) between R13 & R14 (see Fig. 7) was +21.7°± 2°, although between R14 & R15 it was not significantly different from zero. For the case D17, D15, D13 (see Fig. 5) we found that the phase shift (12.2 Hz) between D17 & D15 was +9.3° ± 3° and that between D15 & D13 was -12.7° ± 1°; (b) as regards the thalamo-cortical phase relationships we found that the pulvinar signal which was responsible for a large decrease in intra-cortical coherences after partialisation presented a phase difference of -101.5° ± 5° in relation to D17 and -68° ± 7° in relation to D15; translated into the time domain these phase values correspond to, respectively, 21.1 ± 2.1 ms and 17.6 ± 2 ms. It is interesting to relate these findings to the phase angle found between the LGN (C23) and the cortical site where the largest coherence (0.56 at 12.1 Hz) was encountered: this was the combination C23, C10 with the last electrode situated in the marginal gyrus: the time delay computed in this case was 13 ± 4 ms.

DISCUSSION

The results of partial coherence analysis obtained by eliminating the influence of thalamic signals within the alpha frequency band, upon cortical areas, are clear in showing that a model based on a point-to-point projection (model I of Fig. 1) cannot be accepted for explaining the organization of thalamo-cortical alpha rhythms. Therefore we conclude that one of the versions of model II (a or b, see Fig. 1) should be preferred. Of course, we have first to eliminate the possibility that the large effects of partialisation might be due to spread of electrical potentials in the brain due to the passive process of volume conduction. We have demonstrated here that such a possibility can be excluded since there are significant phase shifts between signals recorded from different cortical and thalamic sites. In this latter respect the present results confirm and extend those previously published[6]; they allow us to formulate the conclusion that as regards the generation of cortical alpha rhythms in dog, there are, over the *marginal gyrus*, different *epicenters* of alpha activity from which this neural activity spreads in several directions just like water springing from small fountains. The present results allow the extension of this formulation to the alpha rhythms arising in the mesial cortex.

The time delays measured between thalamic sites - pulvinar and LGN - and cortex are reported here for the first time; they correspond most certainly to the transmission of neural activity in the thalamo-cortical pathways. Indeed these time delays have a similar order of magnitude as those found for visual evoked potentials[12]. The difference between pulvinar- and LGN-cortical time delays indicates that as regards the generation of alpha rhyhtms the former corresponds to slower pathways.

Another difference between pulvinar- and LGN-cortical relations was the much larger and more widespread influence of the former on cortico-cortical coherences within the alpha band. In particular the pulvinar's influence was the most pronounced upon the mesial cortex neighbouring the marginal gyrus. This area corresponds in general terms to the area of projection of pulvinar afferents (mainly area 18) as studied in detail in other species than the dog[10,11]. Histological analysis of the afferents of area 18 in squirrel monkeys has led to the conclusion that there is in this area an "overlapping or interweaving of thalamic and cortical afferents"[12]; whether this is also the case for dog is, of course, still in question. However, such a pattern of extensive overlapping of pulvinar-cortical afferents with intracortical connections would provide the structural basis which would explain the results obtained in the present investigation using partial coherence analysis.

Nothwithstanding the effects of thalamic signals on cortico-cortical coherences it should be emphasized that the residual coherences measured be-between signals recorded over the mesial cortex were still very large (Table I and II). This implies that there must be other factors which are responsible for the large intracortical coherence. This is also the reason to reject model I (Fig. 1). The question which should be asked then is which of the two versions of model II (a, b, Fig. 1), should be preferred. Of course, it is possible that there are thalamo-cortical afferents involved in the generation of alpha rhythms which bifurcate and project to patches of cortex[12]; this would be compatible with model IIb. However, it is unlikely that such a system of bifurcation would be able to account for the large cortical areas over which coherent activity in the alpha band was encountered. Therefore, we consider more likely that the alternative model IIa may provide the substrate for explaining the present results. There are several arguments to support this hypothesis: the most important one is the fact that a surface-parallel intracortical system of fibres has been described which may extend over a distance of 6-8 mm together with a pyramidal collateral spread module which extends over about 3 mm[13,14]; another one is that the terminals of subcortical projections of different origins end up in overlapping cortical modules so that distinct although neighbouring cortical modules share common inputs. Moreover, it should be noted that the most extensive ordinary coherence and also the largest effects of partialisation were found on the basis of signals recorded from the cortical *surface*. Signals recorded from *within* the cortex showed in general a less extensive area over which partialisation was significant and smaller coherences (cf. C11, C10, Table I) than signals recorded from the cortical surface from electrodes placed at about the same distance. This leads to the conclusion that model IIa is the most likely in order to account for the present results regarding the generation and the spread of alpha rhythms. The basic structural process underlying model IIa would then be the surface-parallel intracortical fibre system[13]. It appears to be an attractive hypothesis that such a system of surface intracortical connections would play a crucial role in the spread of alpha rhythmic activity. Indeed, if such a system did not exist we would be left with a patchy nonoverlapping assembly of cortical modules with their own type of EEG activity with little intracortical coherence; it would be almost impossible in such a case to record an EEG at a relatively large distance from the cortex. This does not imply that to explain cortical *function* the organization principle should be sought in extensive cortico-cortical connections; the most important aspect for cortical function appears to be the basic unit of operation: the cortical *column* or *module* as recently reviewed by Edelman and Mountcastle[15]. The existence of coherent alpha rhythms over large cortical areas should rather be seen as sign that this electrophysiological phenomenon bears no simple relationship to the basic functional units of the cortex; the alpha rhythm would then be a natural consequence of the structural pattern which subserves the communication between modules. Whether the nonlinear filter properties responsible for alpha rhythm generation[16] are contained within a cortical module or not is at present still an open question. We should add that the methodology used here is only applicable to linear systems, so that our conclusions apply only to the linear domain.

The present results have been obtained by way of a method which we may call a "theoretical thalamic deafferentation"; the advantages offered by this signal-analytical methodology as regards a real experimental deafferentation are certainly of importance, since in this way we can "disconnect" brain areas one from the other without changing, by introducing lesions, the system which we want to study.

SUMMARY

The thalamocortical relationships of the alpha-rhythms have been further studied in dog, using a new methodology: a partial coherence function analysis. The aim of this investigation was to clarify whether the large coherence between alpha-rhythms recorded from different cortical sites (intra- and inter-hemispheric) could still be dependent on a common thalamic site. This type of question could not be solved until recently by means of current signal analysis methods such as coherence and cross correlation functions. However, the development of a computer program, which enables the computation of partial coherence functions, opens up new possibilities. By applying this program to the analysis of alpha-rhythms recorded simultaneously from different cortical and thalamic sites it has been possible to demonstrate that in all cases studied so far the alpha-rhythm of the lateral geniculate nucleus influences only in a moderate way the coherence between different cortical alpha-rhythms. However, we found that in several cases the intracortical coherence decreases in a significant way, when the partial coherence is measured using as third signal the alpha-rhythm recorded from the lateral dorsal thalamus, namely the pulvinar nucleus. This finding leads to new interpretations of the influence of the thalamus on different cortical generators of alpha-activity.

The analysis of the data based on partial coherences together with that of phase spectra, by means of which time delays between thalamus and cortex and/or between different cortical areas were determined, leads to a general model of the generation and the spread of alpha-rhythm in the brain.

APPENDIX

We consider here a theoretical case such as model IIa of Fig. 1. We shall investigate the behaviour of the partial coherence functions depending on the relative strength of different signal components. We assume that there are 3 stochastic signals $\underline{y}_1$, $\underline{y}_2$ and $\underline{y}_3$, corresponding to C_1, C_2 and th_1 of Fig. 1 respectively:

$$\begin{aligned} \underline{y}_1 &= \underline{n}_1 + d_{11}\underline{e}_1 + d_{13}\underline{e}_3 \\ \underline{y}_2 &= \underline{n}_2 + d_{21}\underline{e}_1 + d_{23}\underline{e}_3 \\ \underline{y}_3 &= \underline{n}_3 + d_{33}\underline{e}_3 \end{aligned} \qquad \text{(A1)}$$

Intracortical component C_o of Fig. 1 is represented here by $\underline{e}_1$; $\underline{e}_3$ is a component generated in the thalamus; $\underline{n}_i$ (i=1, 2 or 3) represents local noise components. The coefficients d_{11}, d_{13}, d_{21} and d_{33} represent weighing factors. Furthermore we assume without loss of generality that:

$$\text{var}(\underline{n}_1) = \text{var}(\underline{n}_2) = \text{var}(\underline{n}_3) = 1$$

$$\text{var}(\underline{e}_1) = \text{var}(\underline{e}_2) = \text{var}(\underline{e}_3) = 1$$

$$\left.\begin{aligned} \text{cov}(\underline{n}_i\underline{n}_j) &= 0 \\ \text{cov}(\underline{e}_i\underline{e}_j) &= 0 \\ \text{cov}(\underline{n}_i\underline{e}_j) &= 0 \end{aligned}\right\} \quad \begin{matrix} i, j = 1, 2, 3 \\ i \neq j \end{matrix}$$

The variables $\underline{n}_i$ and $\underline{e}_i$ can be considered as spectral components for a particular frequency. We assume that the correlations can be interpreted as the square root of the coherence. For simplicity we will use in the following considerations correlations instead of coherences. We define the variances of $\underline{y}_1$, $\underline{y}_2$ and $\underline{y}_3$ as follows:

$$\begin{aligned} \phi_{11} &= \text{var}(\underline{y}_1) = 1+d_{11}^2+d_{13}^2 \\ \phi_{22} &= 1+d_{21}^2+d_{23}^2 \\ \phi_{33} &= 1+d_{33}^2 \end{aligned} \qquad \text{(A2)}$$

and $$\phi_i = \sqrt{\phi_{ii}}$$

The covariances between the 3 signals are defined as follows:

$$\phi_{12} = \mathrm{cov}(\underline{y}_1\underline{y}_2) = d_{11}d_{21}+d_{13}d_{23}$$
$$\phi_{13} = d_{13}d_{33} \qquad \text{(A3)}$$
$$\phi_{23} = d_{23}d_{33}$$

Thus the normalized correlation, ψ_{12}, between $\underline{y}_1$ and $\underline{y}_2$, (the equivalent to the square root of the coherence function) is given by:

$$\psi_{12} = \frac{\phi_{12}}{\phi_1\phi_2} = \frac{d_{11}d_{21}+d_{13}d_{23}}{(1+d_{11}^2+d_{13}^2)^{\frac{1}{2}}(1+d_{21}^2+d_{23}^2)^{\frac{1}{2}}} \qquad \text{(A4)}$$

and similarly for ψ_{13} and ψ_{23}.

The partial correlation resulting from partialisation on signal $\underline{y}_3$ is as follows (see e.g. equation 27.5 of ref. 20):

$$\psi_{12,3} = \frac{\psi_{12}-\psi_{13}\psi_{23}}{(1-\psi_{13}^2)^{\frac{1}{2}}(1-\psi_{23}^2)^{\frac{1}{2}}} = \frac{\phi_{12}\phi_{33}-\phi_{13}\phi_{23}}{(\phi_{11}\phi_{33}-\phi_{13}^2)^{\frac{1}{2}}(\phi_{22}\phi_{33}-\phi_{23}^2)^{\frac{1}{2}}} \qquad \text{(A5)}$$

Substituting ϕ by the expression given in (A2) and (A3), we can write

$$\psi_{12,3} = \frac{d_{11}d_{21}+d_{13}d_{23}-d_{13}d_{23}[d_{33}^2/(1+d_{33}^2)]}{\{1+d_{11}^2+d_{13}^2-d_{13}^2\ [d_{33}^2/(1+d_{33}^2)]\}^{\frac{1}{2}}\ \{1+d_{21}^2+d_{23}^2-d_{23}^2[d_{33}^2/(1+d_{33}^2)]\}^{\frac{1}{2}}} \qquad \text{(A5)}$$

We can simplify (A5) by substituting $\lambda^2 = 1/(1+d_{33}^2)$;

$$\psi_{12,3} = \frac{d_{11}d_{21}+\lambda^2 d_{13}d_{23}}{(1+d_{11}^2+\lambda^2 d_{13}^2)^{\frac{1}{2}}(1+d_{21}^2+\lambda^2 d_{23}^2)^{\frac{1}{2}}} \qquad \text{(A6)}$$

The question is to compare ψ_{12} (expression A4) with $\psi_{12,3}$ (expression (A6) i.e., the change in correlation between $\underline{y}_1$ and $\underline{y}_2$ depending on $\underline{y}_3$ or, in other words the effect of partialisation of $\underline{y}_3$. We have done this by making a set of numerical examples summarized in Table III.

TABLE III

	$d_{11}=2$, $d_{21}=1$, $d_{33}=1$		$d_{11}=2$, $d_{21}=1$, $d_{33}=10$
	Correlation $\underline{y}_1\underline{y}_2$ ψ_{12}	Partial correlation $\underline{y}_1\underline{y}_2$ depending on $\underline{y}_3$ $\psi_{12,3}$	idem $\psi_{12,3}$
(a) d_{13}, $d_{23}=0$	0.632	0.632	0.632
(b) $d_{13}=2$ $d_{23}=0$	0.471	0.535	0.630
(c) $d_{13}=2$ $d_{23}=2$	0.816	0.756	0.636
(d) $d_{13}=10$ $d_{23}=10$	0.986	0.972	0.707

In case (a) it can be seen that if signal $\underline{y}_3$ does not influence either of the two others (i.e. d_{13} and $d_{23}=0$) the effect of partialisation, depending on $\underline{y}_3$ is, of course, nil. In case (b) we have an example of an asymmetric condition where $\underline{y}_3$ only influences one of the other 2 signals, in this case only $\underline{y}_1$. In such a case the partial correlation is *larger* than the ordinary correlation as it is also the case in the experimental findings for the pairs D11, D13 (partialisation on D20), R18, R19 (partialisation on R2) and C11, C10 (partialisation on C23). In all these cases the ordinary correlation between the cortical signals and the thalamus was much larger for one of the two electrode sites. In these cases partialisation is equivalent to the elimination of a noise term as regards the relationship between both cortical signals.

In cases (c) and (d) we can see the most common finding, i.e. that partialisation leads to a *decrease* of the correlation between the remaining signals. The ordinary and partial correlations are, of course larger for that case where the component $\underline{e}_3$ or $\underline{y}_3$ has a larger coefficient on both $\underline{y}_1$ and $\underline{y}_2$. The last column of table III shows that all effects mentioned occur much more clearly if d_{33} is larger, that is if a larger part of $\underline{y}_3$ is able to influence $\underline{y}_1$ and $\underline{y}_2$.

It should be mentioned, however, that for more complex models it may not be so easy to interpret the effect of partialisation as in the cases described above. This is particularly the case whenever between 3 signals there are feedback loops involving all three.

REFERENCES

1. Lanoir, J. et Cordeau, J.P. (1970) Rhythmes et fuseaux spontanés de l'aire visuelle du chat au cours des différents stades de la veille et du sommeil. J. Physiol. (Pari 62, suppl. 3, 399-400.

2. Lanoir, J. (1972) Etude électrocorticographique de la veille et du sommeil chez le chat. Organisation du cycle nycthéméral. Rôle du thalamus. Thèse Doct. Sci. Marseille. Centre régional de Recherche et de Documentation pédagogique, Marseille, 520 pp.

3. Rougeul, A., Corvisier, J. et Letalle, A. (1974) Rhythmes électrocorticaux caractéristiques de l'installation du sommeil naturel chez le chat. Leurs rapports avec le comportement moteur. Electroenceph. clin. Neurophysiol., 37, 41-57.

4. Storm van Leeuwen, W., Kamp, A., Kok, M.L., de Quartel, F.W., Lopes da Silva, F.H. and Tielen, A.M. (1967) Rélations entre les activités cérébrales du chien, son comportement et sa direction d'attention. Actuel Neurophysiol. (Paris), 7, 167-186.

5. Lopes da Silva, F.H., van Lierop, T.H.M.T., Schrijver, C.F. and Storm van Leeuwen, W. (1973) Organization of thalamic and cortical alpha rhythms: spectra and coherences. Electroenceph. clin. Neurophysiol., 35, 627-639.

6. Lopes da Silva, F.H. and Storm van Leeuwen, W. (1978) The cortical alpha rhythm in dog: depth and surface profile of phase. In: Architectonics of the Cerebral Cortex (Eds. M.A.B. Brazier and H. Petsche), IBRO Monograph series, Vol. 3, Raven Press, New York, pp. 319-333.

7. Gersch, W. and Goddard, G.V. (1969) Epileptic focus location: spectral analysis method. Science, 970, 701-702.

8. Vos, J.E., Lammertsma, A.A. and van Eykeren, L.A. (1977) Ordinary and partial coherences of bipolar and quasi-unipolar derivations of infant electroencephalograms. In: Random Signal Analysis, IEE Conference Publication nr. 159, London, pp. 154-160.

9. Rappelsberger, P., Petsche, H., Vollmer, R. and Lapins, R. (1979) Rhythmicity in seizure patterns - Intracortical aspects. In: Origin of cerebral field potentials (Eds. E.J. Speckmann and H. Caspers), G. Thieme Publications, Stuttgart, pp. 80-97.

10. Dixon, W.J. and Massey, F.J. (1957) Introduction to statistical analysis, McGraw Hill, New York, 488 pp.

11. Storm van Leeuwen, W. and Kamp, A. (1969) Radiotelemetry of EEG and other biological variables in man and dog. Proc. Royal Soc. Med., 62, 451-435.

12. Lopes da Silva, F.H., van Rotterdam, A., Storm van Leeuwen, W. and Tielen, A.M. (1970) Dynamic characteristics of visual evoked potentials in the dog. I. Cortical and subcortical potentials evoked by sine wave modulated light. Electroenceph. clin. Neurophysiol., 29, 246-259.

13. Garcy, L.J. and Powell, T.P.S. (1967) The projection of the lateral geniculate nucleus upon the cortex in the cat. Proc. Roy. Soc. Lond. (Biol.), 167, 107-126.

14. Ogren, M.P. (1977) Evidence for a projection from Pulvinar to striate cortex in the squirrel monkey (saimiri sciureus). Exp. Neurol., 54, 622-625.

15. Curcio, C.A. and Harting, J.K. (1978) Organization of pulvinar afferents to area 18 in the squirrel minkey" evidence of stripes. Brain Res., 143, 147-1.

16. Szentágothai, J. (1978) The local neuronal apparatus of the cerebral cortex. In: Cerebral correlates of conscious experience. INSERM, Symp. no. 6 (Eds. P. Buser and A. Rouguel-Buser), Elsevier, Amsterdam, pp. 131-138.

17. Creutzfeldt, O.D. (1978) The neocortical link: thoughts on the generality of structure and function of the neocortex. In: Architectonics of the Cerebral Cortex (Eds. M.A.B. Brazier and H. Petsche), IBRO Monograph Series, Vol. 3, Raven Press, New York, pp. 357-384.

18. Edelman, G.M. and Mountcastle, V.B. (1978) The mindful brain, The MIT Press, Cambridge, Mass., 100 pp.

19. Lopes da Silva, F.H., van Rotterdam, A., Barts, P., van Heusden, E., and Burr, W. (1976) Models of neuronal populations: the basic mechanisms of rhythmicity. Progr. in Brain Res., 45, 281-308.

20. Kendall, M.G. and Stuart, A. (1951) The advanced theory of statistics, Vol. 2, Griffin & Co., London, 690 pp.

EEG RHYTHMS IN BEHAVIOURAL SITUATIONS

Rhythmic EEG Activities and Cortical Functioning
G. Pfurtscheller et al., editors

A THALAMO-CORTICAL RHYTHMIC SYSTEM ACCOMPANYING HIGH VIGILANCE LEVELS IN THE CAT

JEAN-JACQUES BOUYER, MARIE-FRANCOISE MONTARON, ARLETTE ROUGEUL-BUSER and PIERRE BUSER
Laboratoire de Neurophysiologie comparée, Université Pierre et Marie Curie, 4 place Jussieu, Paris, 75230 Cêdex 05 (France)

INTRODUCTION

In the conventional human EEG, the most commonly identified "spontaneous" rhythms, alpha and mu, are by necessity rather poorly localized and thus tend to provide a fairly holistic view of the brain mechanisms that they are presumed to underlie. It is more surprising that animal studies have seldom been oriented toward the localisation of spontaneous electrocortical (ECoG) activities, despite the possibilities of topographical mapping that these studies may offer. One of the reasons probably lies in the "digital" distinction that is currently made between rhythmic "spindles" accompanying slow sleep and the desynchronized state characterizing wakefulness. When going through the literature, three partially contradictory beliefs are frequently encountered, regarding spindles: (i) that they are indices of sleep as well as drowsiness; (ii) that they are widely distributed over the neocortex; (iii) that they may be considered homologues of the human alpha rhythm.

Only a few authors, including our group, have tried to challenge these points of view by showing that (i) rhythmic activities quite distinct in shape and distribution from sleep spindles may develop over limited cortical zones; (ii) some of these rhythms may accompany specific behavioural situations not necessarily belonging to the family of "sleep states"; (iii) some of these rhythms may be homologues of well-identified activities recorded from the human scalp[1-6].

In our studies over the last 10 years, we have more precisely tried to concentrate on the types of localized rhythms that develop on the sensorimotor cortex and in its vicinity during various behavioural states (in cat and monkey) while the animal is immobile[7]. In the present report, we shall particularly deal with the subset of these rhythms that accompanies an intense level of oriented attentiveness in the cat, and that displays frequencies in the 35 - 45 Hz range. This subset will be designated herein as "fast somato-parietal rhythms".

MATERIALS AND METHODS

To describe briefly the salient aspects of our techniques, cats were implanted with an array of 20 electrodes placed epidurally over the somatosensory (SI), the somatomotor (MI) and the anterior parietal (i.e. anterior suprasylvian, SSA) areas, and a reference electrode was fixed over the frontal sinus. After recovery, the animal was placed, for recording, in a limited (2 m^2), dimly lit, low noise (but not sound-proof) space where it could move. In a preliminary recording session, all electrodes were checked and the ones that displayed the largest fast frequency rhythms (35 - 45 Hz) were selected. From then on, all further records from the successive sessions were taken using the same montage. All derivations were "monopolar", i.e. against the frontal reference electrode.

The animal was either left in the recording chamber with no particular stimulus ("neutral" conditions), or in other cases a mouse was also placed in the room, protected from the cat by a perspex box; the cat could see it, but could not catch it.

The observation time was in the order of 90 min as a rule. Sixteen ECoG channels were recorded, and the animal's behaviour was controlled with a closed TV circuit during that period. ECoG data were stored on tape and further processes on a programmable computer. To perform spectral analysis using the FFT algorithm, 10 successive 5 sec epochs were analyzed with a sampling frequency of 102.4 Hz and a resolution Δf of 0.2 Hz. The individual spectra (μV^2) as a function of F, with F between 0 and 51.2 Hz) were then added and each of the resulting curves thus corresponded to about 1 min of recording. Spectral data were then displayed using the now widely adopted methods of evolutive spectra, presenting the 90 min recording in a condensed form. Finally, coherence functions were also calculated to allow quantitative comparison between two different channels by means of the classical formula (and a resolution of 0,2 Hz).

Thalamic and brainstem lesions were performed electrolytically using either a DC or a high frequency (45 kHz) current.

Two drugs were used in the present study: DOPA was injected IM at a dose of 30 mg/kg; 8 mg/kg apomorphine were administered to normal animals and 4 mg/kg to others after brainstem lesions.

Upon completion of the experimental period, animals were killed with an overdose of anesthetic and perfused with formalin. All depth recording sites and lesions were localized on frozen-sectioned, Nissl-stained 100 μm histological sections.

RESULTS

1. Patterns of somato-parietal rhythms in the normal cat. In the normal, awake cat, three types of rhythms can be recorded from the cortical areas MI, SI and SSA (see Methods); all three types develop only under complete immobility of the animal (except for eye movements): (i) fast rhythms (35- 45 Hz), (ii) rhythms in the middle frequency range (12 - 18 Hz) and (iii) slow rhythms (6 - 8 Hz). Each type displays a particular localization in a "mini-zone" of the anterior convexity (see below). Also worth noting is the fact that each rhythm develops while the animal's behaviour corresponds to a particular state of vigilance: intense and oriented alertness for the fast type (i), quiet wakefulness with no overt sign of attentiveness for type (ii), and drowsiness for type (iii). Moreover, such rhythms, especially those of types (i) and (ii) can develop while the rest of the cortex is fully "desynchronized". Finally, in view of their localization and conditions of occurrence (i.e. immobility), all three types may be regarded as homologues of the human mu or "wicket" rhythm described by several authors[8-12]. More precisely, the fast type of rhythms may be equivalent to the human beta rhythm as first observed in ECoG by Jasper and Penfield[13]. At any rate, these activities are quite distinct from the alpha rhythm, located more posteriorly and developing in very different conditions.

To help interpret the following study with the accompanying evolutive spectra, let us briefly consider the different spectral patterns most commonly encountered when recording is made from one area or another of the anterior cortex (MI, SI, SSA), and correlations are established with the animal's behaviour (Fig. 1).

Six standard features of the power spectrum (summed over 1 min, as explained above) can thus be distinguished. In A, the animal is awake, but displays a considerable amount of movement. Therefore, the only dominant peak is in the lowest frequency range (0 - 3 Hz), corresponding to movement artifacts. In B, the animal is intensely awake but motionless and attentive; in this case, a spectral peak develops in the 35 - 45 Hz range if recording is made from MI and SSA. In C, the animal is still awake, but sitting quietly and this time paying little attention to the environment; concomitantly, rhythms in the middle range (12- 18 Hz) now predominate, but almost exclusively on SI. In D, the animal clearly indicates drowsiness, and a new peak appears in the slowest range of somatic (SI) rhythms (5 - 8 Hz) in addition to the middle range-ones that usually persist. In E, typical slow sleep develops, with spindles (sharp peak at 14 Hz) accompanied by delta (2 - 5 Hz) activity. Spindles are most prominent on MI. Lastly, in F, no rhythms develop at any frequency in any anterior derivation while the animal is immobile; such a pattern indicates paradoxical sleep.

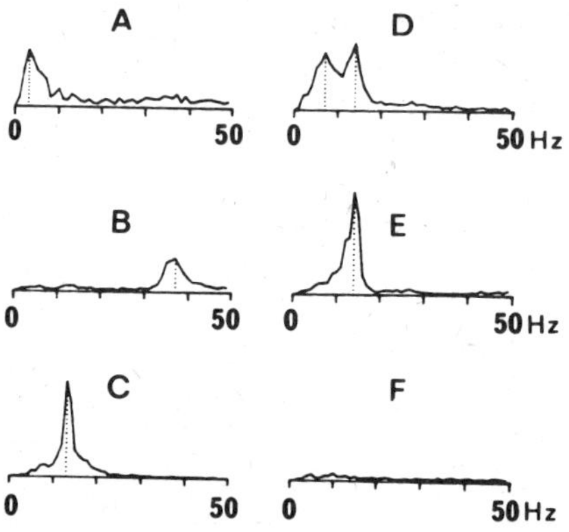

Fig. 1. Typical patterns of power spectra computed from derivations on the anterior cortex, and corresponding to various behavioural states.
Abscissae: frequency in Hz; ordinates: energy in μV^2.
Each curve is the sum of 10 successive spectra taken during the same minute, each one over a 5 sec epoch (sampling frequency: 102.4 Hz; analysis performed between 0 and 51.2 Hz; resolution: 2 Hz).

A. Dominance of very low frequencies (0 - 3 Hz) indicates movement artifacts. No ECoG rhythms are visible, the animal being constantly moving.

B. Fast somato-parietal rhythms, with peak at 35 - 45 Hz (PC or PA derivations on Fig. 3).

C. Rhythms of "quiet vigilance" (peak at 14 Hz) on SI (Fig. 3).

D. Animal in drowsy state. Coexistence of activities at 14 Hz and at 6 Hz in SI.

E. Slow sleep. Spindle peak at 14 Hz accompanied by delta activity on left side of peak (derivation on MI).

F. Paradoxical sleep. No rhythm can be noticed in any range, but the animal is immobile (absence of type A peak).

2. Fast somato-parietal rhythms: neurophysiological exploration. We shall now concentrate on the fastest rhythms (i above) that develop in conditions of intense alertness (Fig. 2). Some remarkable features of these activities will now be considered in more detail.

(i) As already mentioned, they are concentrated in two foci: one is located around the cruciate sulcus, presumably in area MI (pericruciate, PC focus), and the second, on the anterior aspect of the suprasylvian cortex (SSA) in the vicinity of the ansate sulcus (periansate, PA focus) (Fig. 3). Contrasting with this distribution, the middle range (12- 18 Hz) and drowsiness rhythms (6 - 8 Hz) are only present in the SI forearm area. Fig. 3 also indicates that

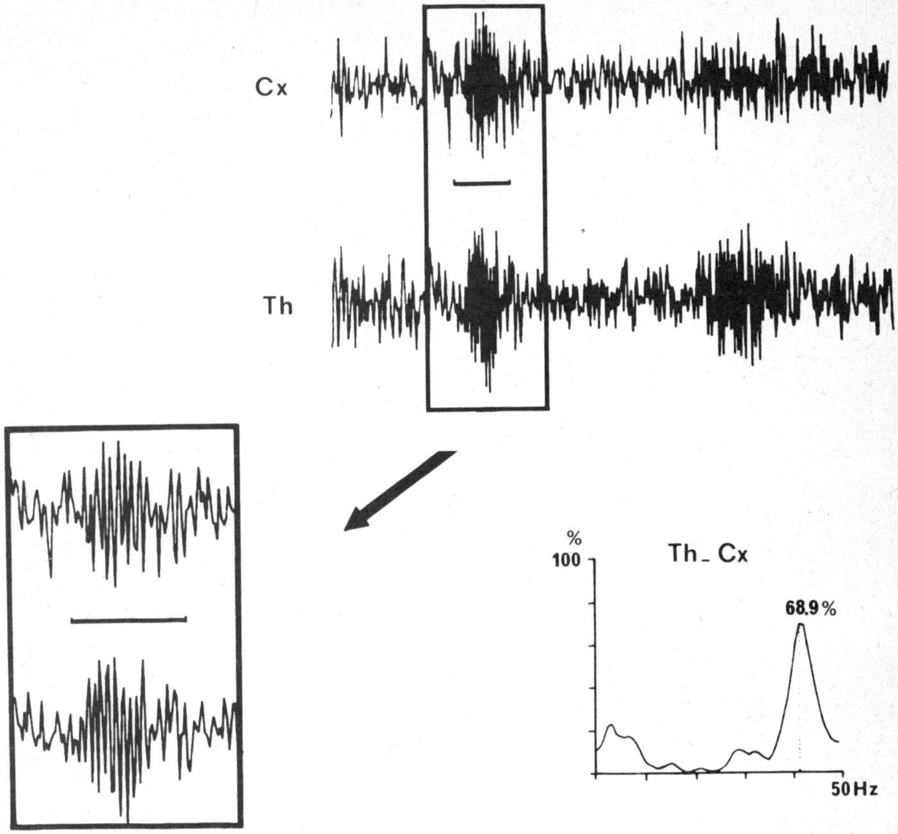

Fig. 2.Fast hypervigilance rhythms simultaneously recorded from the anterior suprasylvian cortex (Cx) and from the posterior group of the thalamus (Th). The close correspondence between the two activities is illustrated: (i) by the lower left record which is an enlargement of the upper two traces in the frame, at twice the recording speed (calibration mark, 500 msec); (ii) by the coherence analysis between cortex and thalamus, performed over 40 records of 1.25 sec each (sampling frequency 102.4 Hz; analysis performed between 0 and 51.2 Hz; resolution: 0.8 Hz). Notice high coherence values in the frequency band 35 - 45 Hz.

typical sleep spindles (sharp peak at 14 Hz) are well developed only in the anterior PC focus.

(ii) Their thalamic origin has now been identified. A site was discovered in the thalamus, that belongs to the "posterior group"[7] (Fig. 4) and displays activities occurring simultaneously and at the same frequency as the fast cortical rhythms (Fig. 2), with a high level of the coherence function calculated between the two loci, thalamic and cortical parietal.

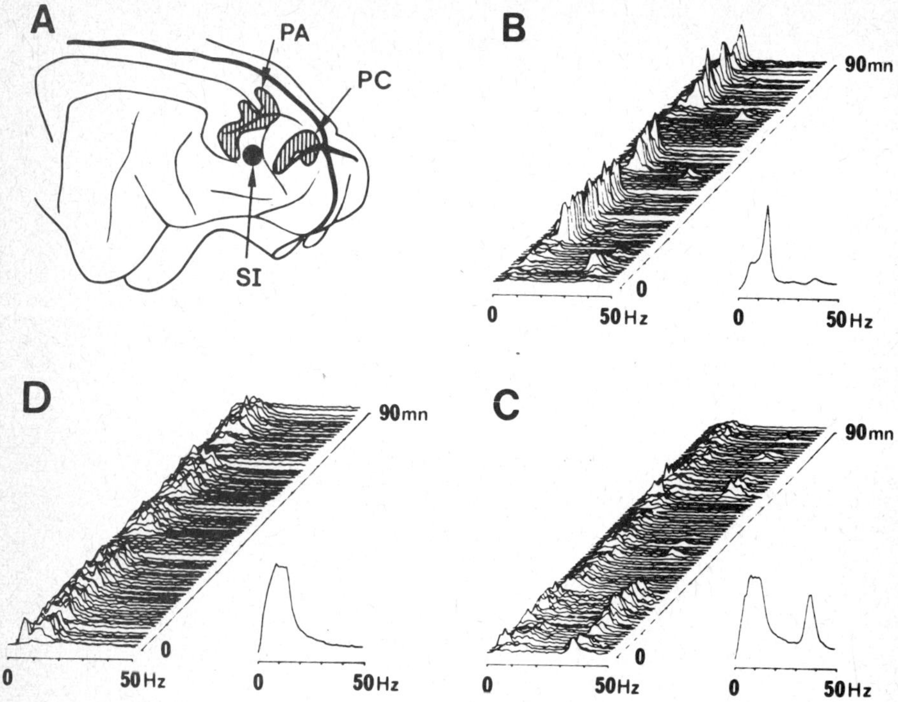

Fig. 3. Cortical distribution of fast hypervigilance rhythms.
A: pooled topographical data over 20 subjects. Notice the existence of 2 foci for fast rhythms, one anterior pericruciate (PC), the second one around the ansate sulcus (periansate, PA) in the anterior suprasylvian cortex.
B: Evolutive spectra taken, min after min, of the ECoG recorded from the pericruciate (anterior PC) focus,with a total recording time of 90 min. Short episodes of fast rhythms can be seen, that accompany high vigilance and immobility (as B on Fig. 1); they alternate with peaks corresponding to sleep spindles (14 Hz, type E pattern on Fig. 1).
C: Activities recorded from the periansate parietal (posterior PA) focus. Notice abundance of high frequency peaks, alternating with rhythms in the lower ranges, corresponding either to quiet vigilance, or to drowsiness or to sleep. On the other hand, no spindles develop in this area (for further details, see Fig.7).
D: ECoG from the somatic cortex (SI, in A), In this case no rhythms can ever be observed in the high frequency range. The only rhythms are those of "quiet vigilance" at 14 Hz and those accompanying drowsiness (14 Hz + 6 Hz), with some shifts toward slow sleep, indicated by the delta activity (3 to 5 Hz). Added to each set of evolutive spectra is the average spectrum computed over the 90 min recording time.

(iii) The fact that a unilateral lesion of the same PO area (Fig. 5) suppresses these rhythms on the same cortical side, leaving the other rhythms unaffected, confirms these data. This suppression was observed throughout the animal's one week follow-up period after lesion.

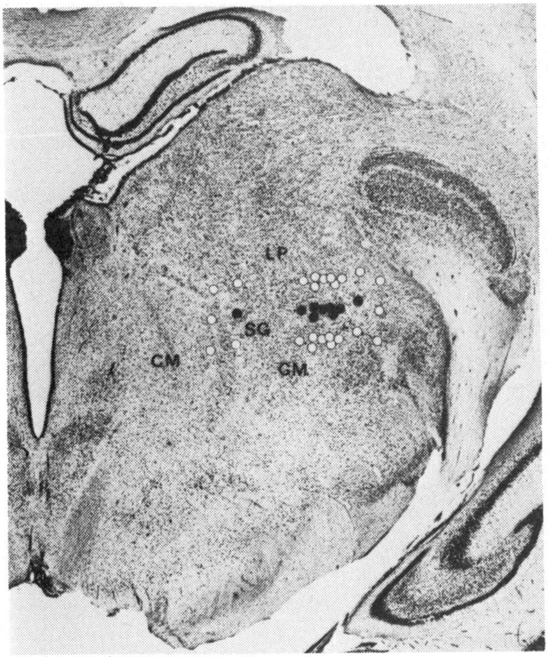

Fig. 4. Thalamic sites from which fast hypervigilance rhythms could be recorded (as on Fig. 2) are indicated by black dots. White dots mark negative points. All recording sites (corresponding to 11 cats) were located in the frontal stereotaxic plane A 6,5. No active points were encountered at 5 mm more anteriorly or more posteriorly from that level.
Abbreviations:
CM, n. centrum medianum;
SG, n. suprageniculatus;
GM, n. geniculatus medialis;
LP, n. lateralis posterior.

(iv) One of the difficulties posed by these rhythms was their similarity to those that are known to develop in the neighbouring olfactory bulb and that roughly display the same frequencies. A series of coherence studies was therefore undertaken to show that no coherence values over 20 % could be obtained when comparing the fast hypervigilance rhythms (at their neocortical or

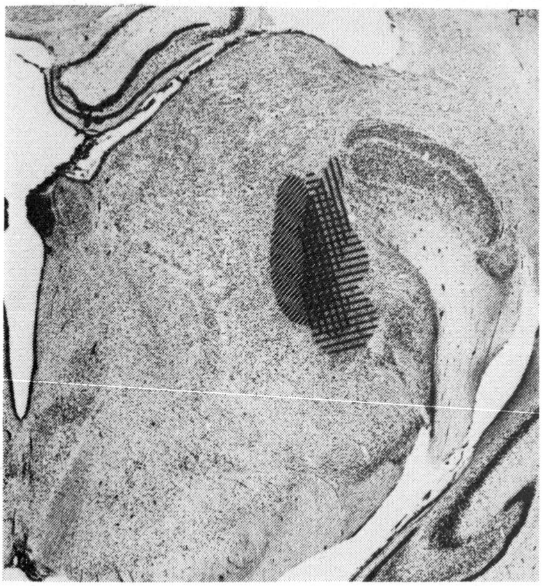

Fig. 5. Lesions performed in the PO thalamic area of 3 animals, that were effective to suppress ipsilaterally the fast somato-parietal rhythms. Lesions are outlined in the frontal stereotaxic plane A 6,5. Each lesion had about the same size in the anterior-posterior direction, than laterally in the frontal plane.

thalamic site) to those in the olfactory bulb, recorded with an extra pair of adequately placed electrodes (Fig. 6).

(v) The most interesting feature is no doubt the behavioural correlates that we were able to detect for the fast somato-parietal rhythms. When the animal was left alone in the recording room with no particular intervening stimulus to draw its attention (neutral conditions), a succession of various types of activities were noticed (Fig. 7, I). During the first few minutes, the animal seemed to be highly attentive, either moving or remaining immobile in a "starring" attitude. In the latter case, fast rhythms immediately developed. Later on the frontal activities shifted towards slower rhythms of quiet wakefulness, and still later, to those of drowsiness with the corresponding changes in behaviour. The cat finally fell asleep, with spindles and slow waves. This pattern of evolution over 90 min of observation was fairly constant when similar sessions were repeated with the same subject. In particular, the initial phase with fast rhythms and immobile attention did not disappear with repetition, as one would have expected.

(vi) On the other hand, when a (perspex protected) mouse was placed in the recording room within sight, the pattern of recorded activity became quite different. In such conditions, the cat usually remained quasi-immobile for about one hour, watching the mouse. Throughout that time of a monotonous type of behaviour, the spectrum remained almost stable, with a high density of fast rhythms presented in the tested parietal derivation (Fig. 7, II). The same also occurred for the anterior PC rhythms, but in this case, our study was restricted to the PA, posterior activities.

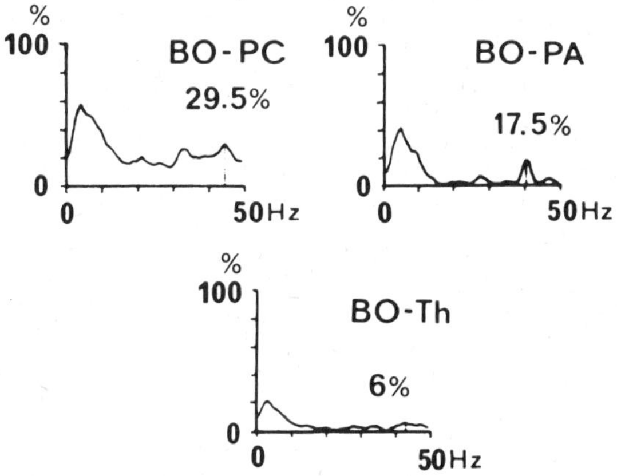

Fig. 6. Coherence analyses between activities recorded from the olfactory bulb (BO) and three sites from which fast hypervigilance rhythms could be maximally obtained: pericruciate (PC) cortex, parietal periansate (PA) suprasylvian cortex and thalamic posterior group (Th). Notice, in all 3 cases, low values of coherence within the 35 - 45 Hz range.

3. Fast somato-parietal rhythms: neuropharmacological and neurochemical data. We then tried to identify the neurochemical systems that could control the development of the fast somato-parietal rhythms and in all probability also the accompanying behaviour of immobility and oriented attentiveness. We were guided in this investigation by previous findings[14] indicating that, in the rat, lesions of the ventral mesencephalic tegmental area (VMT) profoundly altered the "attentive" behaviour of the animal, as it became restless and hyperactive, seemingly unable to "fix" its attention. Moreover, other well accepted data are available, showing that the VMT contains the somas of the A 10 dopaminergic pathway[15]. It was therefore tempting to try to correlate these data to our own observations in the cat by performing the following tests:(i) injection of substances known to stimulate the metabolism of dopamine, i.e. DOPA as a precursor and apomorphine as a stimulus of the dopaminergic receptors, and follow-

ing both the animal's behaviour and its parietal ECoG spectrum in our particular conditions;(ii) lesion of the VMT area and again noting the resulting changes in ECoG and behaviour; (iii) finally, administration of DOPA and apomorphine to the lesioned animals:

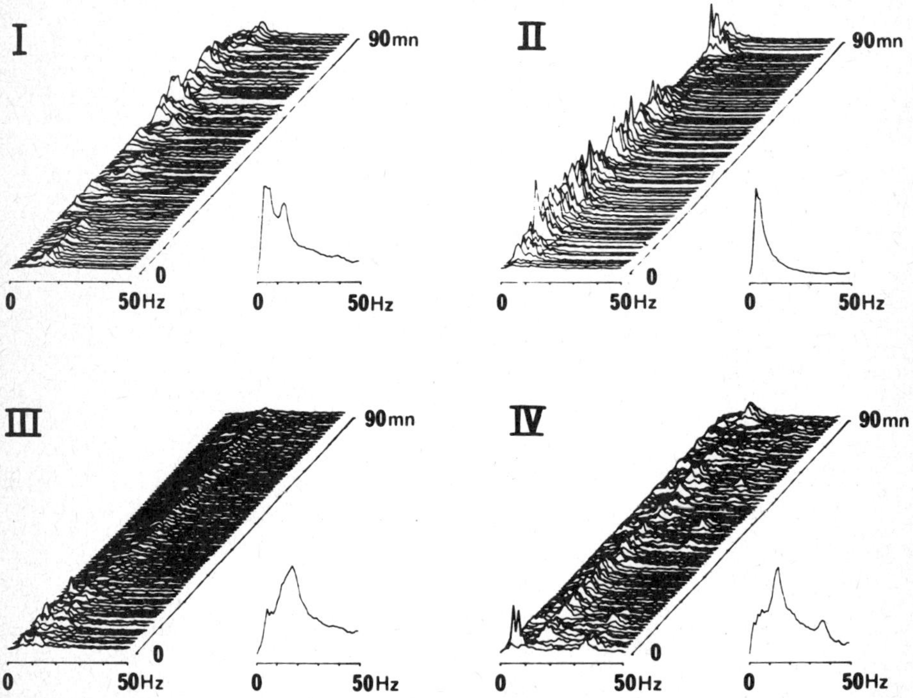

Fig. 7. Evolutive spectra from the periansate (PA) focus in a normal cat,
I: animal in "neutral conditions" (absence of characteristic source of attention). At first, a phase of exploration occurred, with movement artifacts, alternating with immobility and high frequency rhythms. Later on, the animal became drowsy (Fig. 1, D.), then fell asleep (Fig. 1, E). Transient returns to hypervigilance could also be noticed.
II: animal placed in the presence of a mouse. Notice clear prolongation, in this case, of the fast rhythms, for almost 60 min, interspersed with phases of movements. It is only very late that the animal became disinterested. Correspondingly, the hypervigilance peak disappeared, and drowsiness activity developed. A new phase of interest followed, etc....
III: animal as in I, but after administration of DOPA. In this case, notice quasi-permanence of the 35 - 45 Hz peak.
IV: as in I and III, but after injection of apomorphine. Notice coexistence of hypervigilance rhythms and movement artifacts. In fact, the animal was either immobile or developed the "stereotyped" behaviour well known to occur under this drug. After about 80 min, the drug ceased to act and animal became quiet.

(i) Administration of DOPA to the normal cat elicited a drastic change of its behaviour in the recording chamber (Fig. 7, III). Even when no particular item was present to draw its attention, the animal remained immobile, apparently "attentive" and developed fast rhythms throughout the duration of this drug's action (>90 min).

Injection of apomorphine also produced a behaviour of "intense alertness". However, the animal now underwent frequent phases of stereotyped hyperactivity, alternating with some brief periods of immobility. This explains why the recorded spectra most frequently displayed two peaks, one in the range of movement artifacts, the other in the fast rhythm band (Fig. 7, IV).

(ii) We then performed bilateral lesions of the VMT in 4 cats. Throughout the period of observation, i.e. about 2 months, these animals displayed a very characteristic attitude in the recording room. When they are left alone with no overt object to draw their attention, they displayed a very particular locomotor hyperactivity with no sign of any real exploratory or attentive attitude. Correspondingly, their ECoGs remained fully desynchronized, merely with movement artifacts (Fig. 8, I). Contrasting with this pattern of attention, other aspects of the behavioural repertoire of the animal, i.e. quiet wakefulness with middle-range rhythms and slow activities accompanying drowsiness, could occur as in the normal subject. Finally, sleep was qualitatively and quantitatively normal.

The next step was to put the lesioned animal in the presence of a mouse, just as we did with normal cats. The subject now behaved quite differently, circlinguninterruptedly around the perspex box containing the mouse. In this case too, the ECoG remained fully desynchronized, with no sign of fast rhythms, and the evolutive spectra now showed a great abundance of movement artifacts (Fig. 8, II).

(iii) Administration of DOPA to a lesioned cat had quite a different effect than that in the normal subject. This time, the drugged animal adopted a quiet waking attitude and, rather surprisingly, the accompanying rhythms, although barely visible on the parietal leads, could be clearly distinguished as a prolonged ripple throughout the evolutive spectrum, in the middle range frequencies (Fig. 8, III). At any rate, no fast rhythms comparable to those reinforced by DOPA in the non-lesioned cats, could be noticed.

Finally, apomorphine was injected at 4 mg/kg (half the dose administered to normal subjects, see Discussion). In VMT lesioned animals this drug induced the reappearance of brief periods of immobility, accompanied by a transient development of fast parietal rhythms (Fig. 8, IV).

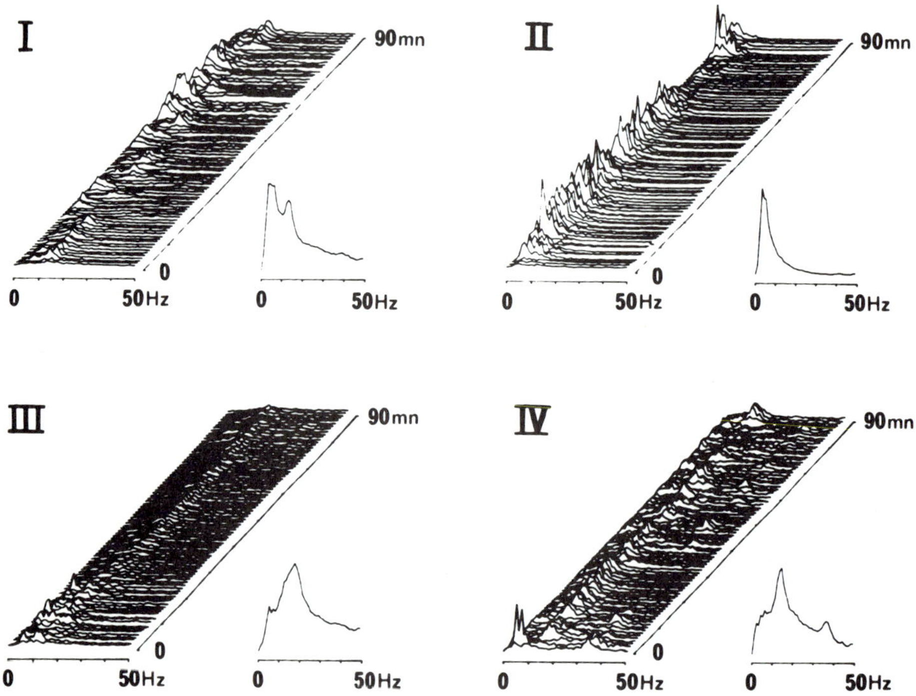

Fig. 8. Evolutive spectra computed as in Fig. 7, in a VMT lesioned animal (PA, anterior suprasylvian derivation).
I: cat was in "neutral" conditions. No peak now occurred in the fast frequency range. Compare to Fig. 7, I.
II: animal put in the presence of a mouse. Cat developed restless behaviour (as indicated by movement artifacts, see Fig. 1, A). No phase of immobility could be observed, nor any fast rhythm.
III: cat after administration of DOPA (same "neutral" condition as in I). Contrasting with pattern III on Fig. 7, no consistent fast rhythms are visible. On the other hand, a peak develops around 14 Hz correlated with "quiet vigilance".
IV: cat after administration of apomorphine. Some fast rhythms now reappeared although not as frequently as in the normal subject.

DISCUSSION

One of the striking features of the present observations is the concomitance that is seen between the sudden development of fast somato-parietal (35 - 45 Hz) rhythms and the characteristic attitude of the animal, i.e. a motionless gaze toward a given (not necessarily identifiable) target in its surroundings, suggesting what we have termed intense vigilance or attentiveness[16]. Undoubtedly, this particular behaviour belongs to the normal repertoire of the animal, especially to its hunting and prey-catching strategies.

In other, fairly distinct environmental conditions, that is, when the animal is quietly sitting in a familiar environment with no overt signs of oriented vigilance, rhythms can be recorded that are quite distinct from the fast ones, in both frequency range and localization (their main focus now being on SI). In both cases, however, the beginning of the rhythmic activity is closely related to corresponding characteristic behaviour. Although we did not specifically concentrate on the problem of chronology between ECoG and behaviour, we think it sensible to consider these activities as event-related phenomena.

Evidently, the test used to evaluate "attention" in its qualitative and quantitative aspects (see[17]) is far from satisfactory. We (as many others) are now faced with the dilemma of either accepting certain behavioural features as indices of attention or rejecting attention itself as a psychological entity, like behaviourists did in the classical period. Since in our observations the only material clues were changes in the "attentive" behaviour of the animal, we definitely decided to accept the validity of this type of criteria, despite the difficulties and criticisms they could feasibly raise. We shall not discuss here whether the simultaneity between fast rhythms and a state of immobile high vigilance is only a sign of a correlation or if a causal link exists between the two phenomena. Let us, however, briefly speculate on the finality of the high vigilance immobility. Two inexclusive issues are possible: (i) immobility may represent a kind of suppression of phasic information from the somatic domain, leaving only steady-state messages from the tonic receptors and facilitating "concentration" upon non-somatic, e.g. visual and/or auditory, stimuli; (ii) immobility may contribute to a "blockade" of the motor system that may be a prerequisite for programming the adequate motor action.

Our next point concerns the effects of drugs upon the dopaminergic metabolism and those of VMT lesions. As mentioned above, our lesion data confirm some previous studies performed in the rat[18,19]. In particular, Galey et al.[14] were able to observe that after VMT lesions, animals were unable to avoid shocks, probably because of their high level of motor activity. "Deterioration of attentiveness" was also noted as a consequence of these lesions. Finally, it was found that the VMT may contribute to the inhibition of motor activity accomanying an intense vigilance level.

Combining these observations with pharmacological data, one may deduce the following: (i) the thalamo-cortical system displays rhythms at a given frequency, with the PO acting as a pace-maker; the transmitters involved remain unknown; (ii) the VMT acts as a permissive (gating) system for this thalamocortical minizone; (iii) from the effects of DOPA, one may conclude that this gating action has a dopaminergic link; (iv) after destruction of the VMT, no further synchronization of that thalamo-cortical sector can occur, at the same time as the attentive immobility state disappears; (v) the fact that apomorphine is still relatively active after VMT exclusion may indicate that dopaminergic receptors are still present, even with the hypersensitivity presumably due to denervation, since half the normal dose is now sufficient to produce some reappearance of high frequency rhythms. At the same time, this action of apomorphine reinforces the assumption that VMT does not act as a pacemaker, but only as a gate, for even in its absence, rhythmicity can occur in this frequency range; (iv) finally, the location of the dopaminergic receptors and, hence, that of the apomorphine action remain to be determined. Data are available which indicate that the A 10 dopaminergic neurons project on the "frontal" cortex[20,21] and also on the striatum[22]. Whether a long-loop system thus exists from VMT to, say, the anterior orbito-frontal cortex and back to the PO thalamic level, is, at the present state of knowledge, pure speculation.

ACKNOWLEDGEMENTS

This study was supported by the following contracts: DRET, Nr 78/1083; CNRS, ERA 411; INSERM, ATP 80 - 79 - 112. The technical assistance of Mrs C. Durand is gratefully acknowledged. Computing facilities were provided by Laboratoires Fournier (Dijon); apomorphine was kindly furnished by Laboratoire Aguettant (Lyons).

REFERENCES

1. Roth, S.R., Sterman, M.B. and Clemente, C.D. (1967) Electroenceph. clin. Neurophysiol., 23, 509-520.
2. Sterman, M.B. and Wyrwicka, W. (1967) Brain Res., 6, 143-163.
3. Lanoir, J. and Cordeau, J.P. (1970) J. Physiol., Paris, 62, suppl. 3 399-400.
4. Rougeul, A., Letalle, A and Corvisier, J. (1972) Electroenceph. clin. Neurophysiol., 33, 23-39.
5. Lopes da Silva, F.H., van Lierop, T.H.M.T., Schrijer, C.F. and Storm van Leeuwen, W. (1973) Rev. Electroenceph. clin. Neurophysiol., 35, 641-645.
6. Rougeul, A., Corvisier, J. and Letalle, A. (1974) Electroenceph. clin. Neurophysiol., 37, 41-57.
7. Rougeul-Buser, A., Boyer, J.J. and Buser, P. (1975) Acta Neurobiol. exp., 35, 805-819.

8. Gastaut, H., Terzian, H. and Gastaut, Y. (1952) Marseille méd., 89, 296-310.

9. Gastaut, H., Jus, A., Jus, C., Morrell, F., Storm van Leeuwen, W., Dongier, S., Naquet, R., Regis, H., Roger, A., Bekkering, D., Kamp, A. and Werre, J. (1957) Electroenceph. clin. Neurophysiol., 9, 1-34.

10. Chatrian, G.E., Petersen, M.C. and Lazarte, J.A. (1959) Electroenceph. clin. Neurophysiol., 11, 497-510.

11. Kuhlman, W.N. (1978) Electroenceph. clin. Neurophysiol., 44, 83-93.

12. Pfurtscheller, G. and Aranibar, A. (1980) Int. Symp. Event-related changes in cortical rhythmic activities. Behavioural correlates. Schladming, Austria.

13. Jasper, H.H. and Penfield, W. (1949) Arch. Psychiat. Z. Neurol., 183, 163-174.

14. Galey, D., Simon, H. and Le Moal, M. (1977) Brain Res., 124, 83-97.

15. Dahlström, A. and Fuxe, K. (1964) Acta Physiol. scand., suppl. 232, 1-31.

16. Mostofsky, D. I. (1970) Attention: Contemporary theory and Analysis, D. Mostofsky ed., Appleton Century Crafts, 9-24.

17. Berlyne, D.E. (1970) Attention: Contemporary theory and Analysis, D.Mostofsky ed., Appleton Century Crafts, N.Y., 1-447.

18. Thierry, A.M., Tassin, J.P., Blanc, C. and Glowinski, J. (1978) In: Adv. Biochem. Psychopharmacol., 19, 205-216.

19. Tassin, J.P., Stinus, L., Simon-Blanc, G., Thierry, A.M., Le Moal, M., Cardo, B. and Glowinski, J. (1978) Brain Res., 141, 267-282.

20. Berger, B., Tassin, J.P., Blanc, G., Moyne, M.A. and Thierry, M. (1974) Brain Res., 81, 332-337.

21. Lindvall, O., Björtelund, A., Morre, R.Y. and Stenevi, U. (1974) Brain Res., 81, 325-331.

22. Beckstead, R.M., Domesick, V.B. and Nauta, W.J.H. (1979) Brain Res., 175, 191-217.

© 1980 Elsevier/North-Holland Biomedical Press
Rhythmic EEG Activities and Cortical Functioning
G. Pfurtscheller et al., editors

COGNITION, ACTION, AND MECHANISMS OF EEG ASYMMETRY

CHARLES D. YINGLING
Langley Porter Institute, University of California, San Francisco, CA 94143

Since the first description of the EEG by Berger (1929)[1], there has been the hope that it can somehow be shown to reflect mental processes and provide as Berger termed it, a "window on the mind". This concept received impetus when Adrian and Matthews (1934)[2] demonstrated that the occipital alpha rhythm could be blocked, not only by opening the eyes and engaging the cortex with processing visual input, but also by performing mental arithmetic with eyes closed. This observation led to the well-known concept of alpha as an "idling" rhythm, manifested primarily during states of inactivity or decreased arousal and tending to disappear during active or attentive mental states.

During the ensuing decades, most research relating the EEG to mental activities focussed on this concept of arousal or activation, which was generally viewed as a unidimensional phenomenon, with a continuum of states of awareness ranging from deep sleep through lighter sleep, drowsiness, relaxed alertness, and finally highly activated states of attention. These states were shown to be accompanied by generally decreasing amounts of synchronous activity in the EEG as the behavioral state moved toward greater arousal or attentiveness.

In 1949, Moruzzi and Magoun[3] demonstrated that the basic phenomena of synchronization and desynchronization of the EEG appeared to be regulated by the reticular formation of the brainstem. The concept of the so-called "reticular activating system" fit in beautifully with the emphasis on arousal level as the main dimension of mental activity affecting the EEG, and became the foundation of a view of brain-behavior relationships which dominated the field until well into the 1960's. During this era, the thought that the kind of mental activity performed might have a significant influence on the EEG was rarely encountered; the emphasis was on "how much" rather than "what kind".

During the 1960's, however, the emphasis changed dramatically with the experiments performed by Sperry and his colleagues on a series of patients in whom the corpus callosum had been severed for treatment of intractable epilepsy. (See Sperry (1974)[4] for review). The knowledge that the two cerebral hemispheres were not functionally identical has existed at least since the time of Broca and Wernicke. However, most thinking emphasized the superiority of the left hemisphere (in right handers) for speech and motor control, and relegated the right hemisphere to a subsidiary role. The demonstration that the "minor" hemisphere was in fact superior in the performance of tasks involving spatial relationships led to a major shift in prevailing views of the brain-mind relationship. The concept of cerebral dominance gave way to that of complementary specialization, generally expressed as a dichotomy between two separate and perhaps incompatible modes of thought, variously categorized as verbal-spatial, analytic-intuitive, sequential-simultaneous, propositional- appositional, etc. This idea rapidly captured the attention of scientists, philosophers, and the general public; today it is hard to read a newspaper or magazine without finding the left/right dichotomy mentioned in some way.

Not surprisingly, EEG researchers were quick to see the implications of this concept, and the emphasis shifted from arousal per se to the more complex notion of differential activation of the two hemispheres by tasks which would engage one more than the other. Galin and his associates (Galin et al., 1971[5]; Galin and Ornstein, 1972[6]) measured whole-band (1-35 Hz) power from temporal and parietal leads while subjects performed verbal (writing a letter, mentally composing a letter) and spatial (Kohs blocks, Minnesota Paper Form Board) tasks. The results confirmed the expected relative suppression of activity on the right during spatial tasks in 8 out of 9 subjects (parietal leads). A subsequent study (Doyle, Ornstein, and Galin, 1974[7]), using Fourier spectral analysis, confirmed these results and demonstrated the greatest task-dependent asymmetry to occur in the alpha band. Morgan, Mc Donald, and Macdonald (1971[8]) measured the relative proportion of alpha activity at left and right occipital leads while subjects performed analytic (ie, mental arithmetic, defining terms) or spatial (imagining a scene) tasks. They noted a relatively lower proportion of alpha over the right hemisphere during the spatial condition in 16 out of 20 subjects.

Since these pioneering studies, a substantial body of literature has emerged generally supporting the conception that differential engagement of the hemispheres by appropriately chosen tasks can be demonstrated with the EEG (McKee et al., 1973[9]; Butler and Glass, 1974[10]; Callaway and Harris, 1974[11]; Morgan et al., 1974[12]; Dumas and Morgan, 1975[13]; Davidson et al., 1976[14]; Davidson and Schwartz, 1977[15]; McLeod and Peacock, 1977[16]; Yingling, 1977[17]; Amochaev and Salamy, 1979[18]; Ehrlichman and Wiener, 1979[19]; Herron et al., 1979[20]; Ornstein et al., 1979[21,22]).

In addition to these studies, which all focussed on measures of ongoing

EEG activity during various tasks, a number of studies using event-related potential (ERP) methods have also reported asymmetries as a function of linguistic/non-linguistic aspects of the stimuli or task demands (Wood, Goff, and Day, 1971[23]; Morrell and Salamy, 1971[24]; Matsumiya, Tagliasco, Lombroso, and Goodglass, 1972[25]; Brown, Marsh, and Smith, 1973[26]; Neville, 1974[27]; Molfese, Freeman, and Palermo, 1975[28]; Roemer and Teyler, 1977[29]; Marsh and Brown, 1977[30]; Wood, 1977[31]).

However, this point of view has not remained unanimous or unquestioned. Friedman, Simson, Ritter, and Rapin (1975[32]) and Galambos, Benson, Smith, Schulman-Galambos, and Osier (1975[33]), in two studies using ERP techniques, failed to demonstrate any asymmetries of the ERPs even though their paradigms involved linguistic judgements and were designed to elicit such asymmetries. Both of these reports suggested that ERP methods were not suitable for demonstrating functional asymmetries, despite earlier claims to the contrary.

Recently, Donchin, Kutas, and McCarthy (1977[34]), in a well-written and thoughtful review, discussed methodological problems which they feel have led to uncertainties and confusions making it "premature to advocate the use of the EEG and ERP parameters as indices of hemispheric utilization..." They listed the following possibilities as reasons why, in some previous studies, expected differences had not been unequivocally demonstrated:

1. The experimental design was not sufficiently sensitive to allow detection of the differences or was inadequate to support the conclusions.
2. The tasks assigned the subject may not in fact have differentially engaged the hemispheres.
3. The effects were range-restricted and the values of the independent variables were out of the relevant range.
4. Subject individual-difference variables were not considered.
5. The parameters of the EEG used as dependent variables were unwisely selected.
6. The measurement techniques used to obtain the parameters were inappropriate.
7. The data were improperly quantified and were inappropriately or insufficiently analyzed.

Despite these methodological problems, Donchin and his colleagues still note that "within the welter of conflicting claims and apparent inconsistencies there is a thread of positive results that indicate the promise of the approach." They state, however, that "If one point emerges with clarity from the studies reviewed, it is this: If there are any differences between the electrocortical activity of the two hemispheres, they will be minute." In the light of the striking functional differences between hemispheres revealed by the split-brain studies, not to mention an enormous body of literature cataloging the different deficits produced by unilateral lesions to one side or the other, this controversy about the demonstrability of clear-cut electrophysiological asymmetries is rather surprising.

An even more skeptical position was taken in a recent report by Gevins, Zeitlin, Doyle, Yingling, Schaffer, Callaway, and Yeager (1979[35]). They argued that "it has never been established that asymmetries or other EEG features are directly related to cognitive activities", pointing out that most tasks which have been used to produce different cognitive demands also differ in "stimulus characteristics, efferent activities (limb and eye movements), and performance-related factors (task demands and a subject's ability and effort)." They presented results which, it was argued, supported their contention that when these noncognitive sources of variance were controlled, "no evidence for lateralization of different cognitive functions was found in the EEG."

I would like to suggest that an alternative explanation of these studies is possible, one which is more consistent both with the existing literature on hemispheric specialization and with recent physiological evidence concerning the mechanisms of generation and regulation of synchronous activities in the brain (Yingling and Skinner, 1975[36], 1976[37], 1977[38]; Skinner and Yingling, 1976[39], 1977[40]). The essence of this view is that the separation of task characteristics into "cognitive" and "motor" components is artificial and misleading. The adaptive significance of a brain to an organism is not that it allows cognition, in the sense of purely mental activity which is not involved with the environment. Nor does it lie in perceptual capacities, though an increasing ability to perceive ever finer distinctions in the sensory world is certainly a concomitant of the evolution of the brain. It is rather in the ability to coordinate and direct activity, patterns of behavior which allow the organism to interact effectively with its environment and communicate with other members of its social group. (This concept of "action" should be distinguished from simple "movement", which does not necessarily imply coordinated, goal-directed patterns of motor output.) Perception and cognition are thus seen as tools in the service of action, rather than as end products of brain activity.

Given this view, it is inappropriate to view the movement-related cortical activity in a task such as writing from memory as an artifact, obscuring the "cognitive" activity which one wishes to study. It is rather part and parcel of the coordinated perceptual-cognitive-motor activity which is the primary function of the brain. Isolated perceptual or "cognitive" activities might more properly be seen as artifactual, representing as they do an isolated subset of the normal functional activity of the brain.

This view of brain-behavior relationships leads to a straightforward prediction about the effects of different cognitive activities on measurable indices of brain function: purely perceptual or "cognitive" tasks will have much less of an effect than those coupled to overt behaviours. It is perhaps analogous to the fact that racing an automobile engine will produce relatively little observable difference unless the car is in gear and the clutch engaged.

With this in mind, let us examine the results obtained in experiment 1 of the Gevins et al. (1979[35]) report, since this experiment compared motor and non-motor tasks in the same subjects. Alpha band (8-13 Hz) spectral intensity will be used for illustrative purposes. In this experiment, 23 right handed subjects performed the following series of tasks: eyes closed, attention to breathing; eyes open, attention to breathing; reading a semi-technical article; writing a summary of the reading passage from memory; constructing a pattern with Kohs blocks from memory, using only the right hand; mental paper folding (after Shepard and Feng, 1972[41]), in which subjects were shown a flat representation of a cube slit along the edges and unfolded, and asked to mentally refold the pattern and determine whether or not two arrows marked on the pattern would meet. In addition, 18 of the 23 subjects manipulated the blocks without trying to create a pattern; and all 23 repetitiously wrote either abcabc or small loops. Finally, 15 subjects performed mental arithmetic, usually serial subtraction of 7's starting from a given number. EEG was recorded from bilateral frontal, central, parietal, and occipital leads, referenced to ipsilateral ears. After editing for artifact, Fourier spectral analysis was computed and banded into standard frequency bands.

It was anticipated that, regardless of lateralization or other topographic effects, performance of any of the tasks would be associated with a general across-the-head reduction of alpha power levels. Surprisingly, this was not always the case. All 23 subjects showed a substantial drop in alpha levels between eyes closed and eyes open conditions. However, mental arithmetic was accompanied by an overall <u>increase</u> in alpha power in 6 out of 15 subjects, a decrease in 7, and no change in 2, when compared with the eyes open condition. This enhancement of alpha during mental arithmetic has been noted previously (Kreitman and Shaw, 1965[42]; Creutzfeldt, Grunewald, Simonova, and Schmitz, 1969[43]; Legewie, Simonova, and Creutzfeldt, 1969[44]). Even more surprising was the fact that 5 out of 23 subjects showed higher alpha levels during reading than in the resting eyes open condition. This observation is certainly hard to reconcile with the classical view of alpha as an "idling" rhythm, blocked by active attention to the visual environment. In some subjects, at least, a complex visuo-perceptive task such as reading can be accompanied by substantial amounts of alpha activity.

The tasks involving active movements, on the other hand, produced very low alpha levels in general, with 20 of 23 subjects showing less alpha during either writing or Kohs blocks than during eyes open fixation. The other motor tasks, random block manipulation and repetitious scribbling, also produced suppression of alpha levels in 15 out of 18 and 19 out of 23 subjects respectively.

In general, then, it appears that tasks involving active manipulation of objects produce more suppression of alpha than do purely perceptual-cognitive tasks, even though they may seem to involve less effort. This effect also extends to lateralized suppression effects produced by tasks expected to differentially engage the hemispheres. For example, Kohs blocks were expected to produce relatively greater alpha suppression over the right hemisphere than writing,and mental paper folding to produce relatively more right suppression than reading. Ratios of R/L alpha power were compared between pairs of tasks, a procedure which is relatively insensitive to absolute differences in spectral intensity between tasks (Yingling, 1977[17]). The blocks vs. writing comparison conformed to the prediction, R/L ratios being higher for writing than blocks for all four pairs of homologous leads. (Table I). On the other hand, the R/L ratios were higher for reading than folding only at the frontal leads, and were reversed (ie, folding reading) at the parietals. For the writing vs. blocks comparison, 18 of 23 subjects showed higher ratios for writing than blocks at central, parietal, and occipital placements, and 15 out of 23 subjects at the frontal leads, but in only 11, 8, and 11 at central, parietal, and occipital leads respectively.

TABLE I

R/L RATIOS OF ALPHA POWER (8-13 Hz)

Mean of 23 subjects (S.D. in parentheses)

	Writing from Memory	Kohs blocks	Reading	Mental Paper Folding
F_4/F_3	1.20 (.26)	1.07 (.15)	1.16 (.21)	1.09 (.15)
C_4/C_3	1.15 (.25)	0.97 (.25)	0.96 (.21)	0.99 (.23)
P_4/P_3	1.18 (.27)	1.00 (.22)	0.97 (.12)	1.07 (.19)
O_2/O_1	1.12 (.33)	0.98 (.23)	1.01 (.29)	1.04 (.22)

In other words, writing and Kohs blocks produce effects on EEG alpha levels which are consistent with predictions based on their differential sensitivity to left and right hemisphere lesions, whereas reading and mental paper folding do not. I do not believe that this necessarily indicates the latter tasks to be any less lateralized than the former. I would rather argue that the lack of active motor involvement eliminates the necessity for much of the cortex to be involved, and thus the scalp EEG does not show a clear pattern of lateralized activity. In contrast, when the lateralized cognitive activities are coupled to cortical mechanisms for motor activity, a larger proportion of the brain is involved, and the resultant patterns are more clearly evident at the scalp. This is true even though the final motor pathways, involving fine finger movements of the right hand, are the same for both tasks. It is not a simple case of coupling, for example, a left or right parietal "cognitive" area to a left central "motor" area. Rather, I suggest that the requirement for planning and coordinating motor output involves a far greater degree of "cognitive" processing, which is ultimately inseparable from the "motor" component. Coordinated action _is_ cognition, and represents the primary product of brain activity.

That greater task-dependent asymmetries are produced by motor than non-motor tasks has been reported in several previous studies. Galin and Ornstein (1972[6]) compared writing a letter (motor), mentally composing a letter (non-motor), Kohs blocks (motor), and a modified Minnesota Paper Form Board test (non-motor). Both temporal and parietal lead pairs showed greater differences between writing and Kohs blocks than between the mental letter and form board tasks, in R/L ratios of 1-35 Hz integrated power. These results were confirmed in a subsequent study (Doyle, Ornstein, and Galin, 1974[7]), which extended the finding to several new tasks, in which the same pattern emerged: motor tasks engaged the hemispheres more asymmetrically. Bennett and Trinder (1977[45]) measured percent time of alpha at temporal leads during letter writing, mental arithmetic, tonal awareness, and Kohs blocks. Again, the motor tasks produced greater differences in asymmetry than the non-motor tasks. Ehrlichman and Wiener (1979[19]) used two verbal-motor (controlled associations, word arrangements), two verbal-nonmotor (nonsense syllogisms, extended range vocabulary), two spatial-motor (puzzle construction, human figure drawing), and two spatial-nonmotor (cube comparisons, surface development) tasks. Integrated alpha levels from left and right temporo-parietal leads were obtained, and log L/R ratios computed for each task from two recording sessions a minimum of one week apart. Both recording sessions produced the same pattern of results: the verbal-motor tasks showed the greatest relative left alpha suppression, with verbal non-motor, spatial non-motor, and spatial motor tasks producing progressively more suppression on the right. Finally, Rebert (1978[46]) observed that playing video tennis (PONG™) produced greater central and temporal alpha asymmetry than simply watching the game, although no difference between playing and watching was noted at the parietal leads.

These studies all pointed to the fact that task-dependent EEG asymmetries are likely to be more robust in active cognitive tasks than in those not involving motor activity. It seems unlikely that these results can all be explained away as resulting from noncognitive motor activities, since all the larger differences reported were appropriately lateralized despite varying motor demands and hand usages. For example, Herron et al.(1979[20]) found alpha power at C_3 to be lower than at C_4 during writing, and higher than at C_4 during block design even though both tasks were performed exclusively by the right hand.

It is crucial to note that in the Gevins et al. study random block manipulation and repetitious scribbling could not be differentiated by any combination of EEG features, whereas actual block design and writing were distinguishable at the $p<10^{-9}$ level. (Gevins et al.[47]).

A position similar to that proposed here has recently been taken by Gazzaniga and LeDoux (1978[48]), who noted that the unique abilities of the right hemisphere in commissurotomized patients were best demonstrated with tasks involving active manipulation. They characterized the right hemisphere superiority as being for "manipulo-spatial", rather than simply spatial activities. And of course, it is well known that the behavior most sensitive to left hemisphere lesions is one involving an incredibly complex motor output: speech.

How is this emphasis on motor activities reflected in the organization of the brain? In a recent series of papers (Yingling and Skinner, 1975[36], 1976[37], 1977[38]; Skinner and Yingling, 1976[39], 1977[40]) it has been suggested that regulation of thalamocortical activity by the thalamic reticular nucleus underlies such phenomena as synchronization and desynchronization of the EEG, cortical and subcortical slow potential shifts, and control of sensory input to the cortex as indexed by enhancement of evoked potentials to attended stimuli. This thalamic gating mechanism is in turn subject to a dual regulation from other brain centers. The mesencephalic reticular formation provides a generalized control, opening all the thalamic gates when activated and thus effecting a general decrease in synchronous EEG (which is produced by cyclic long-duration thalamic IPSP's originating in the thalamic reticular nucleus) and an overall increase in sensory inflow to the cortex. This mechanism is probably responsible for the continuum of arousal states reflected in the EEG, as discussed earlier.

On the other hand, differential patterns of thalamo-cortical activity, including selective enhancement of relevant sensory input and possibly lateralized patterns of activation, are thought to involve highly selective, patterned control of this same thalamic gating mechanism by the cerebral cortex. And it is not, as might be expected, the high-order sensory association areas in the parietal regions which effect this control, but rather the frontal cortex, anatomically and functionally more related to mechanisms for planning and controlling motor behaviors. The cortex does indeed control its own sensory input, but it is the "motor" systems, not the sensory systems, that regulate the influx. This is, of course, quite consistent with the point I have made repeatedly in this chapter, that the brain is organized to carry out its primary function of behavior, not perception or cognition alone, and that this primacy of action is reflected in the results of EEG studies of functional hemispheric asymmetry.

I would like to close with a quotation from Roger Sperry (1952[49]), who presaged many of the ideas in this chapter nearly thirty years earlier: "Perceptions and ideas are found, upon analysis, to have their factual significance and meaning in terms ultimately of overt operation. Their meaning derives from the potential effect, that is, the difference they make or may make in behavior. In both its phylogenetic and ontogenetic histories, mental activity develops out of, and in reference to, overt action."

ACKNOWLEDGEMENTS

This work was supported by National Institutes of Mental Health Interdisciplinary Training Program MH-7082, USPHS Grant NS10471, USPHS Biomedical Research Support Grant RR05755, and Office of Naval Research Contract N00014-75-C-0398. The author wishes to thank Alan S. Gevins and David Galin for many fruitful discussions concerning these issues.

REFERENCES

1. Berger, H., (1929). Über das Elektrenkephalogramm des Menschen. Arch. Psychiat. Nervenkr. 87: 527-570.
2. Adrian, E.D. and Matthews, B.H.C., (1934). The Berger rhythm: potential changes from the occipital lobes in man. Brain 57: 355-385.
3. Moruzzi,G., and Magoun, H.W., (1949). Brain stem reticular formation and activation of the EEG. Electroencephalography and Clinical Neurophysiology 1: 455-473.
4. Sperry, R.W., (1974). Lateral specialization in the surgically separated hemispheres. In F.O. Schmitt & F. Worden, (eds.), The Neurosciences.

Third Study Program. Cambridge, Mass: The MIT Press.

5. Galin, D., Ornstein, R., Kocel, K., and Merrin, E., (1971). Hemispheric localization of cognitive mode by EEG. Psychophysiology 8: 246-247.
6. Galin, D., and Ornstein, R., (1972). Lateral specialization of cognitive mode: An EEG study. Psychophysiology 9: 412-418.
7. Doyle, J.C., Ornstein, R., and Galin, D., (1974). Lateral specialization of the cognitive mode II: EEG frequency analysis. Psychophysiology 11: 567-578.
8. Morgan, A.H., McDonald, P.J., and Macdonald, H., (1971). Differences in bilateral alpha activity as a function of experimental task with a note on lateral eye movements and hypnotizability. Neuropsychologia 9: 459-469.
9. McKee, G., Humphrey, B., and McAdam, D.W., (1973). Scaled lateralization of alpha activity during linguistic and musical tasks. Psychophysiology 10: 441-443.
10. Butler, S.R., and Glass, A. (1974). Asymmetries in the electroencephalograph associated with cerebral dominance. Electroenceph. Clin. Neuropysiol. 36: 481-491.
11. Callaway, E., and Harris, P., (1974). Coupling between cortical potentials from different areas. Science 183: 873-875.
12. Morgan, A.H., MacDonald, H. & Hilgard, E.R., (1974). EEG alpha: Lateral asymmetry related to task and hypnotizability. Psychophysiology 11: 275-282.
13. Dumas, R., and Morgan, A., (1975). EEG asymmetry as a function of occupation, task, and task difficulty. Neuropsychologia 13: 219-228.
14. Davidson, R.J., Schwartz, G.E., Pugash, E. and Bromfield, E., (1976). Sex difference in patterns of EEG asymmetry. Biological Psychology 4: 119-138.
15. Davidson, R.J. and Schwartz, G.E., (1977). The influence of musical training of patterns of EEG asymmetry during musical and non-musical self-generation tasks. Psychophysiology, 14: 58-63.
16. McLeod, S.S. and Peacock, L.J., (1977). Task-related EEG symmetry: effects of age and ability. Psychophysiology 14: 308-311.
17. Yingling, C.D., (1977). Lateralization of cortical coupling during complex verbal and spatial behaviors. In J. Desmedt, (ed.). Language and Hemispheric Specialization in Man: Cerebral Event-Related Potentials, 151-160. Basel, S. Karger.
18. Amochaev, A., and Salamy, A., (1979). Stability of EEG laterality effects. Psychophysiology 16: 242-246.
19. Ehrlichman, H. and Wiener, M., (1979). Consistency of task-related EEG asymmetries. Psychophysiology 16: 247-252.
20. Herron, J., Galin, D., Johnstone, J., and Ornstein, R., (1979). Cerebral specialization, writing posture, and motor control of writing in left handers. Science 205: 1285-1289.
21. Ornstein, R., Herron, J., Johnstone, J., and Swencionis, C., (1979). Differential right hemisphere involvement in two reading tasks. Psychopsychopsysiology 16: 398-401.
22. Ornstein, R., Johnstone, J., Herron, J., and Swencionis, C. Differential right hemisphere engagement in visuospatial tasks. Neuropsychologia, In

Press.

23. Wood, C.C., Goff, W.R., and Day, R.S., (1971). Auditory evoked potentials during speech perception. Science 173: 1248-1251.

24. Morrell, L.K. and Salamy, J.G., (1971). Hemispheric asymmetry of electrocortical response to speech stimuli. Science 174: 164-166.

25. Matsumiya, Y., Tagiliasco, V., Lombroso, C.T. and Goodglass, H., (1972). Auditory evoked response: meaningfulness of stimuli and interhemispheric asymmetry. Science 175: 790-792.

26. Brown, W.S., Marsh, J.T., and Smith, J.C., (1973). Contextual meaning effects on speech-evoked potentials. Behavioral Biology 9: 755-761.

27. Neville, H. (1974). Electrographic correlates of lateral asymmetry in the processing of verbal and nonverbal auditory stimuli. Journal of Psychological Research 3: 151-163.

28. Molfese, D.L., Freeman, R.B., and Palermo, D.S., (1975). The ontogeny of brain lateralization for speech and nonspeech stimuli. Brain and Language 2: 356-368.

29. Roemer, R.A. and Teyler, T.J., (1977). Auditory evoked potential asymmetries related to word meaning, in Desmedt, J.E. (ed.), Prog. clin. Neurophysiol 3: 48-59.

30. Marsh, J.T. and Brown, W.S., (1977). Evoked potential correlates of meaning in the perception of language, in Desmedt, J.E. (ed.), Prog. clin. Neurophysiol 3: 60-72.

31. Wood, C.C., (1977). Average evoked potentials and phonetic processing in speech perception, in Desmedt, J.E. (ed.), Prog. clin. Neurophysiol 3: 73-86.

32. Friedman, D., Simson, R., Ritter, W., and Rapin, I., (1975). Cortical evoked potentials elicited by real speech words and human sounds. Electroencephalography and Clinical Neurophysiology 38: 13-19.

33. Galambos, R., Benson, P., Smith, T.S., Shulman-Galambos, C., and Osier, H., (1975). On hemispheric differences in evoked potentials to speech stimuli. Electroencephalography and Clinical Neurophysiology 39: 279-283.

34. Donchin, E., Kutas, M., and McCarthy, G., (1977). Electrocortical indices of hemispheric utilization. In Harnad, S., Doty, R.W., Goldstein, L., Jaynes, J., and Krauthamer, G., (eds.), Lateralization in the Nervous System, 339-384. New York: Academic Press.

35. Gevins, A.S., Zeitlin, G.M., Doyle, J.C., Yingling, C.D., Schaffer, R.E., Callaway, E., and Yeager, C.L., (1979). Electroencephalogram correlates of higher cortical functions. Science 203: 665-668.

36. Yingling, C.D., and Skinner, J.E., (1975). Regulation of unit activity in nucleus reticularis thalami by the mesencephalic reticular formation and the frontal granular cortex. Electroencephalography and Clinical Neurophysiology 39: 653-642.

37. Yingling, C.D., and Skinner, J.E., (1976). Selective regulation of thalamic sensory relay nuclei by nucleus reticularis thalami. Electroencephalography and Clinical Neurophysiology 41: 476-482.

38. Yingling, C.D., and Skinner, J.E., (1977). Gating of thalamic input to cerebral cortex by nucleus reticularis thalami. In J.E. Desmedt, (ed.). Attention, Voluntary Contraction and Event-Related Cerebral Potentials, 70-96. Basel, S. Karger.

39. Skinner, J.E., and Yingling, C.D., (1976). Regulation of slow potential shifts in nucleus reticularis thalami by the mesencephalic reticular formation and the frontal granular cortex. Electroencephalography and Clinical Neurophysiology 40: 288-296.

40. Skinner, J.E., and Yingling, C.D., (1977). Central gating mechanisms that regulate event-related potentials and behavior: a neural model for "attention". In Desmedt, J.E. (ed.) Attention, Voluntary Contraction, and Event-Related Cerebral Potentials. Basel: S. Karger

41. Shepard, R.N., and Feng, C., (1972). A chronometric study of mental paper folding. Cognitive Psychology 3: 228-243.

42. Kreitman, N., and Shaw, J., (1964). Experimental enhancement of alpha activity. Electroencephalography and Clinical Neurophysiology 18: 147-155.

43. Creutzfeldt, O., Grünewald, G., Simonova, O. and Schmitz, H., (1969). Changes of the basic rhythms of the EEG during the performance of mental and visuomotor tasks. In C.R. Evans and T.B. Mulholland, (eds.), Attention in Neurophysiology, 148-168. London: Butterworths.

44. Legewie, H., Simonova, O., and Creutzfeldt, O.D., (1969). EEG changes during performance of various tasks under open- and closed-eyed conditions. Electroencephalography and Clinical Neurophysiology 27: 470-479.

45. Bennett, J.E. and Trinder, J., (1977). Hemispheric laterality and cognitive style associated with transcendental meditation. Psychophysiology 14: 293-296.

46. Rebert, C.S., (1978). Neuroelectric measures of lateral specialization in relation to performance. Electroencephalography and Clinical Neurophysiology, suppl., 34: 231-238.

47. Gevins, A.S., Zeitlin, G.M., Yingling, C.D., Doyle, J.C., Dedon, M.F., Schaffer, R.E., Roumasset, J.T., and Yeager, C.L. (1979). EEG patterns during "cognitive" tasks. I. Methodology and analysis of complex behaviors. Electroencephalography and clinical Neurophysiology 47: 693-703.

48. Gazzaniga, M.S., and LeDoux, J.E., (1978). The Integrated Mind. New York: Plenum Press.

49. Sperry, R.W., (1952). Neurology and the mind-brain problem. American Scientist 40: 291-312.

Rhythmic EEG Activities and Cortical Functioning
G. Pfurtscheller et al., editors

THE SPECTRAL PROPERTIES OF HIPPOCAMPAL EEG RELATED TO BEHAVIOUR IN MAN

D.E.A.T. ARNOLDS, F.H. LOPES DA SILVA, J.W. AITINK, A. KAMP and P. BOEIJINGA
Brain Research Department, Institute of Medical Physics TNO, Da Costakade 45, Utrecht (The Netherlands)

INTRODUCTION

The relation between hippocampal EEG and behaviour has been studied in a number of animal species. In most cases the rat and to a lesser extent cat and dog have been investigated in this respect. Some studies exist also on the rabbit, some species of monkeys, mouse and guinea pig.

Reviews of recent literature on this subject can be found in O'Keefe and Nadel[1] and in Lopes da Silva and Arnolds[2]. On the other hand only very few data exist on man (Pagni and Marossero[3], Brazier[4], Halgren et al.[5]).

The most striking feature of the hippocampal EEG in dog and cat is the occurrence of a clear rhythmicity in the range of 3 to 8 c/s. In rodents this range may extend up to 12 c/s. This activity is called the hippocampal theta or rhythmic slow activity (RSA). This RSA is prominent under virtually all behavioural circumstances in the rabbit[6,7], dog[8,9,10] and rat[11]. In the cat[12] and much more so in primates[13], the RSA component is often less clear, upon visual inspection.

Spectral analysis of the hippocampus EEG reveals however, that in these species there exists a significant RSA component under most investigated circumstances as well (Crowne[13], Brazier[4], Moise and Costin[14], Arnolds unpublished observations in cat). In general the peak frequency in primates appears to be considerably lower than in other species (± 3 c/s). Relations have been proposed between the occurrence of RSA and various aspects of behaviour such as learning[15], orientation reactions[16], attention and arousal[17]. The simplest and most generally accepted relationship, however, is between the occurrence of RSA and movement, as first proposed although restricted to the voluntary variant by Vanderwolf[18]. In experiments designed to test the relation between hippocampal EEG and more abstract aspects of behaviour such as learning or attention, motor behaviour is always present but no controls for movement are carried out. Hence the previously mentioned behavioural correlates may be considered from a parsimonious point of view to imply motor behaviour related to RSA as argued in Lopes da Silva and Arnolds[2]. A detailed study of the spectral properties of the hippocampal EEG in dog during a variety of behaviours at different levels of complexity[8,9,10] led us to the following generalization concerning the relation between motor behaviour and RSA: when the same behavioural pattern is performed at different levels of intensity, a higher level of behavioural intensity - as measured in terms of amount of motor activity - is related to a higher peak frequency, amplitude and rhythmicity of the RSA component in the concurrent hippocampal EEG.

It is, however, certainly not so that the spectral properties of the hippocampal EEG are proportional to the amount of motor activity when comparing two different behaviour-patterns in this respect.

It is quite possible that besides motor events, sensory aspects of behaviour are also expressed in the spectral properties of the hippocampal EEG[9,10] but this aspect has been less thoroughly investigated.

From the generalization stated above, it may be inferred that in our view the hippocampal EEG should not be considered in terms of "presence" or "absence of RSA", as is done in most studies. On the contrary, the hippocampal EEG should be considered as a signal that under virtually all behavioural circumstances obtains a significant amount of rhythmical activity in the theta band.This was shown to be the case in dog[8] and in rat[20].

It depends on the spectral properties of the hippocampal EEG, i.e. its peak frequency, amplitude and bandwidth, to what extent the rhythmicity is clearly apparent upon visual inspection of an EEG record. As noted before the behavioural correlates of hippocampal EEG have been scarcely investigated in man. Studies in the freely moving human subject are completely lacking. It was thus of interest to investigate whether the generalization formulated above for dog would hold for man as well. We were able to make a contribution in this respect thanks to the cooperation of a patient with epilepsy who had been implanted with electrodes in various brain structures,among them the hippocampus, in an attempt to localize a possible focus of epileptic activity.

METHODS

The subject* was a 39 year old epileptic female whose intellectual ability was in the normal range, according to a number of psychological tests. The patient was left-handed, but had been taught to write right-handedly. The electrode bundle from which records were made consisted of 7 100 μ stainless

---

*Electrodes were implanted in this patient in order to detect, if possible, the epileptogenic area, with the aim of removing it surgically.The patient suffered from partial complex epileptic seizures which could not be controlled by anti-epileptic drugs.

steel wires twisted together. The insulation of the electrodes was scraped away over a length of 2.5 mm. The tip of the hippocampal electrode was aimed at a point 5 mm behind the frontal plane defined by the posterior edge of the Anterior Commissure, 27 mm lateral from the skull midline, 20 mm above the tentorium. The implantation technique has been described in Lopes da Silva and Van Hulten[21]. Apart from the hippocampus, electrodes were implanted in a variety of other structures. The signals derived from these other electrodes have, however, not been analysed with respect to behavioural reactivity; thus they are not included in the present description.

Recordings were against a frontal intracranial electrode the activity of which was of very low amplitude when measured against external EEG electrodes and showed no apparent rhythmicity. The patient was recorded on two different days (session I and II). She was free to move around since we made use of a radio-telemetry system to transmit EEG signals as well as the output of a small accelerometer, which was used as a movement indicator.

The experiments consisted of 3 behavioural tasks. The first one was designed as follows: the patient was seated in a darkened room and a series of about 25 slides was projected. These consisted of schematic drawings of simple objects (key, chair, hammer, etc.). The slides were projected in random order on a screen about 4 mm away from the subject (viewing time $\sim$ 2 sec). By the side of the projection screen was a blackboard, lit by a small lamp. The subject was asked, as soon as she had recognized the object on the slide to get up and walk to the blackboard, where she should write the name of the object and then walk back and sit down again. After a time interval of 5 to 15 sec the procedure was repeated.

During each experimental session this procedure was repeated twice. The first time the subject was instructed to walk briskly (experiments I.1 and II.1). The second time she was asked to walk at a normal pace (experiments I.2 and II.2). Slides were reshuffled before each experiment.

The second behavioural task consisted of a word association test. In this experiment (II.3) the subject was presented on each of 20 trials with the name of a class of objects (e.g."a vegetable") and asked to name an example of a member of that class (e.g. "cauliflower"). The names of the classes were given verbally by the experimenter and the subject was instructed to name the first example of a member of that class that came to her mind.

In the third task (experiment II.4) the reverse was required, i.e. the experimenter named one specific example (e.g. "Amsterdam") and the subject had to name a class to which it belonged (e.g. "a city"). Preceding each of these tests she was verbally instructed and given 3 examples.

All relevant signals, EEGs, movement transducer output, a voice record of both experimenter and subject, and a spoken record of the subject's behaviour, were recorded on a 14 channel FM tape along with a time code. During recording session I the subject was on 300 mg Sodium Valproate (three times per day) and 200 mg Carmabazapine (four times per day).

During the ensuing 5 days this medication was gradually decreased in order to facilitate the occurrence of seizure activity which was necessary for diagnosis. After having been free of medication for 10 days the subject got 300 mg Sodium Valproate (two times per day). Under these circumstances recording session II took place.

Analysis

Essential to our method of analysis is the quantification of both EEG and behaviour, the sampling of both signals at the same rate and the use of computer averaging and statistical analysis. To this end the hippocampal EEG was submitted to continuous spectral analysis by means of a bank of filters with a 1 Hz bandwidth. The output of this analysis was recorded in the computer 5 times per second. In synchrony with sampling of the spectrally analysed EEG, the momentaneous value of the integrated output of a movement transducer on the subject's back was also stored. In order to obtain data reduction to a form easily accessible to averaging and statistical analysis, three "spectral parameters" were derived from each recorded spectrum. One was a measure of the peak frequency of the RSA component in the EEG (I_{PF}), another one gave an indication of the amplitude (I_{AM}) and the third of the rhythmicity (I_R), i.e. it was related to the bandwidth of the spectral peak and to the relation between the peak amplitude value and the background. The behavioural and spectral parameter, or the spectra themselves, recorded during a number of performances of the same type of behaviour were averaged. The average values of each parameter as a function of time could also be computed in relation to a certain trigger point. The significance of differences between the average values of a parameter during different behavioural states was assessed by means of Wilcoxon's test. The significance of fluctuations in the triggered average was determined by comparing each average value to a reference value by means of the sign test. For a more extensive account of our method of analysis see Arnolds et al.[8].

RESULTS

Experiment I.1, I.2, II.1 and II.2

A general inspection of the signal (see Fig. 1) derived from the right (R) hippocampus in this subject revealed a field potential characterized by an irregular rhythmical component at 3-4 c/s. In comparison to the hippocampal EEG of nonprimates, the amplitude and frequency of this activity was low. Clear-cut changes in this signal related to behaviour could not be discerned easily upon visual inspection. In the traces derived from all structures, pathological phenomena such as epileptiform spikes occurred frequently. Along with the decreased medication the amount of this type of pathological activity clearly increased during session II. In the hippocampal records comparatively few epileptiform spikes were found.

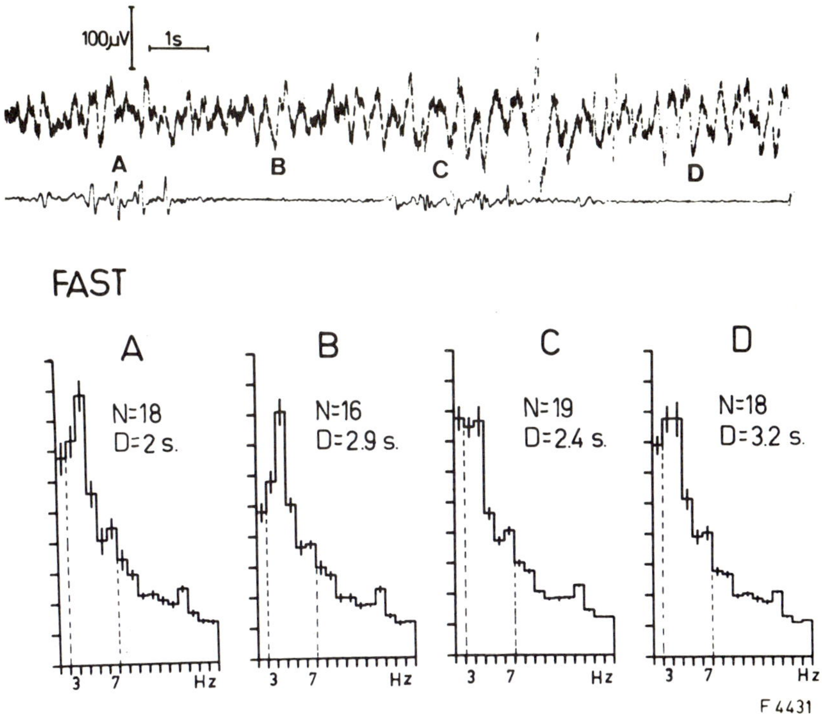

Fig. 1. Above: example of the hippocampal EEG recorded during a single trial of experiment I.1. Below: series of average power spectra of the hippocampal EEG recorded during each of the 4 behavioural states in experiment I.1, i.e. A: walking towards blackboard, B: writing, C: walking back, D: sitting. N is the number of averaged periods. D is the average duration of these periods.

In both sessions there was no obvious relation between the EEG from right (R) and left (L) hippocampal placements. The L hippocampus generally had a smaller amplitude than the R hippocampus, while rhythmical activity was lower in frequency and more variable. The analysis carried out on the EEG from the left side showed little reactivity to behavioural changes. In the following analysis, therefore, we concentrate upon the R hippocampus.

In order to determine the relation between spectral properties of the hippocampal EEG and the behavioural states investigated, we selected from experiments I.1 and I.2 and from II.1 and II.2 a number of periods during which the subject was in a clearcut behavioural state: (a) walking towards the blackboard, (b) writing, (c) walking back and (d) sitting. These periods were selected such that the hippocampal EEG contained neither artifacts nor epileptiform activity. We determined the average value of the spectral parameters and of the integrated output of the movement transducer during these behavioural states (Table I).

TABLE I

For the two experimental sessions the average values of the spectral parameters and the movement parameter (ACC) measured during (a) walking towards the blackboard, (b) writing, (c) walking back, (d) sitting, are tabulated. Columns headed by capital letters refer to the "brisk" version of the behaviour (experiments I.I and II.1), lower case letters to the "moderate" version (experiments I.1 and II.2). When a pair of values is written in *italics*, this indicates a statistically significant (Wilcoxon $p < 0.05$) difference between these measurements (e.g. during session I, I_{PF} was at 3.63 significantly higher while the subject walked briskly towards the blackboard than during the moderate version of the same behaviour).

session I

	A	a	B	b	C	c	D	d
I_{PF}	*3.63*	*3.32*	3.83	3.66	3.16	3.22	3.30	3.50
I_{AM} ▲	6.9	6.6	5.8	6.2	6.7	6.5	6.9	6.7
I_R	*0.441*	*0.364*	*0.466*	*0.418*	0.387	0.363	*0.442*	*0.375*
ACC ▲	*2.12*	*1.46*	*0.58*	*0.62*	*2.00*	*1.37*	*0.57*	*0.50*

session II

I_{PF}	3.40	3.31	*3.95*	*3.64*	3.58	3.50	*3.64*	*3.45*
I_{AM} ▲	5.5	5.4	5.5	5.3	*6.0*	*5.4*	*6.3*	*5.7*
I_R	0.328	0.331	0.363	0.351	0.352	0.368	*0.382*	*0.354*
ACC ▲	*2.86*	*1.76*	*0.86*	*0.74*	*2.22*	*1.40*	*0.660*	*0.659*

As regards behavioural parameters the subject was more active during "1" than during "2" experiments as it was to be expected. However, the fact that a significant difference was found for the output of the movement transducer between the two versions of this task, even while the subject was sitting down, was surprising. The comparison between experiments "1" and "2" as regards the values of the spectral parameters measured during equivalent behaviours showed that whenever a significant difference was observed, the value of the spectral parameter was higher during "the brisk" than during the "moderate" version of the behaviour (see Table I).

Upon comparison of the spectral parameter values associated with different behavioural states, a consistent difference was found between "writing" and the other three behaviours. I_{PF} and I_R during writing were relatively high, while I_{AM} was relatively low (Wilcoxon $p < 0.05$). Other significant differences did occur between behavioural states in this respect, but were not consistent across the two sessions.

As shown in Table I the same trends were visible in both sessions I and II. The rhythmicity of the signal was consistently significantly lower during session II than during session I however (Wilcoxon $p < 0.05$).

Results of experiment II.3 and II.4

The word association tasks used were designed with the objective of having one cue word, spoken by the experimenter and in response a single word uttered by the subject. The subject, however, had often trouble in restricting herself to a single word response and tended to ask for additional clarification. Thus no clearcut trigger point for the subjects response could be defined in most cases. The best defined trigger point was therefore the end of the cue word spoken by the experimenter.

In relation to these triggers changes were found in the spectral parameters. For both experiment II.3 and II.4 these changes were of the same type. The data from both tests were therefore pooled. Fig. 2 shows the spectral parameters as a function of time in relation to the end of the keyword with the pooled data from both experiments II.3 and II.4. The relationship between the spectral parameters and behaviour can be described as follows: the I_{AM} was relatively low during the vocalization of the subject and during the period of silence preceding the presentation of the keyword by the experimenter. It rose during the presentation of the keyword and remained significantly elevated in the period preceding the subject's vocalization. I_{PF} was low in the period of silence preceding the keyword and showed a dip around the moment the

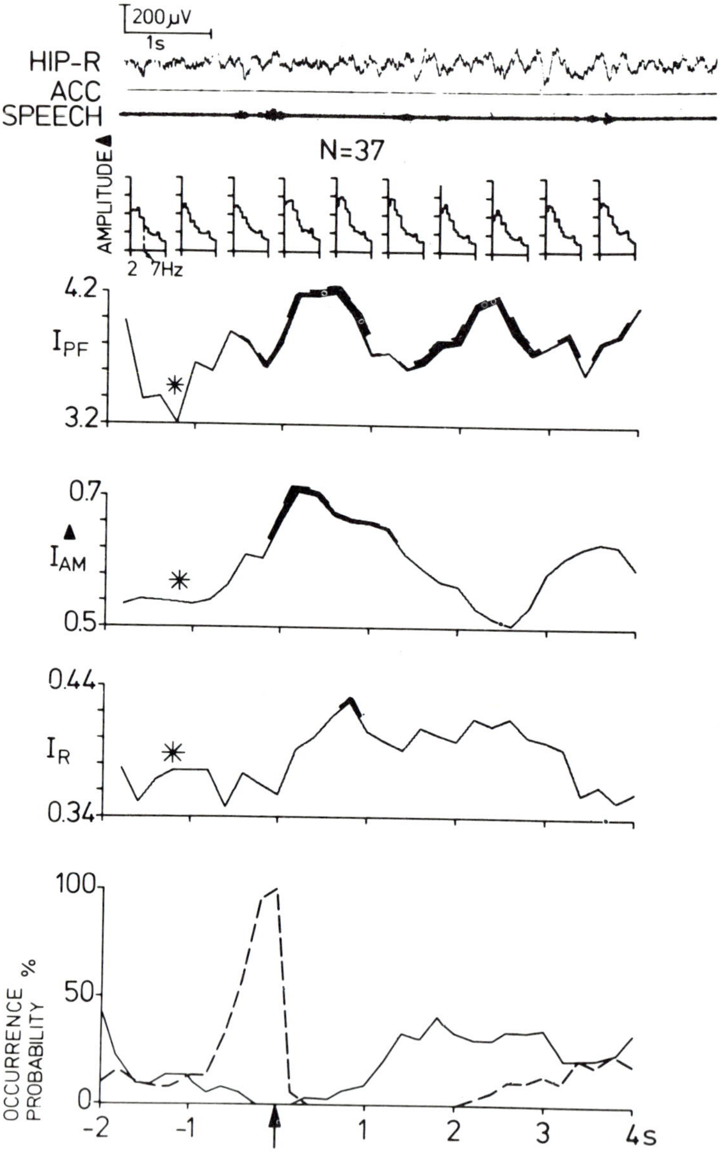
200µV
1s
HIP-R
ACC
SPEECH
AMPLITUDE
N=37
2 7Hz
4.2
I_PF
3.2
0.7
I_AM
0.5
0.44
I_R
0.34
100
50
0
OCCURRENCE PROBABILITY %
-2
-1
1
2
3
4s

response started. I_R also showed a tendency to increase during the period following the presentation of the keyword. It should be added that at the same time the output of the movement transducer showed no significant changes in relation to behaviour.

When some epochs of talking and listening recorded during an informal conversation were compared, I_{AM} appeared to be significantly lower while the subject talked than while she was only listening.

DISCUSSION

An analysis of the hippocampal EEG activity recorded from an epileptic patient under a number of well defined behavioural circumstances showed the existence of a significant rhythmical component between 3 and 5 c/s. The spectral properties of this rhythmical component were significantly different as the subject performed different behaviours. It is remarkable that the values of the spectral parameters of the theta component in the human hippocampus EEG were consistently higher when associated with the more intense performance of a certain behaviour. This outcome is fully in accord with the facts observed in dog[8,9,10].

A strong discrepancy with animal data is, however, the lack of a relationship between gross motor activity (walking vs. sitting) and the spectral properties of the human hippocampal EEG. The variations in the spectral properties appear, on the contrary, to be associated largely with what could be called verbal behaviour, be it writing or speaking. Possibly this finding points indeed in the direction of an involvement of the hippocampus in linguistic processes[1].

Fig. 2. The upper trace gives the hippocampal EEG during a single trial of experiment II.4 (question: "Amsterdam", answer given by the subject: "the capital"). HIP-R = hippocampal EEG, ACC = output of movement transducer on subject, SPEECH = recorded speech of experimenter and subject.

The series of histograms represents the average running spectral analysis of the hippocampal EEG during experiments II.3 and II.4 (data pooled). Each histogram concerns a 200 ms time bin. Inter histogram distance 600 ms. Spectra to be averaged have been aligned with respect to the trigger point: the end of the keyword.

The three following curves show the average I_{PF}, I_{AM} and I_R as a function of time in relation to this trigger point (arrow). Where the curve is thickened there exists a statistically significant difference between that part of the curve and the point indicated by an asterisk, chosen as a reference (sign test). The thicker the line the higher the significance of the statistics (lowest level $p < 0.05$; highest $p < 0.001$). The lowermost curve indicates the probability that the experimenter (stippled line) or the subject (full line) were talking, aligned with respect to the trigger point.

In a previous study of the human hippocampus EEG, Brazier[4] could not find any rhythmicity in the theta range. This might be due to the fact that her spectral analysis consisted of 2 Hz wide bands, which may have been too wide for the phenomenon under consideration.

Recently Halgren et al.[5] described that the hippocampal EEG in man did not exhibit changes correlated with behavioural features in any of 16 psychomotor epileptic patients but one. In this case they reported that the EEG was "desynchronized" in proportion to the apparent difficulty of the task. Neither Brazier nor Halgren employed averaging of EEG spectral parameters in relation to or triggered by behavioural events. This aspect is in our view essential to the discovery of relatively small effects such as those described in this paper.

Of course, our report concerns only a single subject and an epileptic one at that. Nevertheless, the fact that a clearcut reactivity in relation to behaviour was observed implies that, at least in some respects, the hippocampal EEG in man reacts in an analoguous way to that in animal. The fact that writing and verbalization were so clearly correlated with modulation of the hippocampal EEG appears to imply that these types of behaviour are also reflected in neural activity of the human hippocampus, possibly in the same way as motor behaviour, in general, is reflected in the hippocampal EEG of a variety of mammals.

SUMMARY

In an epileptic patient with depth electrodes placed in the hippocampal formation we observed the following:

1) There exists a significant peak in the theta band of the human hippocampal EEG under a number of behavioural circumstances.

2) Four different behaviours were performed at two degrees of intensity. Whenever the values of a certain spectral parameter associated with a low or a high intensity behaviour showed a significant difference, the higher value corresponded with the more intense version of the behaviour.

3) The frequency and rhythmicity of the hippocampal theta component during writing was consistently higher than during sitting or walking. The amplitude was lower however.

4) In a word association task during the period of silence immediately following the question and preceding the answer, the amplitude showed a significant rise as well as frequency and rhythmicity.

REFERENCES

1. O'Keefe, J. and Nadel, L. (1978) The hippocampus as a cognitive map. Clarendon, Oxford, 231-374.
2. Lopes da Silva, F.H. and Arnolds, D.E.A.T. (1978) Physiology of the hippocampus and related structures. Ann. Rev. Physiol., 40, 185-216.
3. Pagni, C.A. and Marossero, F. (1965) Some observations on the human rhinencephalon, a stereo-electroencephalographic study. Electroenceph. clin. Neurophysiol., 18, 260-271.
4. Brazier, M.A.B. (1968) Studies of the EEG activity of limbic structures in man. Electroenceph. clin. Neurophysiol., 25, 309-318.
5. Halgren, E., Babb, T.L. and Crandall, P.H. (1978) Human hippocampal formation EEG desynchronizes during attentiveness and movement. Electroenceph. clin. Neurophysiol., 44, 778-781.
6. Klemm, W.R. (1970) Correlation of hippocampus theta-rhythm, muscle activity and brainstem reticular formation activity. Comm. Behav. Biol., 5, 147-151.
7. Klemm, W.R. (1971) EEG and multiple-unit activity in limbic and motor systems during movement and immobility. Physiol. Behav., 7, 337-343.
8. Arnolds, D.E.A.T., Lopes da Silva, F.H., Aitink, J.W. and Kamp, A. (1979) Hippocampal EEG and behaviour in dog. I. Hippocampal EEG correlates of gross motor behaviour. Electroenceph. clin. Neurophysiol., 46, 552-570.
9. Arnolds, D.E.A.T., Lopes da Silva, F.H., Aitink, J.W. and Kamp, A. (1979) Hippocampal EEG and behaviour in dog. II. Hippocampal EEG correlates with elementary motor acts. Electroenceph. clin. Neurophysiol., 46, 571-580.
10. Arnolds, D.E.A.T., Lopes da Silva, F.H., Aitink, J.W. and Kamp, A. (1979) Hippocampal EEG correlates of stimulus-response tasks and of sexual behaviour. Electroenceph. clin. Neurophysiol., 46, 581-591.
11. Whishaw, I.Q. and Vanderwolf, C.H. (1973) Hippocampal EEG and behaviour: changes in amplitude and frequency of RSA (Theta Rhythm) associated with spontaneous and learned movement patterns in rats and cats. Behav. Biol., 8, 461-484.
12. Kemp, J.R. and Kaada, B.R. (1975) The relation of hippocampal theta activity to arousal, attentive behaviour and somato-motor movements in unrestrained cats. Brain Res., 95, 323-342.
13. Crowne, D.P., Konow, A., Drake, K.J. and Pribram, K.H. (1972) Hippocampal electrical activity in the monkey during delayed alternation problems. Electroenceph. clin. Neurophysiol., 33, 567-577.
14. Moise, S.L. and Costin, A. (1974) Hippocampal, hypothalamic and lateral geniculate activity during visual discrimination in the monkey. Physiol. Behav., 12, 835-841.
15. Adey, W.R. (1967) Hippocampal states and functional relations with cortico-subcortical systems in attention and learning. In: Adey, W.R. and Tokizane, T. (Eds.), Progress in Brain Research. Structure and Function of the Limbic System, Elsevier, Amsterdam, Vol. 27, 228-245.
16. Grastyân, E., Lissâk, K., Madarasz, I. and Donhoffer, H. (1959) Hippocampal Electrical activity during the development of conditioned reflexes. Electroenceph. clin. Neurophysiol., 11, 409-430.

17. Kemp, I.R. and Kaada, B.R. (1975) The relation of hippocampal theta activity to arousal, attentive behaviour and somato-motor movements in unrestrained cats. Brain Res., 95, 323-342.

18. Vanderwolf, C.H. (1969) Hippocampal electrical activity and voluntary movement in the rat. Electroenceph. clin. Neurophysiol., 26, 407-418.

19. Buzsáki, G., Grastyán, E., Tveritskaya, I.N., Czopf, J. (1979) Hippocampal evoked potentials and EEG changes during classical conditioning in rat. Electroenceph. clin. Neurophysiol., 47, 64-74.

20. Coenen, A.M.L. (1975) Frequency analysis of rat hippocampal electrical activity. Physiol. Behav., 14, 391-394.

21. Lopes da Silva, F.H., Van Hulten, K. (1978) Analyse quantitative de l'activité intercritique en EEG et SEEG dans l'épilepsie. Rev. E.E.G. Neurophysiol., 8, 2, 198-204.

MOVEMENT-RELATED CHANGES OF EEG RHYTHMS

© 1980 Elsevier/North-Holland Biomedical Press
Rhythmic EEG Activities and Cortical Functioning
G. Pfurtscheller et al., editors

THE MU RHYTHM: FUNCTIONAL TOPOGRAPHY AND NEURAL ORIGIN

WILLIAM N. KUHLMAN
Neuropsychology Laboratory, Veterans Administration Medical Center, West Haven Connecticut 06516, and Department of Neurology, Yale University, New Haven, Connecticut 06520 (U.S.A.)

INTRODUCTION

This study of the human mu rhythm was somewhat unusual. Whereas the usual sequence of events is to discover animal models for established human EEG rhythms,[1,2] this study was made with the intention of finding a human analog for an animal EEG rhythm.During the past few years,my research has been predominantly the study of EEG feedback training,or operant conditioning,as a possible treatment for epilecptic seizures.[3-6] The rationale for this line of research was derived from studies of a spontaneous EEG rhythm in the cat,termed the sensorimotor rhythm or SMR.[7] This rhythm, with a frequency of 12-18 c/sec, is localized to somatosensory cortex and its main functional characteristic is its occurrence during immobility and its attenuation by movement.[8-11] Based on a finding of delayed seizures in cats conditioned to enhance SMR, conditioning or feedback training of similar frequencies in the human EEG was subsequently applied clinically in human epileptics.[12] A history of EEG feedback training as a therapeutic technique in epilepsy (see Kuhlman and Kaplan,[13] for a review) is not appropriate here, except to point out that there was little evidence to support the assumption of a reliable sensorimotor rhythm in the human EEG - an assumption on which the rationale for feedback training in epileptics was based.

The suggestion that the central mu rhythm in humans may correspond to feline SMR has been made by a number of investigators.[9,10,14,15] The mu rhythm was described by Gastaut[16] as an arch-shaped rhythm within the alpha frequency band recorded from the central or Rolandic area of the scalp. This activity, termed the *rythme rolandique en arceau*[17] or Rolandic wicket rhythm,[18] is functionally similar to the SMR in that it is attenuated or blocked by movement, predominantly by voluntary movement of the contralateral hand, but also by passive or reflex movement and by somatosensory stimulation.[17,18] The term "mu rhythm" was adopted by the International Federation[19] because of the similarity of the shape of this activity to the Greek letter "mu," and this terminology is now commonly used.

Possible relationships between feline SMR and the human mu rhythm have been difficult to investigate, largely because the mu rhythm has been difficult to record under standard conditions. On the basis of visual inspection of EEG records for the characteristic "mu" or "wicket" shape, a mu rhythm was identified in only 18 of 500 recordings by Chatrian et al.,[18] in 18 of 559 recordings by Klass and Bickford,[20] and in one study in only 14 of 10,000 recordings.[21]

Therefore, I began by investigating whether a sensorimotor or mu rhythm could be detected reliably from the human scalp.[22] Power spectral analysis of the EEG was conducted in two small samples of normal subjects and epileptic patients. The occipital alpha rhythm can also be recorded from electrodes placed over central regions, so it was important to devise a method which could differentiate the two rhythms. I would like to elaborate on the main points of this study, and then to speculate on the neural origin of the human mu rhythm. I will propose the hypothesis that the mu rhythm is generated in the posterior bank of the central sulcus, and will present some preliminary data that are consistent with this hypothesis.

FUNCTIONAL TOPOGRAPHY

The method of "functional topography" was used. The EEG was analyzed along three dimensions: (1) *Spatial*, the topographic distribution across the scalp; (2) *Functional*, the reactivity of the EEG to sensory and motor manipulations; and (3) *Temporal*, the characteristic frequency of periodic waveforms.

To obtain a spatial separation of alpha and mu activity, EEG was recorded from two closely spaced bipolar chains of peri-Rolandic and occipito-parietal electrodes, one in a mediolateral direction and the other anteroposterior across the left hemisphere (Fig. 3). In epileptic patients, only the central and most posterior bipolar pairs of the anteroposterior chain were used.

To functionally separate alpha and mu rhythms according to their differential reactivity, a factorial design was used which incorporated interactions of movement and rest with the eyes open and closed (Fig. 1). Movement was voluntary, self-paced, continuous opening and closing of the contralateral (right) hand. This design resulted in four behavioral conditions, each lasting 2.5 minutes, which were replicated twice in random order in each subject. Power spectra (.25 c/sec resolution) were performed on consecutive 4 sec epochs and averaged to yield a spectrum representing 256 secs of activity for each behavioral condition.

Of the eight normal subjects and six epileptic patients studied, all exhibited rhythmic activity represented by dominant peaks in the power spectra in or near the nominal alpha frequency band. These peaks could be classified as mu rhythm or alpha rhythm according to the pattern of variation across the four behavioral conditions and the distribution across the scalp.

The presence of an alpha rhythm was determined in 13 of the 14 subjects. Alpha activity was maximal over posterior scalp with the eyes closed, greatly attenuated with eyes open, but not significantly affected by movement. The presence of a mu rhythm was determined in 7 subjects: in 5 normals and 2 patients. Figure 1 shows the independent variation of mu and alpha activity in the central power spectra. It is useful to examine central spectra obtained from a bipolar pair with an anteroposterior orientation because this derivation also records occipital alpha activity which can be compared with mu rhythm peaks in the same spectra. In this example, an 8 c/sec peak representing the alpha rhythm appears when the eyes are closed but is not greatly affected by movement. An additional peak at 9.5 c/sec representing the mu rhythm is maximal at rest, is abolished by movement of the contralateral hand, but is not greatly affected by opening or closing the eyes. Of all bipolar derivations, mu activity was best recorded over the anterior parietal area (electrodes at C_3 and midway between C_3 and P_3).

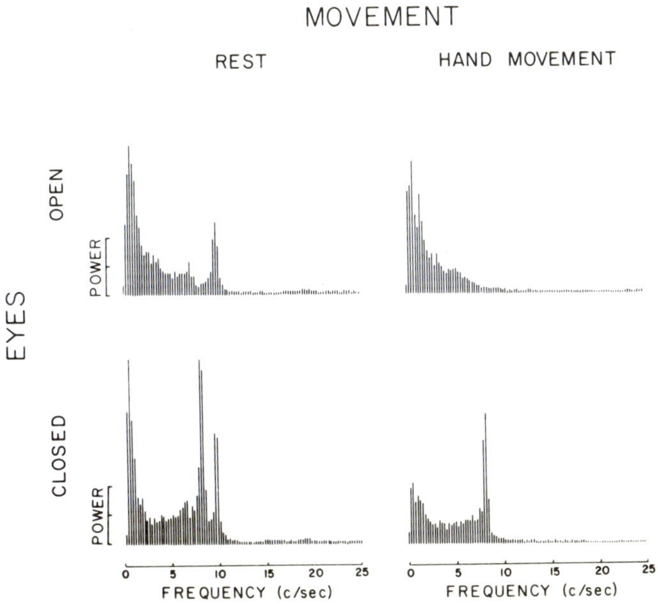

Fig. 1. Example of the independent variation of mu rhythm (9.5 c/sec) and alpha rhythm (8.0 c/sec) in power spectra recorded from the central area (bipolar pair 5, Fig. 3). From Kuhlman (22).

The frequency characteristics of mu and alpha activity obtained in the present sample of subjects is comparable with similar data reported by Storm van Leeuwen et al.[23] Their study utilized the same functional procedure and power spectral analysis. Agreement between the two studies is quite close, with mu rhythms having been detected in about 60% of the subjects, with frequencies in or near the nominal alpha band. Criteria in the present study were not restricted to the usual 8-12 c/sec alpha band since this study was attempting to identify an SMR which might be outside that frequency range. In two subjects the mu rhythm was outside the 8-12 c/sec range, but the majority of subjects exhibited mu activity at frequencies between 9-11 c/sec, illustrating that mu and alpha rhythms occupy the same frequency range. The mu and alpha rhythms were separate in frequency by more than one cycle on the average (1.46 c/sec), with the mu rhythm tending to be slightly higher in frequency (+.29 c/sec).

Figure 2 summarizes the functional topography analysis for the normal subjects in whom a mu rhythm was identified and for whom the full topographic array was used. Values shown represent averaged power of activity at the frequencies identified for mu and alpha rhythms in each subject. Mu rhythm frequencies were determined from bipolar pair 5 (electrodes at C_3 and midway between C_3 and P_3) at rest with the eyes open, and alpha frequencies were determined from bipolar pair 7 (electrodes at P_3 and midway between P_3 and O_1) during the eyes closed resting condition. In the absence of movement, mu activity was maximal at bipolar pair 5. Statistically significant attenuation during movement occurred over the Rolandic area (Fig. 2A). The difference between eyes open and eyes closed conditions was not significant at Rolandic locations, although a significant increase occurred with the eyes closed at the most posterior location. The latter result probably reflects spread of power from adjacent alpha peaks into the frequency of central mu rhythm. In Fig. 2B, the distribution of mu activity at rest relative to the maximum at bipolar pair 5 is compared with percent attenuation by movement. Rolandic derivations in the mediolateral plane showed less mu activity than derivations in the anteroposterior orientation, even though they included a common electrode at C_3. In general, the reactivity of mu rhythm to contralateral hand movement follows closely the distribution of mu activity at rest.

A similar analysis was conducted on alpha activity (Fig. 2C and D). In contrast to mu activity, alpha activity was not significantly attenuated by movement. Maximal power was seen at occipito-parietal locations 6 and 7, where alpha activity was significantly attenuated by opening the eyes. The distribution of alpha activity with eyes closed is comparable to percent

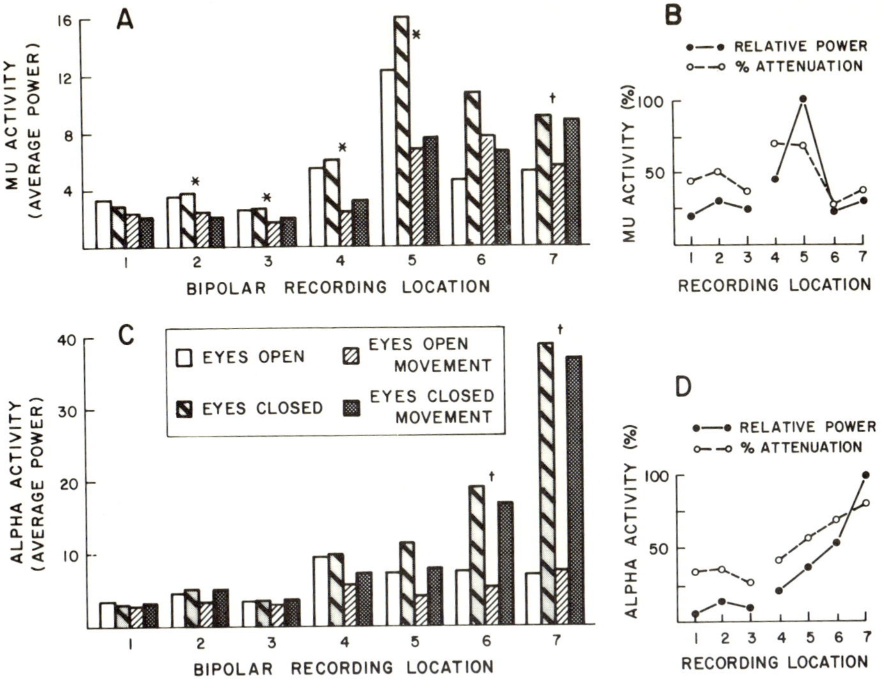

Fig. 2. Summary of functional topography of mu and alpha rhythms. Recording location refers to placement of bipolar pairs (Fig. 3). A and C: average power (arbitrary units) of mu and alpha activity, respectively. Stars indicate significant ($p<.05$) attenuation by movement; crosses indicate significant attenuation by eye opening. B: percent power of mu activity relative to pair 5 (relative power), and percent attenuation by movement (all data with eyes open). D: percent power of alpha activity relative to pair 7 (relative power), and percent attenuation by eye opening (all data with no movement). Adapted from Kuhlman (22).

attenuation of alpha activity by opening the eyes (Fig. 2D). The maximum power of mu activity was approximately 40 % of that of alpha activity.

Figure 3 more directly compares the reactivity to movement and eye opening for both mu and alpha rhythms. Reactivity is expressed as percent attenuation from the eyes closed resting condition for both rhythms. Attenuation was calculated for each subject separately. If power decreased to zero during movement or eye opening, attenuation would be 100%. If no attenuation occurred, a score of 0% was recorded. Attenuation was then averaged across subjects. Movement was with the eyes closed. This analysis further illustrates the functional and spatial distinction between mu and alpha activity, with mu activity showing maximal attenuation over the central areas during movement, predominantly at bipolar pair 5, and alpha activity showing

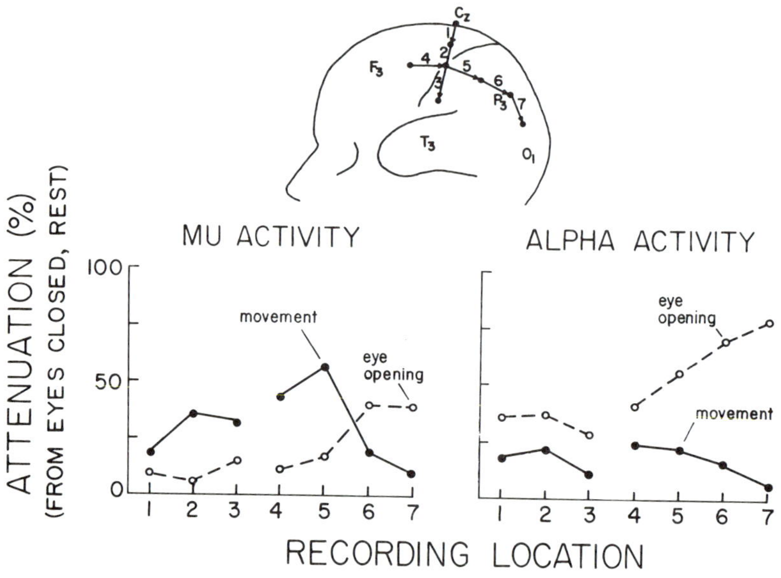

Fig. 3. Reactivity of mu and alpha activity to contralateral hand movement and eye opening expressed as per cent attenuation from the eyes-closed, no-movement condition. If activity disappears, attenuation = 100%. Location of bipolar electrode pairs is shown at top.

attenuation toward the posterior scalp with the eyes open. As was previously noted, mu rhythm is minimal at posterior scalp (Fig. 2B) such that attenuation at these locations (6 and 7) by eye opening is most likely due to the influence of alpha rhythm.

The reactivity of alpha rhythm to eye opening (maximum attenuation of 79%) was greater than that of mu rhythm to movement (maximum attenuation of 56% with eyes closed and 69% with eyes open). This might be accounted for by assuming that the eyes open condition is a more constant influence across the 2.5 minute periods used here than is phasic fist clenching. An analysis of mu activity conducted during the immediate fist clench over short epochs, for example by the methods reported by Pfurtscheller and Aranibar,[24] might be expected to yield greater attenuation. Nevertheless, the analyses were adequate to identify mu rhythm and to distinguish the mu rhythm from the alpha rhythm spatially, functionally, and temporally. The identification of mu rhythm in a much higher percentage of subjects than in previous studies may be attributed to the specific methods used. Mu rhythm was identified in previous studies primarily by retrospective visual inspection of the EEG records for the characteristic "wicket" shape.[17,18,20] In addition, subjects in previous studies were predominantly from clinical populations. There is some indication, for example, that mu activity may be less frequently detected in epileptic patients than in normal subjects.[6,22]

The available evidence from the study of human subjects is consistent with the idea that the human mu rhythm is analogous to the 12-18 c/sec SMR of the cat, as previously suggested by several investigators.[9,10,14,15] Cortical mapping studies in the cat have shown that the sensorimotor rhythm is primarily a somatosensory rhythm rather than a motor rhythm. SMR has been localized cortically to the primary somatosensory area SI, the posterior sigmoid gyrus of the postcruciate region, and subcortically to the ventral posterolateral nucleus (VPL) of the thalamus.[9-11]

Kuhlman[22] reported that the frequency of the mu rhythm remained stable across 25 days of recording over a 2-3 month period. Those data were collected in conjunction with another study in which subjects were given feedback training to enhance mu activity.[3] Subjects showed reliable enhancement of mu rhythm across sessions. This long term increase above resting levels has yet to be demonstrated in feedback training studies of the alpha rhythm,[13] and provides further support for the analogy between the mu rhythm and the feline SMR which has also been operantly conditioned.[15] The feedback signals were visual; the task required sustained visual attention, whereas drowsiness inhibited the mu rhythm. These behavioral correlates are generally consistent with the observations of behavioral correlates of sensorimotor rhythms reported in animals by Rougeul et al.[25] in the baboon and squirrel monkey and by Bouyer et al.[26] in the cat.

NEURAL ORIGIN

Having reviewed some available evidence regarding the distribution of the mu rhythm, its similarity to the sensorimotor rhythm, and the distinction between the mu rhythm and the alpha rhythm, I now turn to consideration of the neural origin of human mu activity. The analogy between the mu rhythm and the feline SMR and the localization of SMR to somatosensory cortex suggests that the mu rhythm might also be generated in somatosensory cortex. As previously mentioned, the mu rhythm is attenuated not only by voluntary movement of the contralateral hand, but also by passive movement and somatosensory stimulation.[18,27] The latter result, which was verified in 2 subjects by Kuhlman[22] in referential recordings during contralateral median nerve stimulation, indicates that efferent motor commands are not essential for the disruption of the rhythm. Kuhlman[22] also noted in intracranial depth probe recordings in one epileptic patient in whom a mu rhythm was identified, a correspondence between the location of mu activity and the location of the largest amplitude somatosensory evoked potential (SEP).

The specific inferences one can draw about the origin of mu rhythm from comparisons of such evoked potential data and the EEG obviously depends upon the specificity with which particular SEP components can be attributed to activity in particular anatomical structures. In the somatosensory evoked potential, the initial neural activity in primary cortex is recorded over parietal scalp as a biphasic negative-positive sequence of components at about 20 and 30 msec, the "N20-P30 complex".[28,29] Based on cortical surface recordings of SEPs in epileptic patients, Broughton[30] and Allison et al.[29] found that N20-P30 inverted in polarity across the central sulcus, but did not invert from surface to depth. This has been verified in our laboratory in every patient in whom recordings from the pial surface of the hand area were available. These data have been interpreted as indicating that the primary parietal components, as well as their precentral polarity-inverted reflections, are produced by an equivalent dipole generator oriented perpendicular to the central sulcus in Brodmann's area 3b of the postcentral gyrus.[29,30]

In bipolar scalp recordings of the mu rhythm, in subjects whose mu rhythm showed a clear "mu" or "wicket" shape, an apparent polarity inversion was sometimes noted.[22] Bipolar recordings involving common electrodes, however, are difficult to interpret as to which electrode of a particular pair is in fact the "active" electrode. Referential recording using a distant reference generally is preferable, but such recordings rarely show clear mu activity. Most subjects showed little mu activity anterior to C_3, and in subjects with occipital alpha activity, the mu rhythm tended to exhibit a sinusoidal waveform, making polarity determinations difficult.

In one normal subject with no detectable alpha rhythm, the mu rhythm was clearly visible in referential recordings to interconnected ears (Fig. 4). This activity was determined to be mu rhythm in the power spectra by the functional procedure previously described. Scalp recorded mu activity at 11 c/sec (eyes open, rest) was seen at postcentral electrodes, and was less well developed at precentral locations (Fig. 4A). Comparison of the polarity of the mu rhythm across the estimated location of the central sulcus[31] indicates an inversion from a negative sharp wave and positive slower wave posteriorly (a "mu shape"), to a positive sharp wave and negative slower wave anteriorly, (a "wicket" shape, Fig. 4B). A similar polarity inversion was seen in the primary cortical response of the somatosensory evoked potential recorded from the same electrodes in the same subject (Fig. 4C). The polarity of components occurring at approximately 19 and 32 msec (the N20-P30 complex) clearly inverted from a

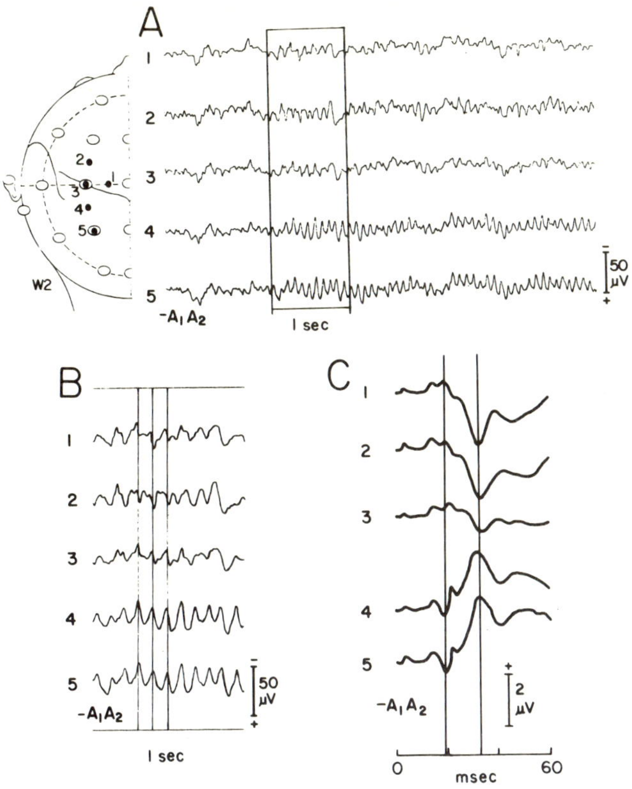

Fig. 4. Referential recordings of mu rhythm (A) and somatosensory evoked potentials (C) in the same subject. B: expanded section of the record shown in A (box). An apparent polarity inversion of mu rhythm was noted. A similar reversal was seen in the somatosensory evoked potential. Stimulation was a shock to the median nerve of the contralateral (right) wrist at 3 ma above thumb twitch threshold. Each trace is the average of 192 responses.

negative-positive sequence posteriorly to a positive-negative sequence anterior to the central sulcus. For both spontaneous and evoked activity a transitional waveform was recorded at the electrode located approximately over the central sulcus (C_3) which had some characteristics of both precentral and postcentral waveforms.

Two quantitative methods were used to evaluate the impression of a polarity reversal of mu rhythm across the central sulcus. The first, which may be called "wave averaging," is illustrated in Fig. 5. This procedure is similar to averaging evoked potentials, but instead of time locking the average to a sensory stimulus, an arbitrary but unambiguous point on each wave is used. In the case of the mu rhythm, the negative sharp wave or "spike" of the wave was used. First, precentral and postcentral channels of EEG were digitized simultaneously in 4 sec epochs with 4 msec resolution. The postcentral recording from an electrode midway between C_3 and P_3, referred to interconnected ears, showed the clearest mu activity and was used to synchronize the averaging process. Representative time points for averaging are indicated by dots (Fig. 5, top). An average waveform of the mu rhythm was then obtained by averaging forward 160 msec and backward 40 msec from each spike. Using the same time points derived from the postcentral channel, an average waveform was similarly obtained for the precentral channel (electrode midway between C_3 and F_3). In Fig. 5, bottom, the average of 40 individual waves is shown. The dot represents the synchronization point of the average. The average precentral mu wave is an approximate mirror image of the mu wave exhibited posterior to the central sulcus. The period of this activity is 92 msec, or a frequency of about 11 c/sec.

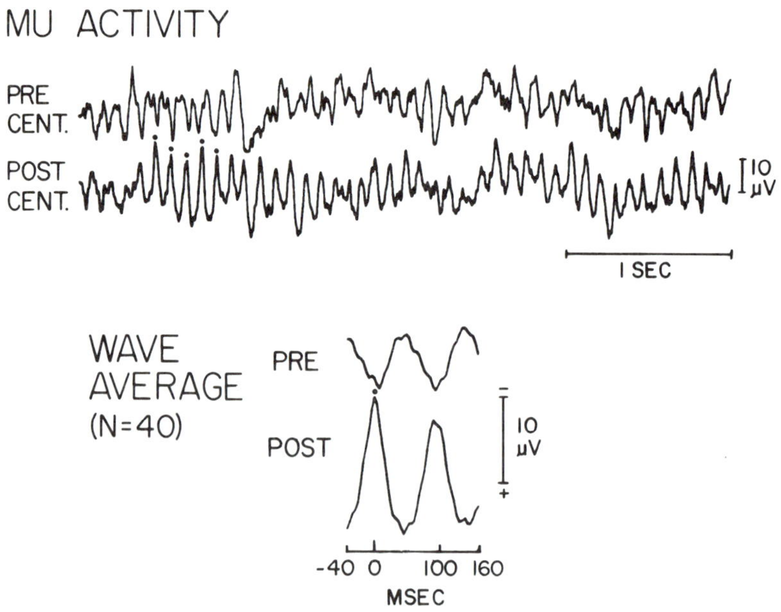

Fig. 5. "Wave averaging" of mu rhythm. Top: pre- and postcentral recordings (electrodes 2 and 4, Fig. 4) referred to interconnected ears. Dots illustrate representative synchronization time points for averaging. Bottom: average mu waves calculated from top. Forty waves were averaged from the sample. Same subject as in Fig. 4.

Crosscorrelation functions were performed on the data of Fig. 5 (top) to compare the phase relationship between pre and postcentrally recorded mu activity. A crosscorrelation function plots the degree of correlation between two signals as one signal is shifted in time to the other. Figure 6 first presents autocorrelation functions for each channel, in which each signal is time shifted relative to itself. These functions show positive peaks at 92 msec (11 c/sec) corresponding to the 92 msec period evident in the wave average in Fig. 5. Correlations were +.64, and +.30 in the post- and precentral recordings, respectively. The crosscorrelation function (Fig. 6, bottom) shows a negative correlation at zero time delay and negative values at leads and lags corresponding to the period of mu activity (92 msec). At zero time delay, the coefficient is negative in sign, although small in magnitude due to the presence of other activity in phase in both channels (see Fig. 5). The symmetrical values of about -.5 as the signals are shifted forward and backward by one cycle of the mu rhythm indicate that the pre- and postcentral recordings are out of phase by approximately one cycle, consistent with a polarity reversal across the central sulcus.

Figure 7 illustrates one possible interpretation of the site of origin of the human mu rhythm that is consistent with the polarity inversion just described. The drawing is of the central area of the left hemisphere of a human brain cut saggitally through the hand area approximately 7 cm from the

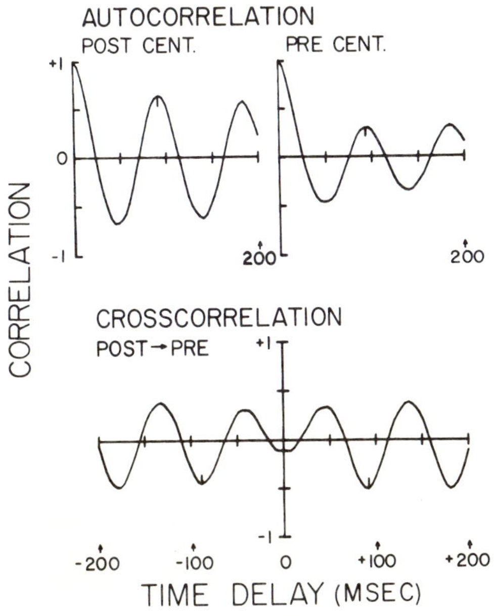

Fig. 6. Autocorrelation and crosscorrelation analyses performed on the data of Fig. 5, top. Tics indicate 92 msec time points (frequency of mu rhythm = 11 c/sec).

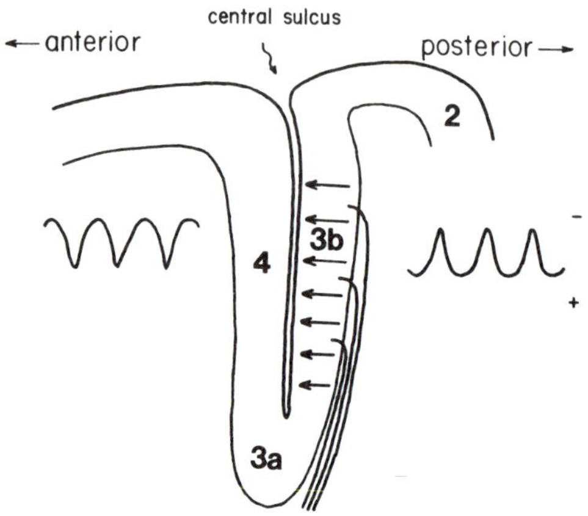

Fig. 7. Drawing of saggital section of human sensorimotor cortex of the left hemisphere. A hypothetical dipole generator of mu activity in the posterior bank of the central sulcus (somatosensory SI, Brodmann's area 3b) is assumed to produce a polarity inversion of the mu rhythm across the central sulcus.

midline. Precentral gyrus is to the left; postcentral gyrus is to the right. Somatosensory afferents from the hand are represented within the cortex of the postcentral gyrus. Note that most of the postcentral gyrus is buried in the central sulcus. Heavy thalamic afferent projections from VPL to Brodmann's area 3b, which is infolded to form the posterior bank of the central sulcus, have been shown in rhesus monkey[32] and may be assumed to exist in man. A horizontal equivalent dipole generator of mu rhythm in somatosensory SI, oriented perpendicular to the central sulcus in Brodmann's area 3b and parallel to the surface, is assumed to produce a polarity reversal of the mu rhythm across the central sulcus. According to this model, a different polarity of the mu rhythm would be recorded depending upon which "side" of the equivalent dipole one records from. Thus, the mu rhythm would be recorded as a "wicket" shape, with a positive spike and negative slower wave, from the "surface" or precentral side of the dipole, and as a "mu" shape, with a negative spike and positive wave, from the "depth" or postcentral side of the equivalent dipole. A similar explanation has also been advanced by Allison et al.[29] to account for the polarity reversal of some primary SEP components.

It should be recognized that the evidence reported above suggesting a polarity reversal of mu rhythm across the central sulcus was based on a single subject. Because clear mu activity is not typically observable in referential recordings in most subjects, the existence of such a polarity inversion will be difficult to verify. The data from this subject may not be representative of mu activity in the general population. On the other hand, mu activity in more typical subjects may not be essentially different from that shown for this particular subject, but may simply be masked by the presence of other activity. Recordings from the cortical suface in monkey and, when possible, in man are desirable to assess this question. While clearly speculative, the model presented is consistent with several findings obtained in bipolar recordings. The orientation of this proposed generator could account for the predominant distribution of mu rhythm in the anteroposterior plane. In the bipolar arrays of the present study, mu activity was largest in the anteroposterior chain and was reduced in the mediolateral chain across C_3. Due to the polarity reversal, one would also expect that a trans-Rolandic bipolar derivation with electrodes anterior and posterior to C_3 would augment recording of the mu rhythm as Kuhlman[22] has reported. Two independent generators of mu activity would be assumed to exist, one in each postcentral gyrus of each cerebral hemisphere. Storm van Leeuwen et al.[23] found by analyzing coherence functions that mu rhythms were generally bilaterally asymmetrical and not bilaterally coherent.

CONCLUSION AND SUMMARY

The spatial, functional, and temporal distinctions between the mu rhythm and the occipital alpha rhythm would seem to eliminate the possibility that rhythmic activity within the alpha band arises from only one source. In Berger's[33] original reports of electrical activity recorded from the human scalp, he believed that alpha activity arose from the entire cerebral cortex. Later, Adrian and colleagues[34,35] demonstrated that the alpha rhythm originated from occipital regions and was related to functions of the visual modality. The present study serves to extend these classical observations. The "alpha rhythms" may be considered a class of resting rhythms which are modality specific. Specific sensory modalities may independently generate specific rhythmic activity while at rest or while not receiving or processing input. Thus the occipital alpha rhythm may be considered a "visual alpha rhythm" and the mu rhythm may be considered a "somatosensory alpha rhythm." A similar hypothesis has previously been advanced by Chase[15] based on studies of the cat. This notion would predict that other sensory modalities would also generate specific rhythmic activities. The auditory modality would seem most amenable to further investigations by scalp recordings.

While it is clear that somatic stimulation may disrupt mu activity in the absence of movement, a conclusion that only afferent input from the periphery may block the mu rhythm would be contradicted by one important finding, namely, the blocking of mu activity in advance of movement, as reported by several investigators (see Chatrian,[36] for a review). Assuming that the mu rhythm is a reflection of "resting" somatosensory cortex, blocking preceding motor activity may reflect preparation of somatic cortex for receipt of the impending inflow of afferent feedback produced by the movement. This preparatory mechanism would favor the rapid evaluation of the outcome of the movement such that subsequent motor corrections could be applied without delay. It is clear that input from sources other than the periphery may disrupt mu activity.

To summarize, I have presented data which indicate that the mu rhythm is a normal resting EEG rhythm which can be frequently detected from the scalp by power spectral analysis and functional testing. I have suggested that the mu rhythm may be generated in somatosensory cortex, based on its scalp topography, its similarity to sensorimotor rhythms in animals, and its relationship to the distribution of somatosensory evoked potentials. I have hypothesized that the mu rhythm is generated in the posterior bank of the central sulcus in an area corresponding to area 3b and have presented some preliminary data to illustrate this hypothesis. It should be emphasized that this hypothesis, while a reasonable one, is not necessarily the correct one. Clearly, evidence in additional subjects regarding the phase and polarity relationship of the mu rhythm is needed in order to evaluate its neural origin. In addition, the description by Rougeul et al.[25] and Bouyer et al.[26] of multiple foci and multiple frequencies of sensorimotor rhythms in the monkey and cat suggests that, in addition to the class of "alpha rhythms," of which the mu rhythm is a part, we may also think in terms of "mu rhythms." Thus, there appear to be more animal rhythms for which we may discover a human analog.

ACKNOWLEDGMENTS

Supported by the Medical Research Service of the Veterans Administration and by NIMH Grant MH-05286. I thank T. Allison, W.R. Goff and C.C. Wood for advice and discussion and R. Bartozzi for preparing the manuscript.

REFERENCES

1. Andersen, P. and Andersson, S. (1968) Physiological Basis of the Alpha Rhythm, Appleton-Century-Crofts, New York.
2. Lopes da Silva, F.H., van Lierop, T.H.M.T., Schrijer, C.F. and Storm van Leeuwen, W. (1973) Organization of thalamic and cortical alpha rhythms: spectra and coherences. Electroenceph. clin. Neurophysiol., 35, 627-639.
3. Kuhlman, W.N. (1978) EEG feedback training: enhancement of somatosensory cortical activity. Electroenceph. clin. Neurophysiol., 45, 290-294.
4. Kuhlman, W.N. (1978) EEG feedback training of epileptic patients: clinical and electroencephalographic analysis. Electroenceph. clin. Neurophysiol., 45, 699-710.
5. Kuhlman, W.N. and Allison, T. (1977) EEG feedback training in the treatment of epilepsy: some questions and some answers. Pavlov. J. Biol. Sci., 12, 112-122.

6. Kuhlman, W.N. and Mattson, R.H. (1979) EEG feedback training in the treatment of seizures: mechanisms and maintenance of effect. Biofeedback Self-Regulation, 4, 267.

7. Roth, S.R., Sterman, M.B. and Clemente, C.D. (1967) Comparison of EEG correlates of reinforcement, inhibition, and sleep. Electroenceph. clin. Neurophysiol., 23, 509-520.

8. Brazier, M.A.B. (1963) The problem of periodicity in electroencephalogram: studies in the cat. Electroenceph. clin. Neurophysiol., 15, 287-298.

9. Howe, R.C. and Sterman, M.B. (1972) Cortical-subcortical EEG correlates of suppressed motor behavior during sleep and waking in the cat. Electroenceph. clin. Neurophysiol., 32, 681-695.

10. Rougeul, A., Letalle, A. and Corviser, J. (1972) Activité rythmique du cortex somethésique primaire en relation avec l'immobilité chez le chat libre éveillé. Electroenceph. clin. Neurophysiol., 33, 23-39.

11. Waller, H.J. and Feldman, S.M. (1973) Correlations between somatosensory thalamic activity and cortical rhythms. Brain Res., 57, 417-441.

12. Sterman, M.B., MacDonald, L.R. and Stone, R.K. (1974) Biofeedback training of sensorimotor EEG in man and its effects on epilepsy. Epilepsia, 15, 395-416.

13. Kuhlman, W.N. and Kaplan, B.J. (1979) Clinical applications of EEG feedback training. In R.J. Gatchel and K.P. Price (Eds.), Clinical Applications of Biofeedback: Appraisal and Status, Pergamon, New York, pp. 65-96.

14. Chase, M.H. and Harper, R.M. (1971) Somatomotor and visceromotor correlates of operantly conditioned 12-14 c/sec sensorimotor cortical activity. Electroenceph. clin. Neurophysiol., 31, 85-92.

15. Chase, M.H. (1974) Research strategies in operant EEG conditioning: an investigation of sensorimotor cortical activity. In M.H. Chase (Ed.), Operant Control of Brain Activity. Perspectives in the Brain Sciences, vol. 2, BIS/BRI, UCLA, Los Angeles, pp. 39-60.

16. Gastaut, H. (1952) Etude électrocorticographie de la réactivité des rythmes rolandiques. Rev. Neurol., 87, 176-182.

17. Gastaut, H., Terzian, H., and Gastaut, Y. (1952) Etude d'une activité électroencéphalographique méconnue: "Le rythme rolandique en arceau." Marseille-med., 89, 296-310.

18. Chatrian, G.E., Petersen, M.C. and Lazarte, J.A. (1959) The blocking of the rolandic wicket rhythm and some central changes related to movement. Electroenceph. clin. Neurophysiol., 11, 497-510.

19. Storm van Leeuwen, W. et al. (1966) Proposal for an EEG terminology by the Terminology Committee of the International Federation for Electroencephalography and Clinical Neurophysiology. Electroenceph. clin. Neurophysiol., 20, 306-310.

20. Klass, D.W. and Bickford, R.G. (1957) Observations on the rolandic arceau rhythm. Electroenceph. clin. Neurophysiol., 9, 570.

21. Brechet, R. and Lecasble, R. (1965) Reactivity of mu rhythm to flicker. Electroenceph. clin. Neurophysiol., 18, 721-722.

22. Kuhlman, W.N. (1978) Functional topography of the human mu rhythm. Electroenceph. clin. Neurophysiol., 44, 83-93.

23. Storm van Leeuwen, W., Wieneke, G., Spoelstra, P. and Versteeg, H. (1978) Lack of bilateral coherence of mu rhythm. Electroenceph. clin. Neurophysiol., 44, 140-146.

24. Pfurtscheller, G. and Aranibar, A. (1979) Evaluation of event-related desynchronization (ERD) preceding and following voluntary self-paced movement. Electroenceph. clin. Neurophysiol., 46, 138-146.

25. Rougeul, A., Bouyer, J.J., Dedet, L. and Debray, O. (1979) Fast somato-parietal rhythms during combined focal attention and immobility in baboon and squirrel monkey. Electroenceph. clin. Neurophysiol., 46, 310-319.

26. Bouyer, J.J., Montaron, M.F., Rougeul-Buser,A. and Buser, P. (1980) A thalamo-cortical rhythmic system accompanying high vigilance levels in the cat. In G. Pfurtscheller et al. (Eds.) Rhythmic EEG activities and cortical functioning. Elsevier/North Holland, Amsterdam.

27. Broughton, R., Regis, H. and Gastaut, H. (1965) Modification of somaesthetic evoked potentials during bursts of mu rhythm and fist clenching. Electroenceph. clin. Neurophysiol., 18, 720.

28. Goff, W.R., Allison, T. and Vaughan, H.G. (1978) The functional neuroanatomy of event related potentials. In E. Callaway, P. Tueting and S.H. Koslow (Eds.), Event-Related Brain Potentials in Man, Academic Press, New York, pp. 1-79.

29. Allison, T., Goff, W.R., Williamson, P.D. and VanGilder, J.C. (1980) On the neural origin of early components of the human somatosensory evoked potential. In J.E. Desmedt (Ed.), Clinical Uses of Cerebral, Brainstem, and Spinal Somatosensory Evoked Potentials, Prog. clin. Neurophysiol., vol. 7, Karger, Basel, pp. 51-68.

30. Broughton, R.J. (1969) Discussion in E. Donchin and D.B. Lindsley (Eds.), Average Evoked Potentials: Methods, Results, and Evaluations, NASA SP-191, Government Printing Office, Washington, D.C., pp. 79-84.

31. Jasper, H.H. (1958) The ten twenty electrode system of the International Federation. Electroenceph. clin. Neurophysiol., 10, 371-375.

32. Jones, E.G. and Powell, T.P.S. (1973) Anatomical organization of the somatosensory cortex. In A. Iggo (Ed.), Handbook of Sensory Physiology, Springer-Verlag, Berlin, 2, 579-620.

33. Berger, H. On the electroencephalogram of man. English translation by P. Gloor. Electroenceph. clin. Neurophysiol., 1969, Suppl. 28.

34. Adrian, E.D. and Matthews, B.H.C. (1934) The Berger rhythm: potential changes from the occipital lobes in man. Brain, 57, 355-385.

35. Adrian, E.D. and Yamagiwa, K. (1935) The origin of the Berger rhythm. Brain, 58, 323-351.

36. Chatrian, G.E. (1976) The mu rhythm. In A. Rémond (Ed.), Handbook of Electroencephalography and Clinical Neurophysiology, vol. 6, Part A, The EEG of the waking adult, Elsevier, Amsterdam, pp. 46-69.

© 1980 Elsevier/North-Holland Biomedical Press
Rhythmic EEG Activities and Cortical Functioning
G. Pfurtscheller et al., editors

COHERENCE ESTIMATES IN DETERMINING CENTRAL ALPHA AND MU ACTIVITIES

M. SCHOPPENHORST, F. BRAUER, G. FREUND and St. KUBICKI
Abteilung klin. Neurophysiologie, Klinikum Charlottenburg
Spandauer Damm 130, D-1000 Berlin 19 (F.R. Germany)

Various figures have been reported for the percentage of mu waves in normal subjects [1-4]. This may be due to several reasons.

While visually evaluating routine EEGs we noted that mu waves were remarkably unstable. During a single recording mu waves which had been present for several seconds would occasionally disappear and then reappear. On closer observation we found that the lability of the mu rhythm could be attributed to slight changes in the degree of vigilance, often hardly discernible from changes in the alpha rhythm.

Another difficulty in differentiating mu waves is often caused by their low amplitude.

Chatrian et al. [5] and Bostem et al. [6] mentioned that mu waves frequently show an amplitude of only 2o to 3o μV and can be reliably identified only in this range. If, in addition, an alpha rhythm projected to fronto-precentral regions is superimposed on mu waves, it is virtually impossible to identify them by means of visual analysis alone.

To overcome these difficulties, we applied the coherence method [7,8,9] for the detection of mu waves in a similar way to Storm van Leeuwen [1o].

If we assume the mu and the alpha rhythm to have different local origin, they should be incoherent; where "coherence" is a function estimated by computer for a pair of signals, indicating the degree of linear relationship between them. It is obtained by comparing the frequency distributions of the two signals by means of their cross-power spectrum, a complex quantity with magnitude and phase[11].

The magnitude is normalized to a range between zero and one; this is called the coherence function. A value close to zero indicates little or no linear dependency between the signals; a large value (close to one) indicates a common component of the two signals in the corresponding frequency band. The phase angle is only reliable if coherence differs significantly from zero.

Blocking the mu rhythm by voluntary movement, while suppressing mu coherence, should not affect the coherence of alpha waves. If both rhythms were present in one lead, with frequencies close to each other, alpha coherence might even increase with the disappearance of mu waves. Therefore the evaluation of changes in coherence, in addition to the visual analysis of the EEG recorded, should prove a useful tool for verifying mu waves even when they are of short duration, of low potential, or superimposed with an alpha rhythm.

In a study on 70 healthy subjects aged between 12 and 70 years, every person had to clench and unclench one fist continuously over a 3 minute period, followed by a 3 minute period of rest. The EEG was recorded from two bipolar longitudinal rows according to the 10-20 system. The results of this study and details of the data processing are reported[12].

To demonstrate the coherence method for determination of mu waves two examples are given.

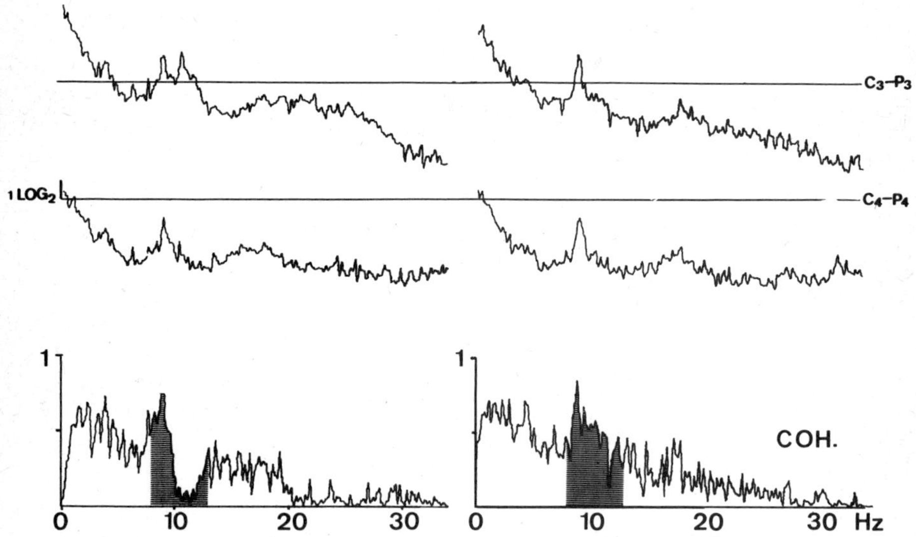

Fig. 1. Logarithmic power spectra and coherence function during rest (left) and fist clenching (right). The coherence within the alpha band is marked. Spectra (upper row) are calculated from left hemisphere (C_3-P_3) and right hemisphere derivation (C_4-P_4)(middle row). Note the unchanged 9c/sec coherence between both states, typical of an alpha rhythm and the increase in the 11c/sec coherence during motor activity indicating a superimposition of alpha and mu rhythms.

Interhemispheric coherence

Fig. 1 shows logarithmic power spectra calculated from the bipolar derivation C_3-P_3 and C_4-P_4, during rest (left on Fig. 1), and motor activity (right). The coherence function is plotted below the spectra; the coherence within the alpha band is marked. The spectra calculated from resting EEG show two peaks, one at 9 and the other at 11 Hz. During fist clenching the 11 Hz activity was attenuated, as shown in the reduced peak on the right-hand side of Fig. 1. The fact that the interhemispheric coherence of the 9 Hz peak was always high and independent of the motor activity indicated an alpha rhythm of 9 Hz, extending at least as far as the electrodes P_3 and P_4.

The 11 Hz peak belongs to mu-activity, since the 11 Hz activity was reduced during motor activity (corresponding to the definition of mu activity). It is of interest that, during rest, the interhemispheric coherence of the 11 Hz activity was low, and increased markedly during movement. This suggests that mu and alpha activities, both of 11 Hz, were superimposed during rest, while the mu activity was suppressed during fist clenching.

Subjects with predominant mu waves over the central region, or with mu waves as the only recognizable activity, always showed a low interhemispheric coherence during rest.

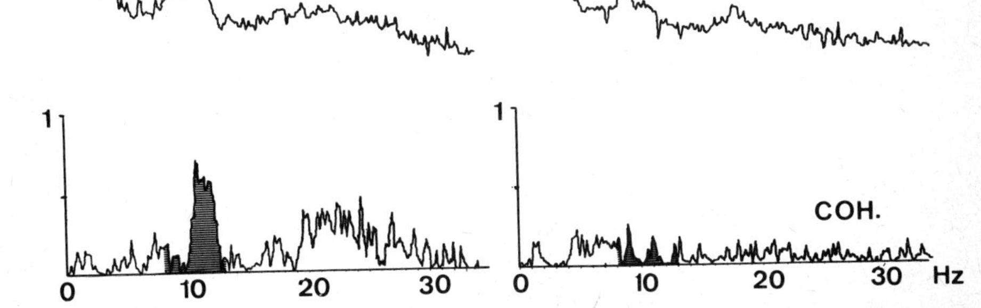

Fig. 2. For explanation see Fig. 1, but now all spectra are taken from left hemisphere. Note the high coherence of the 11 c/sec activity during rest (left) and the disappearance of the coherence during the motor action, indicating a 11 c/sec mu rhythm.

Intrahemispheric coherence

Intrahemispheric coherence was studied using the derivations F_3-C_3 and C_3-P_3. One example is given in Fig. 2 (same subject as in Fig. 1). The spectra during rest (left) indicate one clear peak in the anterior derivation (11 Hz) and a double peak (9 and 11 Hz) in the posterior EEG. During rest the intrahemispheric coherence was high, with the 11 Hz activity. The motor activity dampened the 11 Hz rhythm (no peak in the spectra on the right-hand side) and reduced the coherence to zero. This behaviour of the 9 and 11 Hz activity demonstrated an alpha field reaching to P_3 but not to C_3, and a mu field under C_3. Owing to the latter the intrahemispheric coherence was high during rest and low during movement. The intrahemispheric alpha coherence (9 Hz activity) was always low due to the fact that the alpha field did not reach the electrode C_3. Both examples demonstrate that (1) mu activity can either increase or decrease coherence, depending on the distribution of mu and alpha fields, (2) there is no coherence between mu and alpha waves, (3) the alpha activity shows a high interhemispheric coherence, (4) the mu waves lack interhemispheric coherence and (5) alpha and mu waves influence their coherence function reciprocally.

In this study, 60 % of the subjects were found to be mu waves carriers [12]. This proportion is similar to the one found by Storm van Leeuwen [10].

REFERENCES

1. Gastaut, H., Bacher, F., Bert, J., Blanc-Garin, J., Fessard, A., Fraisse, P., van Goethem, M. and Roger, A. (1959) Relations entre les variables électroencéphalographiques et celles exprimant la personnalité et les fonctions sensomotorices. Résultats d'une enquête effectuée sur une population homogêne de jeune adultes mâles agés de 20 ans. Rev. neurol.101, 320-390.

2. Schnell, R.G. and Klass D.W. (1966) Further observation on the Rolandic arceau rhythm. Electroenceph. clin. Neurophysiol. 20, 95.

3. Niedermeyer, E. and Koshino, Y. (1975) µ-Rhythmus: Vorkommen und klinische Bedeutung. Z.EEG-EMG 6, 69 - 78.

4. Kuhlmann, W.N. (1978) Functional topography of the human µ rhythm. Electroencephal. clin. Neurophysiol. 44, 83-93.

5. Chatrian, G.E., Petersen, M.C. and Lazarte, J.A. (1959) The blocking of the rolandic wicket rhythm and some central changes related to movement. Electroenceph. clin. Neurophysiol. 11, 497-510.

6. Bostem, F., Dongier, M., Demaret, A. and Herzet J.P. (1964) Discussion á propos du rhythme mu. Rev. neurol. 111, 335-337.

7. Brauer, F., Schoppenhorst, M. and Kubicki, St. (1975) Comparative spectral analysis and coherence funtions in the light of activations of sound and damaged parts of the human brain. In: M. Matejcek and G.K. Schenk (Eds.) Quantitative analysis of the EEG. AEG Telefunken EDP-Division, Konstanz, 183-194.

8. Schoppenhorst, M., Brauer, F. and Kubicki St. (1976) Spectra and coherence functions of μ waves and μ wave foci. Electroenceph. clin. Neurophysiol. 41, 659.

9. Schoppenhorst, M., Brauer, F. and Freund, G. (1977) Identifications of μ rhythm by means of spectral analysis and coherence functions. Electroenceph. clin. Neurophysiol. 43, 131.

10. Storm van Leeuwen, W., Wienecke, G., Spoelstra, P. and Versteeg, H. (1978) Lack of bilateral coherence of μ rhythm. Electroenceph. clin. Neurophysiol. 44, 140-146.

11. Jenkins, G.M. and Watts, D.G. (1968) Spectral analysis. Holden-Day, San Francisco

12. Schoppenhorst, M., Brauer, F., Freund, G. and Kubicki, St. (1980) The significance of the coherence estimates in determining central alpha and mu activities. Electroenceph. clin. Neurophysiol. 48, 25-33.

Rhythmic EEG Activities and Cortical Functioning
G. Pfurtscheller et al., editors

BLOCKING DIFFERENCES OF PRECENTRAL ALPHA-RHYTHM AND THEIR RELATION TO CEREBRAL POTENTIALS PRECEDING VOLUNTARY FLEXION AND EXTENSION MOVEMENTS IN MAN

L. DEECKE
Abteilung Neurologie, Universität Ulm, D - 79 Ulm (F.R. Germany)

It has been known now for three decades that periodic cortical activity can be blocked before or at the onset of a voluntary movement. Jasper and Penfield[1] had already noted in 1949 that the precentral beta rhythm was blocked prior to a hand movement. Kruger and Henry[2] reported that the Rolandic beta-rhythm in monkeys was blocked with the onset of movement. While recording through the intact skull, Klass and Bickford[3] and Chatrian et al.[4] showed that the rare Rolandic μ rhythm (Arkadenrhythmus, rhythme en arceau, wicket rhythm) was blocked shortly before the onset of a voluntary movement.

The significance of the particular morphology of the μ rhythm is not known. It is doubtful whether the peculiar en arceau shape has any functional meaning, since the occipital alpha-rhythm can exhibit similar deviations from the ideal sine wave, especially in children. We will, therefore, use the term precentral alpha rhythm instead of μ rhythm and summarize that the precentral alpha rhythm can be blocked or diminished over the precentral region shortly before and during a voluntary hand or finger movement. Blocking is usually more pronounced contralaterally than ipsilaterally to the moving hand. To our knowledge, systematic investigations comparing flexion and extension movements have not been reported so far.

Thirty subjects participated in the study. Except for one relatively ambidextrous subject all were right-handed according to a rating scale[5]. Seventeen were female and 13 were male, average age 22.3 years. Subjects were seated comfortably in an EEG chair in an electrically shielded room. They fixed their gaze on a given spot and performed quick movements of the right forefinger at their own volition. Movements ($\pm 40^{\circ}$ around the metacarpophalangeal joint) were restricted by an apparatus. The indifferent position was determined under complete relaxation prior to the experiment. It usually was a 25° flexed position. The sequence of blocks each consisting of about 100 movements, was randomized according to direction. Four hundred artifact-free trials were averaged for either type of movement. Eye movements and other artifacts were deleted by gaze fixation and on-line editing. This was done by marking artifact - contaminated trials with an electric pulse causing automatic reject of such trials during off-line time-reversed averaging.

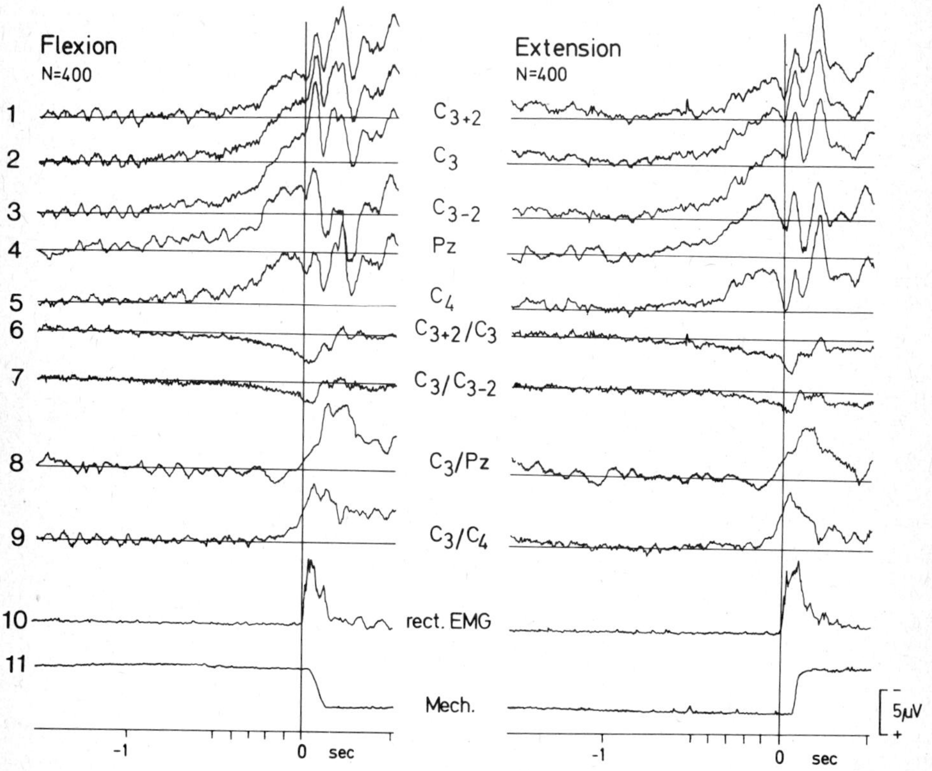

Fig. 1. Cerebral potentials preceding voluntary flexion and extension movements. Averages over 400 movements each. Trace 1 to 5, monopolar recordings versus linked ears; trace 6 to 9 bipolar recordings as indicated. C_{3+2}, two cm anterior, C_{3-2}, two cm posterior to C_3. Trace 10, rectified and averaged EMG from flexor indicis or extensor indicis muscle, respectively. Trace 11, mechanogram (finger position). Note equal readiness potential (BP) preceding flexion and extension (traces 1 to 5) but larger motor potential (MP) preceding extension as compared to flexion (traces 8 and 9). Pre-motion positivity best seen in traces 4 and 5, especially prior to extension.

The EEG was recorded over the left precentral area at C_3, as well as from locations 2 cm anterior (C_{3+2}) and 2 cm posterior (C_{3-2}). Recordings were also made from the mid-parietal area (P_z). All scalp leads were referred to linked ear lobes. Bipolar recordings consisted of C_3/C_4, C_3/C_z, C_{3+2}/C_3 and C_3/C_{3-2}. Movement onset was determined by the EMG of the right m. flexor indicis or m. extensor indicis, respectively. The EMG was rectified and averaged along with the EEG channels. The finger position was monitored by means of a potentiometer and averaged (Fig. 1). In half of the subjects, the alpha rhythm was high enough to permit single trial evaluation of the EEG for precentral alpha-blocking. In order to achieve 400 artefact-free movements for averaging, the subject usually performed 600 to 700 movements. Thus, a total of about 10,000 extensions and about 10,000 flexions were evaluated for precentral alpha blocking. This was defined as a disappearance or significant diminution of the alpha rhythm in the contralateral precentral leads preceding the movement but outlasting the onset of the movement. An example is given in Fig. 2, third movement from left. All other conditions were rated as "unclear blocking" (cf. Fig. 3).

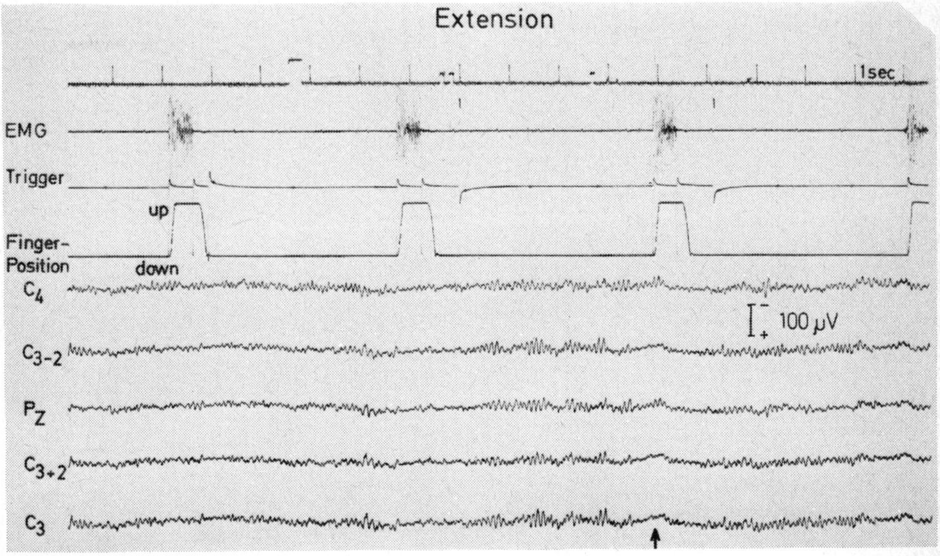

Fig. 2. Original EEG record showing precentral alpha blocking prior to extension (arrow). Blocking occurs unilaterally being restricted to the contralateral precentral leads C_3, C_{3+2} and C_{3-2}, while the readiness potential - here visible in the single trial - occurs bilaterally. No association between precentral alpha blocking and readiness potential (see text).

Blocking of the precentral alpha rhythm was recognized in about half of the extension movements (mean across the 15 subjects 52.4 %, SD 19.6), whereas it was present in only one third of the flexions (34.2 %, SD 11.45). The difference was highly significant ($t = 6.19$, $2p<0.001$). The mid-parietal and ipsilateral precentral recordings could also show contemporaneous blocking. More frequently, they showed no clear blocking as is seen in Fig. 2. Even in the averages over 400 movements, the remaining alpha rhythm (declining not with $\frac{1}{\sqrt{N}}$ but with $\frac{1}{\sqrt{N} \cdot \sqrt{\pi/2}}$ cf. Kornhuber and Deecke[6]) could exhibit these blocking differences (Fig. 4). In this case no blocking occurred prior to flexion. Whereas prior to extension, the contralateral precentral average curves showed a definite blocking in contrast to the ipsilateral precentral region where no significant blocking occurred. The finding of a higher incidence of contralateral precentral alpha blocking prior to extensions as compared to flexions may indicate a higher degree of precentral cortical activation prior to extensions as compared to flexion movements of comparable magnitude.

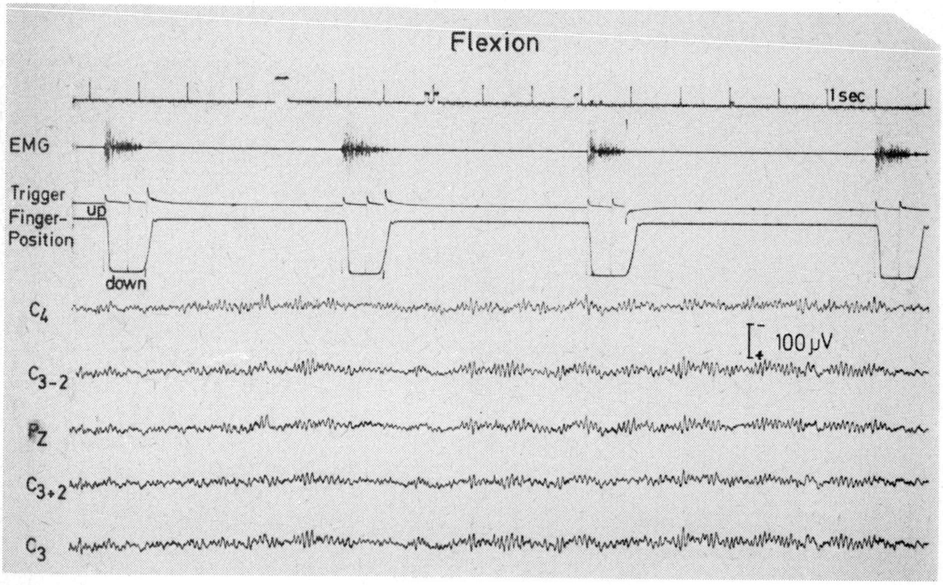

Fig. 3.Original EEG record with flexion movements. No typical blocking of the precentral alpha rhythm is found. Nevertheless, a readiness potential is seen with the third movement from the left.

This difference also corresponds to a significant difference in the cerebral potentials preceding voluntary movement: the motor potential was larger prior to extension movements as compared to flexion. The motor potential (MP) was taken as the additional negativity over the contralateral motor cortex as measured in the bipolar C_3/C_4-derivation occurring on average 50 to 60 msec prior to the onset of EMG activity (cf. Fig. 1). The MP averaged -1.63 μV prior to extension but only -1.38 μV prior to flexion ($2p<0.005$).

The pre-motion positivity (PMP), typically seen in leads P_z and C_4 prior to extension (cf. Fig. 1) averaged +0.77μV (SD1.61) prior to extension and +0.55μV (SD 1.31) prior to flexion as measured at P_z. Thus, it was slightly larger prior to extension. However, the difference was just below statistical significance.

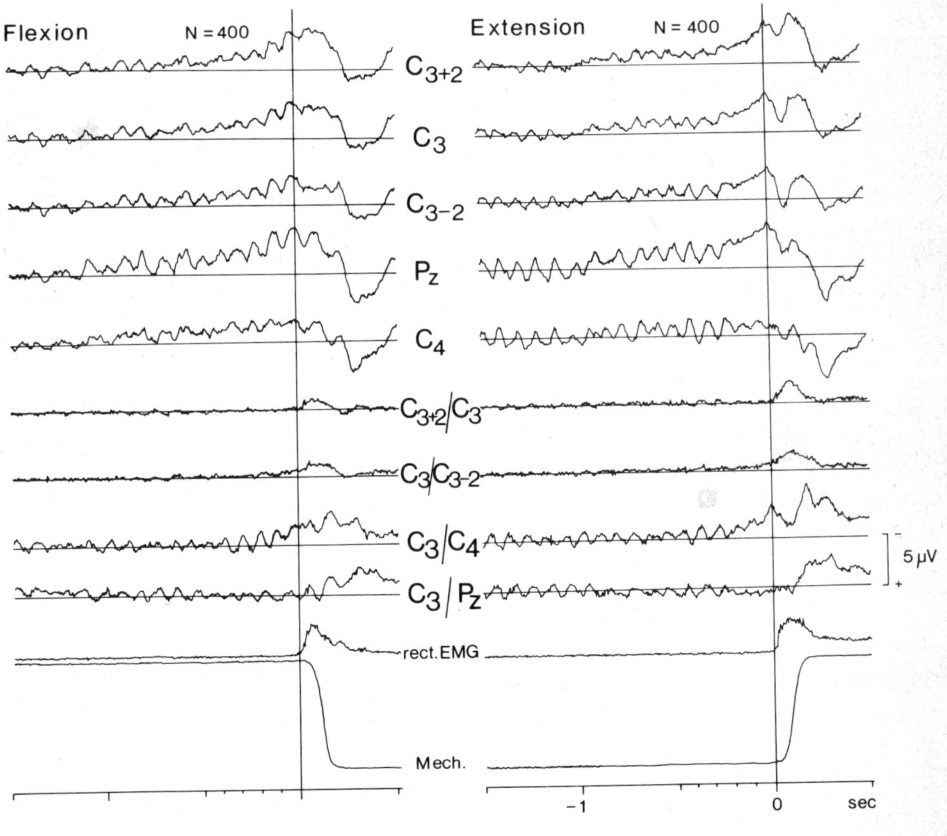

Fig. 4. Differences in precentral alpha blocking between flexion and extension. Even after averaging over 400 trials, the different blocking behaviour between flexion and extension is clearly present. Alpha with its constant frequency is less readily reduced by an averaging procedure than white noise (see text). In this case, no significant blocking occurs prior to flexion. However, blocking occurs prior to extension. It is present in the contralateral precentral leads but not in the ipsilateral precentral lead.

The Bereitschaftspotential (BP or readiness potential) showed no differences between flexion and extension in the contralateral precentral leads. In the ipsilateral and the mid parietal leads there were slight differences. In both leads, extension generated higher BP amplitudes than flexion at all three points of measurement. BP_{-150} at P_z averaged -3.1 μV (SD 1.98) for extension and only -2.6 μV (SD 1.98) for flexion ($2p<0.05$). BP_{-75} at P_z averaged -3.3 μV (SD 2.1) for extension and -2.9 μV (SD 1.89) for flexion ($p<0.05$). BP_{-150} at C_4 averaged -1.95 μV (SD 1.45) for extension and -1.59 μV (SD 1.74) for flexion, which was just below statistical significance. The other amplitude differences were not significant.

In conclusion, we found a significantly higher incidence of blocking of the precentral alpha rhythm prior to extension of the right index finger as compared to flexion of the right index finger, both movements being of comparable force and magnitude. Blocking did not occur on every trial. It occurred in half of the extension movements but only in one third of the flexion movements. Blocking differences could even be present in the curves averaged over four hundred movements. Due to its constant frequency, the alpha rhythm is not reduced through the averaging process with $\frac{1}{\sqrt{N}}$ as is white noise but with $\frac{1}{\sqrt{N}\,\sqrt{\pi/2}}$. Therefore, average curves contain residual alpha modulation even with large numbers of trails and with random triggering. Thus,there is no indication for phase lock or synchronisation between the alpha rhythm and the voluntary movement[6]. Blocking was generally lateralized in that it typically involved only or predominantly the contralateral precentral region. The observed blocking differences seem to be valid although no computer-assisted evaluation of precentral alpha blocking has been used. Such method would have made the results only more pronounced in view of the fact that an event-related decrease of alpha power would then have been found in 100 % and not only in half of the subjects (cf. Pfurtscheller and Aranibar[14]).

These blocking differences between extension and flexion are not reflected by similar differences of the Bereitschaftspotential or readiness potential, which had the same amplitude before extension and flexion in contralateral precentral leads. Over the ipsilateral precentral region and the parietal midline there was somewhat more BP activity before extension as compared to flexion that just reached the 5 % level of significance. There was a suggestion that the PMP is larger preceding extension than preceding flexion, but this difference just failed to reach statistical significance.

A striking difference was found with the motor potential (MP). The MP was 18 % larger prior to extension as compared to flexion. The MP is the only unilateral potential in the family of pre-movement potentials. It is thought to be related to the activation of motor cortical neurons associated with the pyramidal tract volley[7,8]. Thus, it may be very similar to what Craggs[9] called a "motor cortex field potential" that he recorded directly from motor cortex in monkey. The fact that the human MP has some characteristics in common with motor cortex cells as recorded in monkey (e.g. the differences between flexions and extensions) might support our belief that the MP reflects motor cortical activation. Clough et al.[10] found that the mean amplitudes of maximal motoneural EPSPs from motor cortex stimulation are larger for extensor muscles that for flexors. And in single cell recordings in motor cortex of conditioned monkeys, Fetz and Cheney[11] found that the increment in firing rate per increment in force ("motor gain") was about twice as great for extension-related as for flexion-related motor cortex cells.

The present results are in accordance with the importance of extension in the central organization of primate hand- and finger movements. In grasping movements, for example, extension necessarily precedes flexion[12]. Also in primates, the flexor arm muscles are stronger than the extensors so that on the cortical level it may require more cortical activation for the extensor than for the flexor system to perform a movement of equal size.

Precentral alpha blocking shares two important features with the motor potential : asymmetry and the difference between extension and flexion. However for the BP our results suggest rather a dissociation between this slow preparatory potential and precentral alpha blocking: The contralateral BP shows absolutely no difference in amplitude between extension and flexion while precentral alpha blocking does. The BP is rather a bilateral phenomenon and involves both hemispheres, while precentral alpha blocking involves the contralateral hemisphere more than the ipsilateral one. Considering the onset times, blocking occurs much later than the onset of the BP (cf. Fig. 4). Blocking seems to coincide with the time when the contralateral preponderance of negativity[8,13] begins (as seen in the bipolar C_3/C_4 derivation of Fig. 4, right side).

We may conclude from these considerations that the precentral alpha blocking is not associated with the early bilateral preparatory process of the BP, but it may be associated with its late-developing asymmetry. This asymmetry appears, on average 400 to 300 msec prior to the onset of movement and obviously is an expression of additional activity occuring later in the course of motor preparation which involves that particular motor cortex which is contralateral to the intended movement. This additional contralateral activity appears to be accompanied by desynchronisation of the periodic rhythms to such an extent, that it presents as blocking of the precentral alpha rhythm.

REFERENCES

1. Jasper, H., Penfield, W. (1949). Electrocorticograms in man:Effect of voluntary movement upon the electrical activity of the precentral gyrus.Arch. Psychiat. Nervenkr. 183, 163-174
2. Kruger, L., Henry, C. (1957). Electrical activity of rolandic region in unanesthetized monkey. Neurology (Minneap.) 7, 490
3. Klass, D.W., Bickford, R.G. (1957). Observations on the rolandic arceau rhythm. Electrocenceph.clin.Neurophysiol. 9, 570
4. Chatrian, G.M., Petersen, M.C., Lazarte, J.A. (1959). The blocking of the rolandic wicket rhythm and some central changes related to movement. Electroenceph.clin.Neurophysiol. 11, 497-510
5. Ferner, H., Horn, B., Suchenwirth, R. (1970). Die Bestimmung der Händigkeit beim Erwachsenen. Med. Welt 1, 33-35
6. Kornhuber, H.H., Deecke, L. (1965). Hirnpotentialänderungen bei Willkürbewegungen und passiven Bewegungen des Menschen: Bereitschaftspotential und reafferente Potentiale. Pflügers Arch. 284, 1-17
7. Deecke, L., Scheid, P., Kornhuber, H.H. (1969). Distribution of readiness potential, pre-motion positivity and motor potential of the human cerebral cortex preceding voluntary finger movements. Exp. Brain Res. 7, 158-168
8. Deecke. L., Grözinger,B., Kornhuber, H.H. (1976). Voluntary finger movement in man: Cerebral potentials and theory. Biol. Cybernetics 23,99-119
9. Craggs, M.D. (1974). Electrical activity of the motor cortex associated with voluntary movements in the baboon. J. Physiol. (Lond.) 237,12-13
10. Clough, J.F.M., Kernell, D., Phillips, C.G. (1968). The distribution of monosynaptic exitation from the pyramidal tract and from primary spindle afferents to motor neurons of the baboon's hand and forearm. J. Physiol. (Lond.) 198, 145-166
11. Fetz, E.E., Cheney, P.D. (1979).Muscle fields and response properties of primate cortico-motor neuronal cells. In: Granit, R., Pompeiano, O. (Eds.) Reflex control of posture and movement. Prog. Brain Res. 50
12. Phillips, C.G., Porter, R. (1977). Corticospinal neurons. Their role in movement. London, Academic Press
13. Deecke. L., Becker, W., Grözinger, B., Scheid, P., Kornhuber, H.H. (1973). Human brain potentials preceding voluntary limb movements. Electroenceph. clin. Neurophysiol. 33 (Suppl.), 87-94
14. Pfurtscheller,G., Aranibar, A. (1979). Evaluation of the event-related desynchronization (ERD) preceding and following voluntary self-paced movement. Electroenceph. clin. Neurophysiol. 46, 138-146.

© 1980 Elsevier/North-Holland Biomedical Press
Rhythmic EEG Activities and Cortical Functioning
G. Pfurtscheller et al., editors

THE SPATIAL AND TEMPORAL DISTRIBUTIONS OF ALPHA ACTIVITY AND THEIR MODIFICATION DURING MOTOR PREPARATION

P. V. POCOCK
Burden Neurological Institute, Bristol BS16 1QT (England)

INTRODUCTION

Scalp-recorded alpha activity still remains something of an enigma despite the many studies reported over the years. What was originally thought to be a simple rhythm[1] is now known to be a complex and involved phenomenon to the extent that some authors today feel impelled to define their subject of interest as 'rhythmic activity in the alpha band'.[2]

Alpha activity was soon realised to be responsive to changes of state, e.g. eye closure, and it would seem that observation of the changes of such activity under imposed conditions would therefore be one of the most fruitful ways of improving understanding of its mechanism and significance. Before modification can be investigated the baseline condition must be well-defined and modern computing facilities enable this to be done with the necessary greater precision than hitherto.

'Alpha activity', which is here taken to be rhythmic activity recordable on the scalp in the frequency range 8 to 13 Hz, must now be considered as multi-dimensional encompassing, at the very least, the aspects of its relative presence (amplitude), location on the scalp, time relationship at different locations (phase) and frequency.

This study was designed to investigate in detail the distributions of alpha activity over the scalp and, in view of the somewhat inconsistent results reported in the past regarding the so-called 'travelling' phenomenon[3-7], special emphasis was placed on this aspect. The changes in these distributions that occurred in the short interval immediately prior to a simple motor action (button-press) were then examined. The mechanism and generation of the Bereitschaftspotential (BP), the cortical manifestation of the preparedness for motor action, are now becoming well understood[8]. Here, the BP was used simply as an electroneurophysiological indicator of a change of state and no attempt was made to correlate any details of it with alpha activity changes.

METHOD

Fifteen normal right-handed adult males acted as subjects. EEGs were recorded from them whilst reclining in a relaxed state, with eyes closed in a semi-darkened room. In order to observe widespread scalp distributions, a regular array of twelve electrodes was used, located in four rows (occipital, parietal, central, frontal) of three (left-, right-hemisphere, midline), each referred to a mid-forehead (FPz) reference shown previously[9] to be the most appropriate for such recordings. Eye movements were monitored by referring the mid-forehead electrode to linked mastoids. Two further electrodes were located on the right forearm for registering the myogram associated with the button-press.

Amplification was achieved using an Elema Schonander Mingograph instrument set with 5 second time constant and 30 Hz high frequency filters. A PDP 12 computer could be triggered to sample and store 4 second epochs (sampling rate 128 Hz) of the output of the electroencephalograph, 2 seconds of which occurred before the trigger.

To establish the distributions for 'resting' alpha activity, in the first experimental stage (resting) twenty multichannel samples were stored on magnetic tape, the trials being triggered by the experimenter. An on-line visual display enabled artefact contaminated trials, e.g. with large eye movements, to be rejected. In the second experimental stage (active), the subject was asked to press a button with a rapid thumb movement with the right hand, voluntarily (self-paced). Here the button-press acted as the computer trigger and forty such trials were collected. Because time into the experiment was not of primary interest and indeed, effects from it had to be minimised, in the final stage twenty more trials were collected in the resting state and these trials combined with those in the first stage for subsequent analysis. In each stage the intertrial interval was approximately random with an average of about ½ minute.

Figure 1 gives an example of a polygraphic recording in the active stage showing the EMG acting as the trigger.

ANALYSIS AND RESULTS

Bereitschaftspotential

Linked mastoids (2M) is a suitable reference for recording BPs and so the activity on the eye movement channel (FPz-2M) was added to the activity on each other recorded EEG channel (e.g. Cz-FPz) to produce activity as if recorded by each scalp electrode referred to linked mastoids (e.g. Cz-2M). This was done for each trial and all trials added for the passive and active conditions separately. Figure 2 shows the distribution of this BP over the scalp averaged over subjects for the active condition, the arrows indicating the trigger

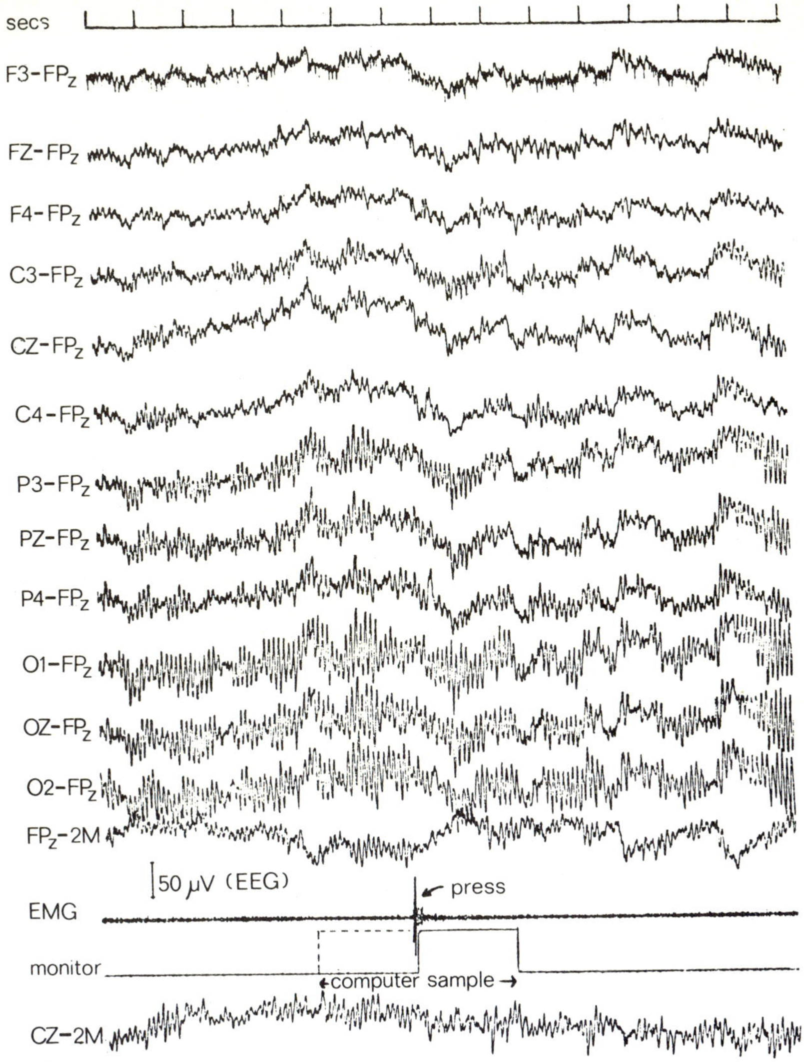

Fig. 1. Sample of EEG record in active condition

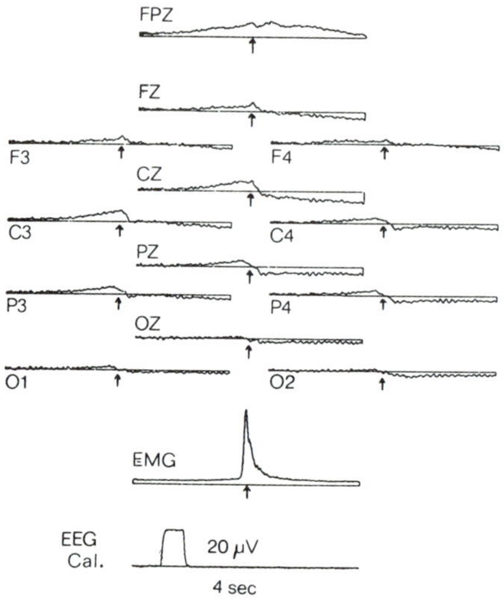

Fig. 2. Distribution of BP averaged over subjects

(button-press). The corresponding traces for the passive condition show no discernible slow potential shifts. This figure demonstrates the well-known BP distribution being maximal at the vertex (average amplitude 4.2 μV over 150-300 msec before the trigger) and is slightly larger contralateral to the operant hand[8]. The amplitudes are statistically significant, using the t-test, at all sites except F_4 and 0_2. The onset of the rise of the BP measured to the start of the EMG is 980 msec.

Alpha activity

To reduce any effects due to drowsiness only early trials (the first ten 2 second epochs) were used in the following analysis.

As already implied, what constitutes an 'alpha rhythm' is a matter of considerable debate. In order to try and minimise the chances of blurring the issue by confounding many different rhythms it was decided to concentrate analysis on single frequency components in individual trials and average at as late a stage in the processing sequence as possible. Because individual frequency components and their phase pattern in particular, were of interest, Fourier Analysis was chosen as the most appropriate basis for the processing. Consequently, for each subject and for each experimental condition, the (11) components in the alpha band of the power spectrum for each (12) channel of data were examined for all (10) trials collectively. Of all these (11x12x10) components, the ten maximum ones were selected (it was only counted as one selection if the same component for the same trial would have been selected from two or more electrodes). In other words, analysis for each subject and condition was focussed on the ten most dominant single alpha-band components e.g. the 9½ Hz component in the 4th trial which occurred maximally at, say, Oz, in the passive condition for subject 1, and so on. In practice, all these maximum components occurred posteriorly.

Once a component had been selected in this way, it was the distribution of precisely that particular component that was investigated at all the electrodes, results being finally averaged over trials and subjects. In the active condition, only the first half of each trial, i.e. the 2 second interval immediately preceding the button press, was considered in the above analysis.

TABLE 1

MEANS (OVER SUBJECTS) AND S.D.s OF AMPLITUDE (IN μV) AND SIGNIFICANCE TESTS* FOR EACH ELECTRODE

Electrode	Resting Mean	Resting (S.D.)	Active Mean	Active (S.D.)	Wilcoxon T Statistic	Signif. of T (n = 15)
F3	3.1	(1.4)	3.0	(2.0)	47	-
FZ	3.6	(1.6)	3.4	(2.3)	33	-
F4	3.4	(1.3)	3.1	(2.1)	28	-
C3	7.0	(3.6)	6.0	(3.8)	28½	-
CZ	7.5	(4.1)	6.2	(4.0)	23½	$p < 0.05$
C4	7.7	(3.8)	6.3	(3.7)	14½	$p < 0.01$
P3	13.2	(7.5)	11.6	(7.2)	23	$p < 0.05$
PZ	15.5	(8.2)	13.3	(7.9)	11½	$p < 0.01$
P4	15.2	(7.2)	13.6	(7.7)	10	$p < 0.01$
01	20.0	(11.0)	19.8	(11.7)	58	-
OZ	22.5	(10.1)	22.3	(10.3)	62	-
02	22.7	(6.5)	23.7	(9.9)	68	-

* For occipital electrodes amplitude values were used in the Wilcoxon matched pairs test. For other electrodes, percentage of the laterally corresponding occipital value were used.

Resting condition. The amplitude of alpha activity showed the familiar progressive attenuation from posterior to anterior regions. The amplitude values averaged over subjects are given with their respective standard deviations in Table 1 for each electrode for the resting condition. Figure 3 shows this effect with left, midline and right values averaged in each row and expressed as the square root of amplitude. It also gives the fitted regression line indicating an approximate adherence of amplitude (over these limited electrode positions) to a square law. A contour plot of the amplitude distribution is given as the uppermost diagram in Figure 4.

The phase characteristics of these components were investigated by plotting the phase of each particular component at each electrode as a vector (Figure 5). In this example a pattern of a progressively changing angle in the longitudinal direction, corresponding to a time shift lagging towards the back, can be

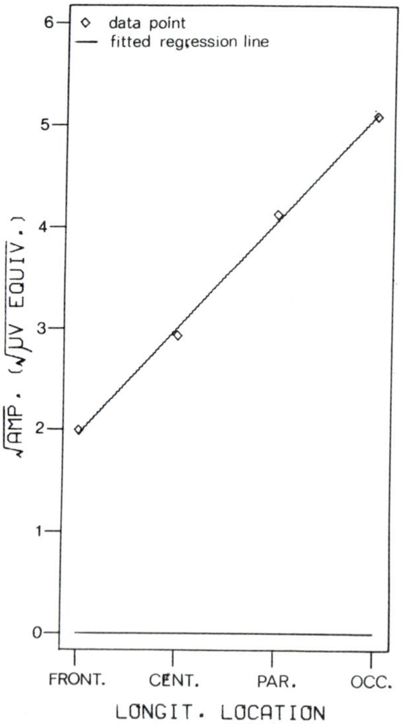

Fig. 3. Average longitudinal distribution of the square root of amplitude in the passive condition (including the "best straight line").

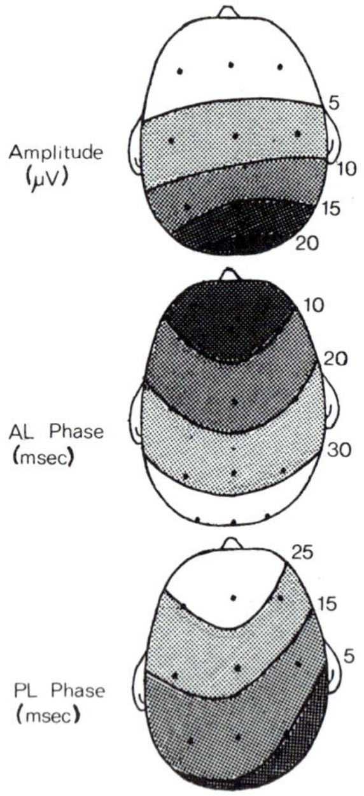

Fig. 4. Average amplitude and phase distributions in passive condition as contour plots.

clearly noticed. Many examples of such a pattern were seen and also many of the reverse, viz. lagging from back to front. For brevity these are termed anterior leading (AL) and posterior leading (PL) components respectively. The relatively few components which couldn't be seen to conform to either of these patterns were subsequently ignored.

Patterned cases were averaged into their two respective types after 'normalising' had been done by rotating their origins to coincide. Figures 6 and 7 give the subject averages of the AL and PL components respectively, averaging over transverse locations. It is this progressive longitudinal phase or time shift that others have referred to as sweeping, spreading, travelling, migration and propagation[3,4,10.11]. It is seen here to be approximately uniform with a velocity apparently in the order of 10 m/sec. Table 2 gives the mean and standard deviation of these time differences for each electrode when averaged over subjects for the resting condition and for both posterior and anterior leading components.

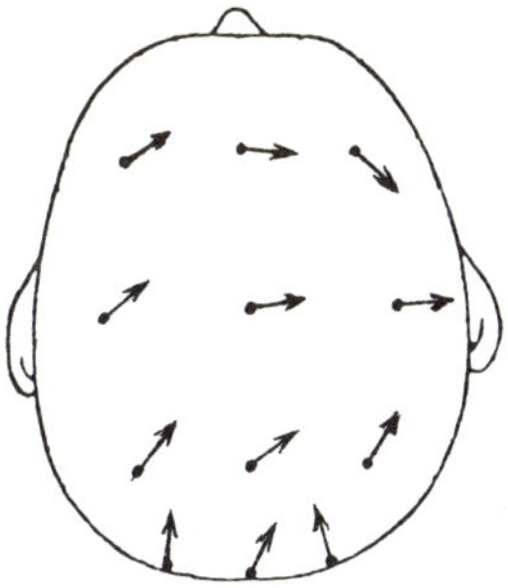

Fig. 5. Example of a phase pattern.

Average interhemispheric differences were about 10 msec but in no significant consistent direction.

The lower two diagrams of Figure 4 show these time shifts and reveal a further finding,namely that for AL components, midline locations 'lead' co-lateral electrodes and 'lag' them for PL components.

A simple gauge of frequency for these components was obtained by averaging together their individual frequency values for each condition and subject.

Modification during active condition. The distributions of amplitude and phase of alpha activity take on the same basic form during preparation for motor action but closer inspection reveals interesting differences.

Table 1 also gives the mean amplitude values (and standard deviations) averaging over subjects for each electrode in the active condition. It can be seen here that in parietal and central rows there is an attenuation of amplitude in the active condition, the appropriate value of Wilcoxon's matched-

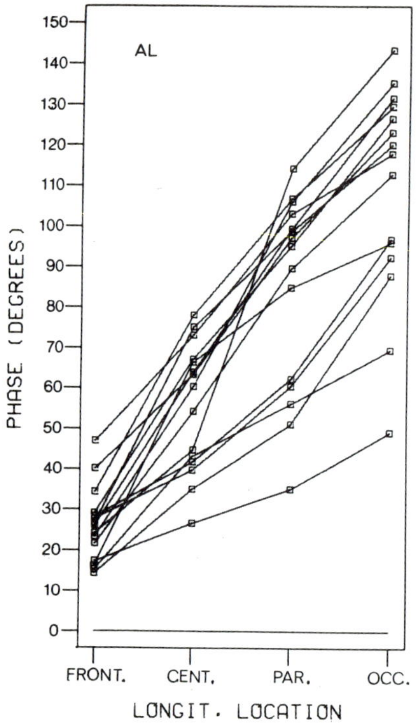

Fig. 6. Longitudinal distribution of phase of AL components in the passive condition.

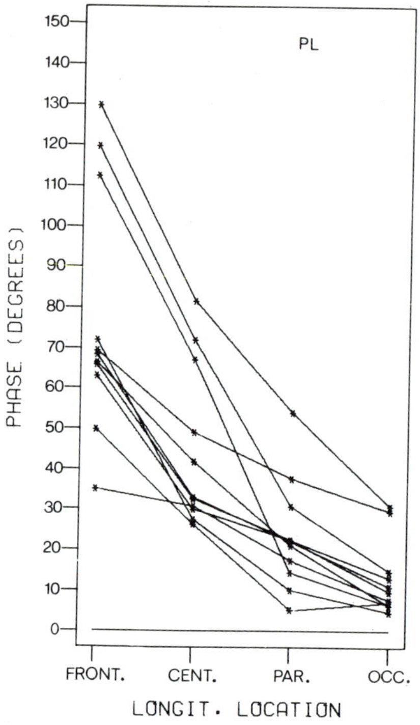

Fig. 7. Longitudinal distribution of phase of PL components in the passive condition.

pairs T statistic and its significance being given in the rightmost columns. No significant differences occur in occipital or frontal areas.

Transverse time differences were significantly shorter in the active condition for AL and PL components combined, averaging over longitudinal channels ($n=21$, $T=51\frac{1}{2}$, $p < 0.05$). Time differences were also shorter in the longitudinal direction (averaging over transverse channels) for PL components though statistical significance was not quite attained. Such differences were inconclusive for AL components. Mean and standard deviation values for these time differences averaged over subjects for each electrode are given in Table 2 for AL and PL components. The inconclusiveness of some of these differences and the difficulty in interpreting them will be appreciated in the light of what follows in the next section.

The average frequency of the selected components was slightly but significantly faster in the active condition (resting f=9.6 Hz, active f=9.8 Hz, n=15, T=20, $p < 0.02$).

Bi-directional Sweeping. The apparent longitudinal movement of alpha activity over the scalp has been often reported, some observing only anterior to posterior sweeping[6,10], some only the reverse[5,12], some both[3,4] and some could not discern any consistent patterns[7]. The velocity of propagation has also been variously reported, with equivalent approximate values of ½ m/sec[5], 3 m/sec[4], 5 m/sec[3], 6 m/sec[6] and 7 m/sec[12]. Animal work, however, has tended to produce slower speeds[7,13] but closer electrode spacing is generally used.

This leads to the hypothesis that the observation of bi-directional sweeping and of widely differing velocities is due to 'spatial aliasing'. This is where, analogous to the familiar temporal aliasing experienced with inadequate sampling

TABLE 2
MEANS (OVER SUBJECTS) AND S.D.s OF PHASE (IN MSEC) FOR EACH ELECTRODE

	Resting				Active			
	Post. Leading		Ant. Leading		Post. Leading		Ant. Leading	
Electrode	Mean	(S.D.)	Mean	(S.D.)	Mean	(S.D.)	Mean	(S.D.)
F3	8.5	(5.1)	21.0	(9.9)	8.7	(7.1)	23.3	(10.7)
FZ	4.7	(3.6)	24.6	(8.1)	4.0	(3.2)	26.2	(14.2)
F4	8.6	(6.3)	19.0	(8.0)	8.3	(5.6)	16.4	(9.7)
C3	16.8	(4.9)	13.1	(6.5)	15.8	(6.6)	15.3	(9.4)
CZ	12.5	(5.7)	16.1	(6.2)	10.7	(5.1)	16.9	(7.6)
C4	17.0	(7.3)	8.0	(5.2)	17.8	(5.8)	9.4	(7.8)
P3	23.9	(6.6)	7.2	(4.1)	22.0	(6.9)	9.6	(6.6)
PZ	21.7	(6.9)	7.8	(4.4)	19.8	(6.2)	9.3	(6.1)
P4	24.1	(7.8)	4.5	(3.1)	24.7	(8.1)	5.0	(3.1)
01	29.7	(7.1)	4.0	(2.8)	29.4	(8.9)	7.6	(6.0)
0Z	29.9	(7.7)	4.3	(3.4)	28.4	(8.9)	6.3	(5.6)
02	30.9	(7.7)	2.4	(3.2)	30.8	(9.6)	2.3	(1.7)

rates in frequency analysis, a wave frequency (seen here as a velocity of spread) is misinterpreted by virtue of an inadequate sampling interval (in space rather than time). This is illustrated in Figure 8 showing a rhythm seen at two

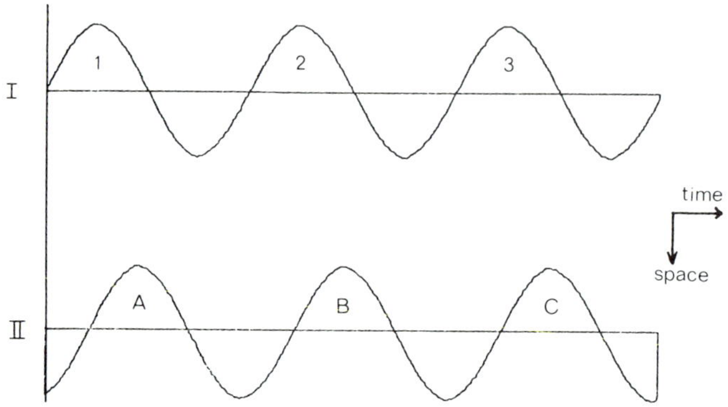

Fig. 8. Spatial aliasing principle

longitudinally separated locations I and II on the scalp. If it is assumed that wave 1 at I corresponds with wave A at II, etc., then a certain speed and direction of propagation is obtained. But if in 'reality' wave 2 corresponded to wave A then a different direction (and velocity) of propagation would be obtained.

If such spatial aliasing did occur, the situation could be as illustrated in Figure 9 where electrodes at the longitudinal sites are shown with uniform phase shifts of a rhythmic component, shaded waves corresponding. The magnitude

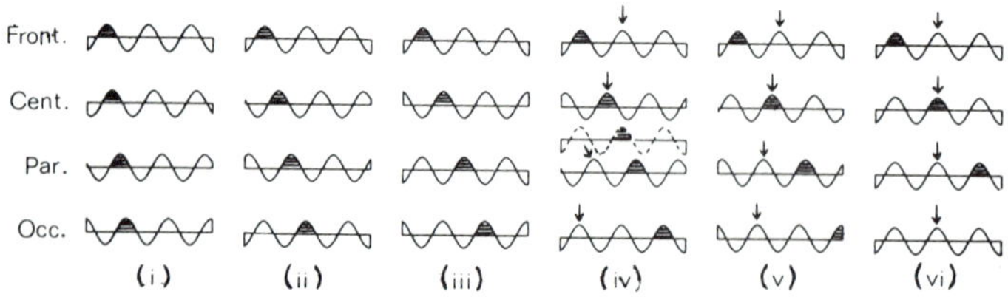

Fig. 9. A/P leading by spatial aliasing. Scheme I.

of phase shifts increases i.e. velocity of spread decreases, from (i) to (vi), each with anterior leading. However, in (iv) and (v) the analysis would assume the waves indicated by the arrows corresponded and hence would interprete these as posterior leading. The whole range of directions and velocities met in practice could have been obtained in this way, by having a constant direction but different velocity of spread.

However, relatively few examples were seen where the pattern would have arisen from a situation similar to (iii). An unusual "U"-shaped distribution of velocities could have accounted for this, but more likely is the extended spatial aliased scheme depicted in Figure 10 where only posterior leading components 'really' exist but with differing velocities, here centred around $\frac{3}{4}$ m/sec.

One implication of this hypothesis is that closer electrode spacing gives slower recorded velocities, a fact verifiable from the literature quoted earlier.

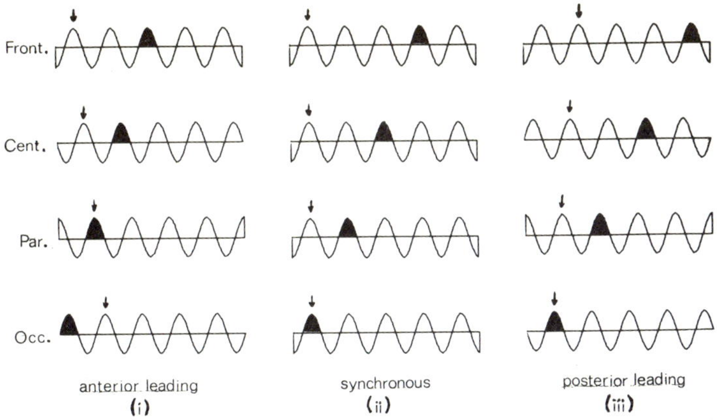

Fig. 10. A/P leading by spatial aliasing. Scheme II.

Secondly, the midline lead/lag phenomenon described previously follows automatically as illustrated in Figure 11. This shows how a midline leading component in one direction (top half of the figure) becomes interpreted as midline lagging in the opposite direction (bottom half) if the spread velocity is slower, both longitudinally and transverse.

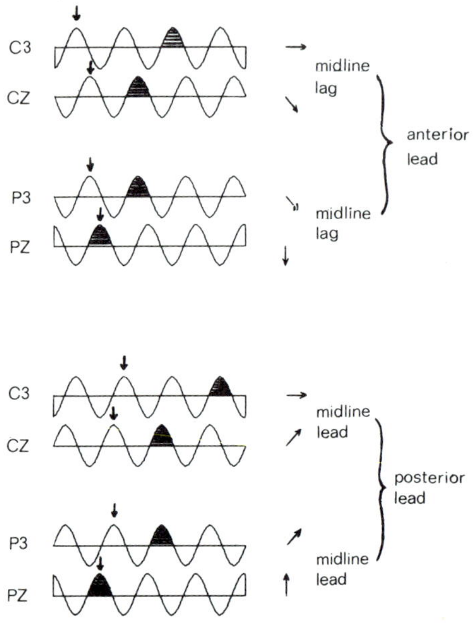

Fig. 11. Midline lead/lag by spatial aliasing

DISCUSSION

It is universally accepted that alpha activity attenuates with external stimulation, the most obvious example being the blocking of occipital rhythms with visual input e.g. on eye opening. Most studies, however, have only been concerned with posterior rhythms whereas here there were no attenuating effects seen at occipital sites during motor preparation (button-pressing condition). Pfurtscheller and colleagues[2], in recordings made in a study of other event-related potentials, have underlined the close relation between attenuation of alpha-band rhythms and specific sensory stimulation. They go on to point out that blocking of central rhythms occurs preceding the imperative stimulus of a CNV paradigm regardless of the modality (visual or auditory) of that stimulus. In this present study, no stimuli were used and the attenuation of alpha-band rhythms was observed during motor preparation only in centro-parietal areas. These localised areas coincide with those of maximal BP thus supporting theories that, as the BP is thought to be related to sensori-motor pre-activation[14], desynchronization of cortical rhythms is related to thalamo-cortical mechanisms. Extension of the argument that local desynchronisation occurs in 'involved' areas might suggest that prominent alpha-band activity is rarely seen anteriorly as these non-specific areas are constantly 'involved'.

Alpha activity could thus be viewed as the manifestation of 'idling' cortex.

Modification of phase (time) relationships have been less extensively reported, increased arousal having been proposed as affecting longitudinal time delays[10]. Here, the precise nature of the modification of the propagation is not clear, confounded, no doubt, by the apparent bi-directional sweeping. The general tendency, though, is for faster spreading during motor preparation, localised effects not being observable. Although here it seems unlikely that the simple undemanding task could significantly affect the level of arousal (an impression confirmed by post-experimental subjective reports) one has to look possibly toward this to account for the effect. This could be manifested by changes in the excitability of the cortex due to the preparation for motor action, as evidenced by the widespread rise in negative potential of the BP, and brought about by generalised thalamic or brain-stem effects.

The apparent increase in frequency during the motor preparation could be accounted for by a suppression of only slower alpha components[15]. However, it is in general agreement with others who report increases during mental tasks[16] and in on-going motor tasks[15], thus reflecting increased cerebral activation.

Central to the question of alpha generation is the relative involvement of cortex and deeper structures but this must be viewed in relation to the three dimensions of alpha activity seen on the scalp, viz. the actual generation of the rhythmic potentials, the pacing or modulation of these rhythms and their spread over the cortex.

Much work over the years points towards the potentials being generated in the cortex[17,18], confirmed recently by definitive work with intra-cortical electrodes[7]. Localised attenuation here in involved cortical areas fits this notion whilst indicating sub-cortical (thalamic ?) influences on the size of synchronous neural masses, perhaps by modification of the dendritic networks in the different cortical layers.

The thalamus has been shown to exhibit rhythmic activity in the alpha band, a strict point-to-point thalamo-cortical relationship having been proposed[19], thalamic rhythms thereby being relayed to the cortex[20]. The brain-stem and other parts of the basal ganglia have also been proposed as having a generalized influence on cortical rhythmicity[21] perhaps by largely controlling thalamic activity[22]. The generation of the BP has been shown to involve the thalamus and brain-stem[23,24] and so the generalised frequency modulation during motor preparation adds weight to this view of the influence of these structures on cortical rhythmicity.

Although some workers propose that subcortical areas are largely responsible for the travelling of alpha rhythms[21], most evidence suggests that the cortex plays the major, if not single-handed, role[5,11] although the precise mechanisms may not be clear. Although faster spreading during the widespread increased cortical activation of the button-pressing condition might support this theory (both implying altered characteristics of dendritic networks and the horizon-horizontal colaterals in particular), further work would have to be done to substantiate this.

In general, the results here help corroborate the notion of localised alpha-band rhythmic activity being generated, under thalamic influences, potentially over the whole cortex at numerous "epicentres"[7] and spreading cortically in all directions.

SUMMARY

In an experiment where samples of alpha activity were collected at rest and during the interval immediately prior to a simple voluntary motor action, the distributions of alpha activity over the scalp in terms of amplitude and phase were investigated.

A hypothesis is proposed to account for the observed bi-directional spreading.

The modifications of the distributions during motor preparation of localised (centro-parietal) attenuation, faster spreading and increased frequency are seen as furthering the understanding of the generation and significance of alpha activity.

REFERENCES

1. Berger, H. (1929) Über das Elektrenkephalogramm des Menschen. Arch. Psychiat. Nervenk., 87, 527-570.
2. Pfurtscheller, G. and Aranibar, A. (1977) Event-related cortical desynchronisation detected by power measurements of the scalp EEG. Electroenceph. clin. Neurophysiol., 42, 817-826.
3. Cooper, R. and Mundy Castle, A.C. (1960) Spatial and temporal characteristics of the alpha rhythm : a toposcopic analysis. Electroenceph. clin. Neurophysiol., 12, 153-165.
4. Rémond, A., Lesèvre, N., Joseph, J.P., Reiger, H. and Lairy, G.C. (1969) The alpha average. I. Methodology and description. Electroenceph. clin. Neurophysiol., 26, 245-265.
5. Nunez, P.L. (1974) Wavelike properties of the Alpha Rhythm. I.E.E.E. Trans. on Biomedical Engineering - B.M.E. 21, 6, 473-482.
6. Suzuki, H. (1974) Phase relationships of alpha rhythm in man. Jap. J. Physiol., 24, 569-586.
7. Lopes da Silva, F.H. and Storm van Leeuwen, W. (1978) The cortical alpha rhythm in dog: the depth and surface profile of phase. In M.A.B.Brazier and H.Petsche (Eds.). "Architectronics of the Cerebral Cortex". Raven Press, N.Y., 319-333.

8. Deecke, L., Grozinger, B. and Kornhuber, H.H. (1976) Voluntary finger movement in man: cerebral potentials and theory. Biol.Cybernetics, 23,99-119.

9. Pocock, P.V. (1976) On the collection and analysis of alpha rhythm data. Electroenceph. clin. Neurophysiol., 40, 542P.

10.Shaw,J.C. and McLachlan, K.R. (1968) The association between alpha rhythm propagation time and level of arousal. Psychophysiology, 4, 307-310.

11.Petsche, H. and Rappelsberger, P. (1970) Influence of cortical incisions on synchronisation pattern and travelling waves. Electroenceph. clin. Neurophysiol., 28, 592-600.

12.Walter, D.O., Rhodes, J.M., Brown, D. and Adey, W.R. (1966) Comprehensive spectral analysis of human EEG generators in posterior cerebral regions. Electroenceph. clin. Neurophysiol., 20, 224-237.

13.Petsche,H. and Šterc,J. (1968) The significance of the cortex for the travelling phenomenon of brain waves. Electroenceph. clin. Neurophysiol., 25, 11-22.

14.Papakostopoulos, D. (1978) Electrical activity of the brain associated with skilled performance. In D.Otto (Ed.). "Multidisciplinary Perspectives in Event-Related Brain Potential Research". U.S.Environmental Protection Agency, Washington D.C., 134-137.

15.Pfurtscheller, G., Maresch, H. and Schuy, S. (1977) Inter- and intra-hemispheric differences in the peak frequency of rhythmic activity within the alpha band. Electroenceph. clin. Neurophysiol., 42, 77-83.

16.Walter,W.G. (1959) Intrinsic rhythms of the brain. In J.Field (Ed.). "Handbook of Physiology, Sec.1. Neurophysiology Vol.1". Amer. Physiol. Soc., Washington D.C., 1, 279-298.

17.Bremer, F. (1953) Some problems in neurophysiology. University of London, Athlone Press.

18.Eccles, J.C. (1953) The Neurophysiological Basis of Mind : the Principles of Neurophysiology. Clarendon Press, Oxford.

19.Andersen,P., Olsen,L., Skrede,K. and Sveen,O. (1969) Mechanisms of thalamo-cortical rhythmic activity with special reference to the motor system. In F.J.Gillingham and I.M.L.Donaldson (Eds.). "Third Symposium of Parkinson's Disease". Livingstone, Edinburgh.

20.Anderson,S.A., Holmgren,E. and Manson,J.R. (1971a) Synchronization and desynchronisation in the thalamus of the unanaesthetized decorticate cat. Electroenceph. clin. Neurophysiol., 31, 335-345.

21.Garoutte,B. and Aird,R.B. (1958) Studies on the cortical pacemaker: synchrony and asynchrony of bilaterally recorded alpha and beta activity. Electroenceph. clin. Neurophysiol., 10, 259-268.

22.Anderson,S.A., Holmgren,E. and Manson,J.R. (1971b) Localised thalamic rhythmicity induced by spinal and cortical lesions. Electroenceph. clin. Neurophysiol., 31, 347-356.

23.Haider,M., Ganglberger,J.A. and Groll-Knapp,E. (1968) Computer analysed thalamic potentials and their relation to expectancy waves in man. Acta Neurol. Latinoamer., 14, 132-137.

24.McCallum,W.C., Papakostopoulos,D. and Griffith,H.B. (1976) Distribution of CNV and other Slow Potential Changes in Human Brainstem Structures. In W.C.McCallum and J.R.Knott (Eds.). "The Responsive Brain". John Wright and Sons Ltd., Bristol, 205-210.

© 1980 Elsevier/North-Holland Biomedical Press
Rhythmic EEG Activities and Cortical Functioning
G. Pfurtscheller et al., editors

VOLUNTARY MOVEMENT ERD: NORMATIVE STUDIES

GERT PFURTSCHELLER and ALBERTO ARANIBAR
Institute of Biomedical Engineering, Technical University of Graz
Inffeldgasse 18, A-8010 Graz (Austria)

Recent studies of human neural systems involved during the execution of voluntary movement, by way of computer analysis of EEG signals derived from scalp electrodes, have reemphasized the role of rolandic or central-parietal areas. In these regions, two types of simultaneous changes can be observed, depending on the electrode positions and methods of analysis being used:

1. Through DC or long time constant recording and averaging, triggered by repetitive, self-paced voluntary hand or finger movements, the Bereitschaftspotential or Readiness Potential, the Premotion and Motor Potentials have been obtained in normal human subjects [1,2,3].
2. Through analysis of the EEG frequency bands, particularly concentrated on the alpha and neighbouring beta bands, a decrease in amplitude or power within these frequency ranges has been observed to occur during voluntary, repetitive hand movements [4,5], which correspond to the classical central mu activity.

Both types of changes can be observed simultaneously through recording from central scalp electrodes in humans and triggering by voluntary, self-paced repetitive finger or thumb movements [6]. With the use of special techniques developed in our laboratory [6,7], the time course of the decrease in alpha power in relation to this voluntary movement could be accurately described. It was shown to consist of a phasic, i.e. short-lasting power decrease which has been designated event-related desynchronization or ERD [8]. Some of the normal features of these two EEG responses to intentional motor behaviour, across the central-vertex areas, have been reported [6].

However, the pattern of finger or thumb movement has proved inadequate to be transposed, as a reference, to the study of patients with motor disorders, particularly hemiparesis or hemiplegia. As is well known, many of these patients, having lost command of individual finger movements, can only attempt - if at all - rough global hand, even whole arm movements.

Therefore, a new experimental movement pattern was studied in our laboratory, requiring whole hand squeezing. The object of the present study is to define quantitatively the normal characteristics of the ERD responses, particularly within the alpha band, during this type of performance, i.e. voluntary hand squeezing by normal human subjects.

We have investigated 33 right-handed neurologically normal human subjects and the aim was to establish normative, statistical criteria that could be used as reference for further quantitative studies.

MATERIAL AND METHODS

Subjects. Two groups of subjects took part in these experiments, a first group of 16 students aged from 20 to 27 and a second, of 17 subjects aged between 46 and 80 years.

For statistical analysis, both groups were either pooled together or treated separately. All subjects were paid.

Recording. EEG was recorded using a bipolar transverse montage, as a coronal chain along the central areas and across the vertex. Six mm diameter disc electrodes were fixed with collodion in the following scalp positions: C_Z-C_4', C_4'-C_4, C_4-T_4', C_Z-C_3', C_3'-C_3 and C_3-T_3', the primed notations indicating intermediate positions between neighbouring electrodes of the 10-20 system. An SLE-8 channel machine was used, with a time constant T=1 sec and an upper cut-off standard frequency at 75 Hz.
The analog data from a channel with the trigger signal given by movement onset, and the data from the six EEG channels were stored on a HP 3968A analog tape recorder. For calibration, series of 10 rectangular 25μV pulses were stored.

Experimental sequence. The following sequence was used:

1. Control situation: EEG recording with subject's eyes closed, without intentional hand movements, for a period of 12 minutes.
2. Right hand movement: voluntary, repetitive, self-paced squeezing of a rubber ball, with eyes closed. Interval between movements was at least 10 seconds or longer. The total period was about 30 minutes.
3. Left hand movement: similar to right hand movement.

Throughout the experiment the subjects were sitting comfortably in a chair.

Data processing. The EEG signals were processed according to methods developed in our laboratory and described in previous reports [6,7,8]. These consisted of analyzing bandpass filtered EEG signals in the 8-10 Hz and 16-20 Hz frequency bands. Sampling epochs were of 6 sec, each including 4 sec preceding and 2 sec following the trigger by movement onset.

The sampling rate was 128 Hz and the central processing was done by a PDP11/45 computer. A 4th order low pass filter with cut-off frequency at 30 Hz was used for the EEG channel to eliminate aliasing errors. Each channel was calibrated individually.

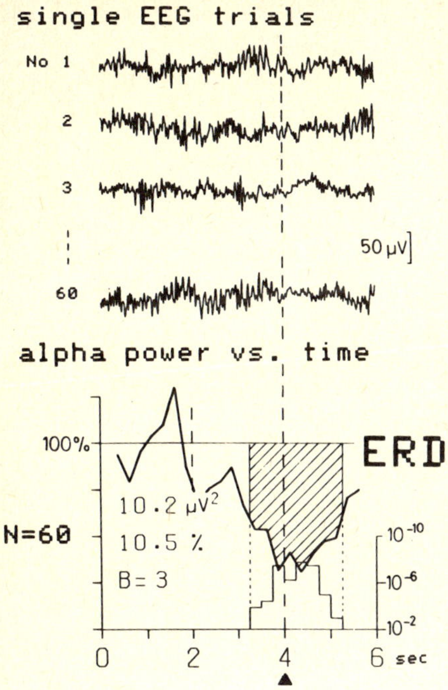

Fig. 1. Above: Single 6-sec EEG epochs are collected, up to N=60, then bandpass filtered. Below: Time course of the corresponding alpha power for 60 trials. The vertical broken line at t=4 sec indicates onset of hand movement. Left vertical scale: alpha power with 100 % as the normalized reference (average of the first two seconds). Before and during movement significant alpha power decrease (hatched area). Statistical significance of this decrease is indicated by the step function with levels to be read on right vertical scale. Reference alpha power: $10.2 \mu V^2$ or 10.5 % of total EEG power.

For each EEG channel a curve was plotted, showing the time course of power within the alpha (Fig. 1) or beta (16-20 Hz) band, with a time resolution of 0.25 sec. Each of these power vs. time graphs represented the average of about 60 individual 6-sec EEG trials.

In addition to the power-time curves, the final graphic output also included the conventional unfiltered average (i.e. without bandpass filtering) of the same 6-sec EEG trials, containing, in case of hand movements, the Bereitschaftspotential. Furthermore, two superimposed logarithmic power spectra were plotted, one correspnding to the first 2 sec of the interval taken as a reference and the second to the 2-sec segment around the trigger time, i.e. 1 sec preceding and 1 sec following movement onset.

An example of the final graphical output for one subject (subject 244) is presented in Fig. 2. In the control situation (Fig. 2a) there is no Bereitschaftspotential (BP) and the superimposed logarithmic power spectra are nearly identical to each other. The 6-sec power-time curves, within the alpha

(8-12 Hz) or beta (16-20 Hz) band, show no statistically significant variations. The reference to judge these variations was the mean power during the first 2 sec of the interval.

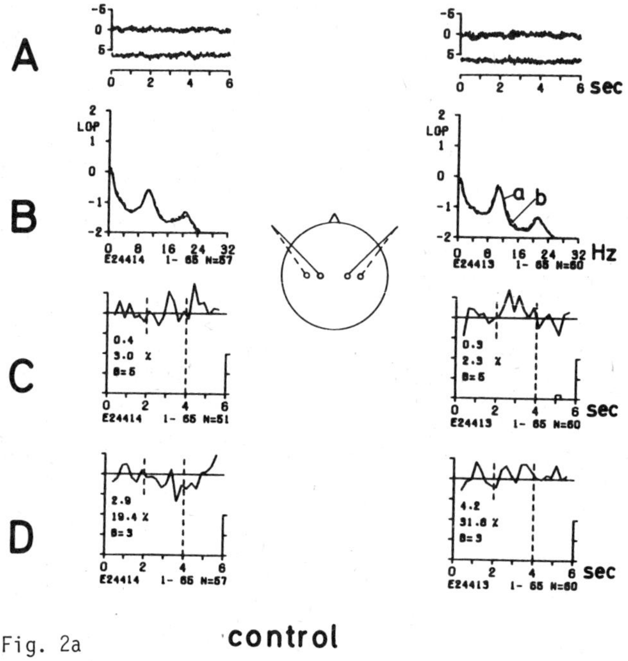

Fig. 2a

Fig. 2. Comparisons between two homologous, midcentral EEG derivations (C_3'-C_3 and C_4'-C_4) from a 24 y.o. subject, under control conditions (Fig. 2a), in trials with right hand (Fig. 2b, left) or left hand (Fig. 2b, right) squeezing. In each panel, from top to bottom: (A) averaged unfiltered EEG (with BP in Fig. 2b) with intertrial S.D. below, ordinates in μV, negative up; (B) superimposed logarithmic 2-sec spectra for the reference period, t=0-2 sec (full line) and for the interval t=3-5 sec (stippled line). In the last two rows power vs. time plottings as in Fig. 1, for 16-20 Hz (C) and 8-12 Hz (D) frequency bands.

In contrast, during repetitive right hand (Fig. 2b, left panel) or left hand (Fig. 2b, right) squeezing there were significant power changes within the beta (16-20 Hz) and particularly the alpha bands (8-12 Hz) as shown by the hatched areas. The criteria for establishing statistical significance of these changes are explained below.

Another version of graphical output, for the power within the alpha band, is displayed in Fig. 3, where the topographical distribution along the central areas and across the vertex is emphasized.

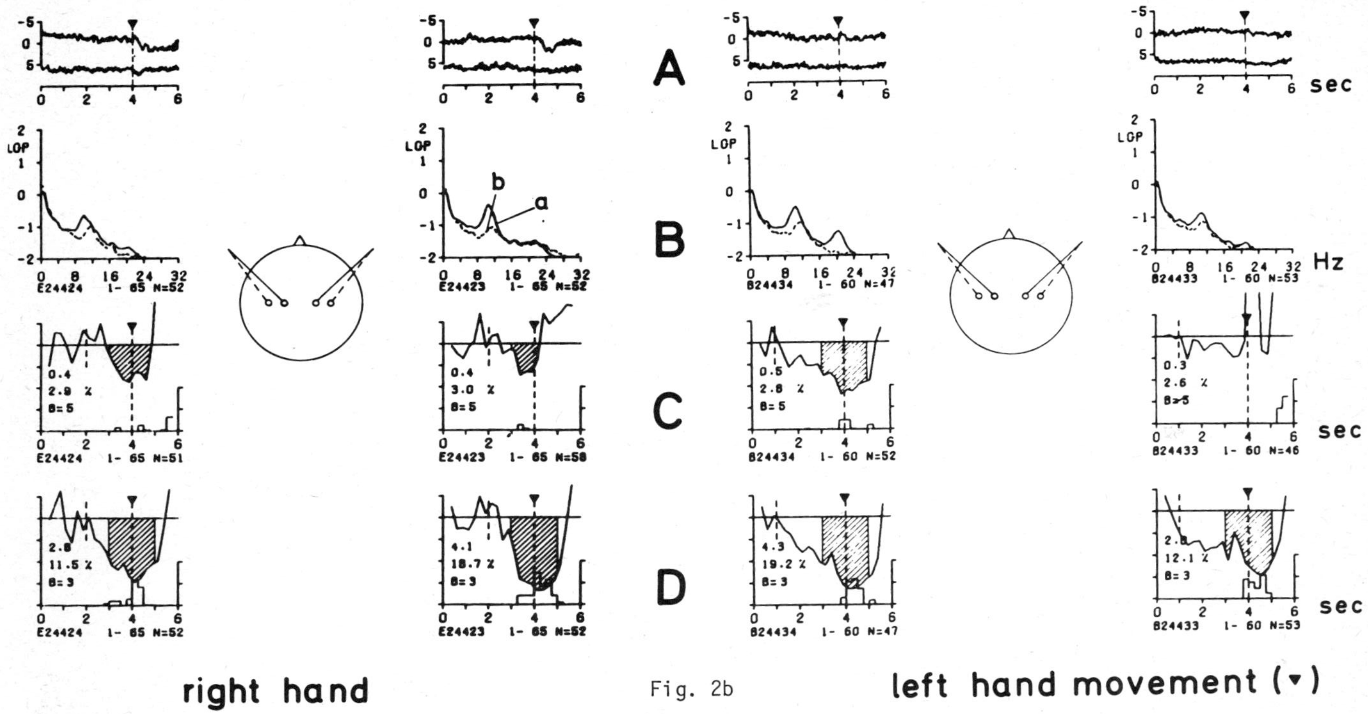

Fig. 2b

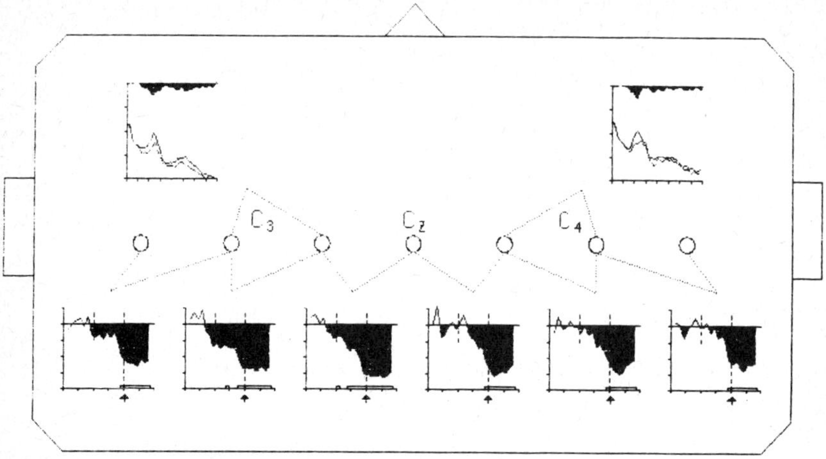

Fig. 3. Topographical display of EEG alpha power vs. time curves taken from central derivation, in a 46 y.o. subject.2-sec spectra for midcentral regions are also plotted. Scales and notation as in Figs. 1 and 2. Short lines parallel to abscissa below indicate 1 % level of significance.

Statistical analysis. For the statistical analysis of the changes in alpha band power, the reference was taken as the mean power during the first two seconds of the interval, i.e. from 4 to 2 seconds preceding movement onset. This was expressed either as an absolute power (μV^2) or as a relative reference, i.e. normalized and considered as 100 %.

The changes occurring in the 2 second segment, which included 1 second preceding and 1 second following movement onset, were quantitatively determined. The statistical significance of the changes within this time series segment was evaluated by the nonparametric sign test, taking as reference the power-time series of the first 2 seconds. These power changes related to movement were also quantitatively expressed as percentages of the reference alpha power. Reference alpha powers or significant power decreases, i.e. event-related desynchronizations (ERDs), or peak frequency within the analyzed bands, were compared between homologous areas of both hemispheres.Criteria of symmetry were established through linear or linearized regression analysis, with confidence and tolerance limits determined by using the corresponding F-values. The quality of fitting was evaluated by Pearson's correlation coefficients and t-tests of significance.

Similar statistical analyses were made to scrutinize dependence relationships between reference power and ERD, as well as between other parameters such as absolute and relative reference power and age vs. alpha peak.

RESULTS

The results reported here are limited to changes occurring in the activity within the alpha band in relation to voluntary hand squeezing.

1. Reference Alpha Power

As explained above, the reference measurements of alpha power corresponded to the first interval, from 4 to 2 seconds preceding movement onset, within each 6-sec trial.

Alpha Peak. In 5 subjects no clearly definable frequency peak could be detected within the alpha band in any of the used EEG derivations. Therefore, the results presented below correspond to the 28 subjects who displayed clear alpha peaks.

The mean frequency peaks within the alpha band, for the reference interval, are listed in Table 1. In all experimental situations, during control or before hand squeezing, the mean alpha peaks showed an essentially uniform distribution, i.e. minimal, nonsignificant differences, including for its topography.

TABLE 1

ALPHA FREQUENCY PEAKS

Mean frequency alpha peaks ($\pm$S.D.), in Hz, in the reference interval preceding hand movement in 28 normal subjects.

EEG DERI-	Central-vertex		Midcentral		Central-temporal	
VATION	Right	Left	Right	Left	Right	Left
CONTROL	9.8$\pm$1.5	9.8$\pm$1.5	9.7$\pm$1.5	9.7$\pm$1.5	9.5$\pm$1.4	9.7$\pm$1.5
RIGHT HAND	9.7$\pm$1.5	9.6$\pm$1.3	9.6$\pm$1.5	9.5$\pm$1.3	9.5$\pm$1.4	9.5$\pm$1.3
LEFT HAND	9.6$\pm$1.4	i.6$\pm$1.4	9.4$\pm$1.5	9.5$\pm$1.2	9.4$\pm$1.3	9.6$\pm$1.4

TABLE 2

ALPHA FREQUENCY PEAKS, INTERHEMISPHERIC CORRELATION

Mean slopes ($\pm$ S.D.) of regression lines (S) of mean reference alpha peak values of left on right homologous scalp EEG derivations and correlation coefficients (C) from 28 normal subjects, preceding hand movement.

PAIRS OF EEG DERIVATIONS		Central-vertex	Midcentral	Central-temporal
	S	0.963 ± 0.026	0.979 ± 0.035	1.012 ± 0.066
CONTROL	C	0.990	0.984	0.948
	p<	0.001	0.001	0.001
	S	0.878 ± 0.048	0.820 ± 0.060	0.916 ± 0.037
RIGHT HAND	C	0.964	0.936	0.979
	p<	0.001	0.001	0.001
	S	0.888 ± 0.062	0.829 ± 0.058	1.013 ± 0.044
LEFT HAND	C	0.943	0.942	0.976
	p<	0.001	0.001	0.001

The symmetry of the mean reference alpha peak values corresponding to homologous areas of both hemispheres was tested by linear regression and correlation analysis. The results are presented in Table 2. In general, as indicated by the mean slope values, the symmetry was greatest during the control condition and decreased a little, though significantly enough (F-value confidence limits, $p < 0.05$) during the reference periods preceding hand movements, in the upper and midcentral areas but not significantly in the central-temporal regions.

An example of this type of symmetry analysis is illustrated in Fig. 4, for the central-temporal areas (C_3-T_3' vs. C_4-T_4') in control conditions. The 5 % confidence and tolerance elliptic borders show that the paired values distribute themselves in a random, bivariate gaussian manner. This was true for all sets of paired values analyzed.

A weak but significant linear dependence was verified and was confirmed by the correlation analysis of the mean reference alpha peaks on the age of the subjects when all 28 subjects (mean age $\pm$ S.D.: 42 ± 19 years) were considered together. This is shown in Fig. 5 for the right central-vertex (C_Z - C_4') region during the control period. However, as clearly shown in this figure, two age subgroups could be separated, which were significantly different ($t = 11.846$, $p < 0.001$), the younger (14 subjects) with mean age ($\pm$S.D.) of 24 ± 2 years and the older group (14 subjects) with mean age 60 ± 11. Separate analysis of these two age groups and corresponding alpha peaks did not confirm the above

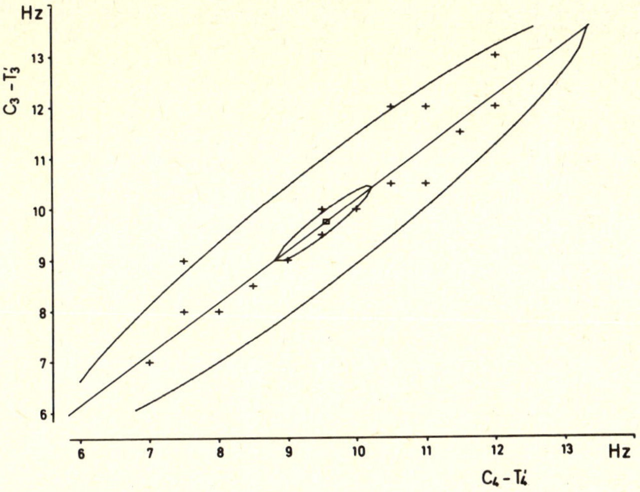

Fig. 4. Regression line of the alpha frequency peak values (Hz) between left central-temporal region (C_3-T_3') as y, vs. corresponding values of the homologous contralateral region (C_4-T_4') as x, from 28 normal subjects, during control conditions. $y = 0.081\ (\pm 0.639) + 1.012\ (\pm 0.066)x$, with correlation coefficient $r = 0.948$ ($p<0.001$). Inner ellipse: 5 % confidence contour around mean and other ellipse: tolerance contours. Some of the points (crosses) are identical for two or more subjects.

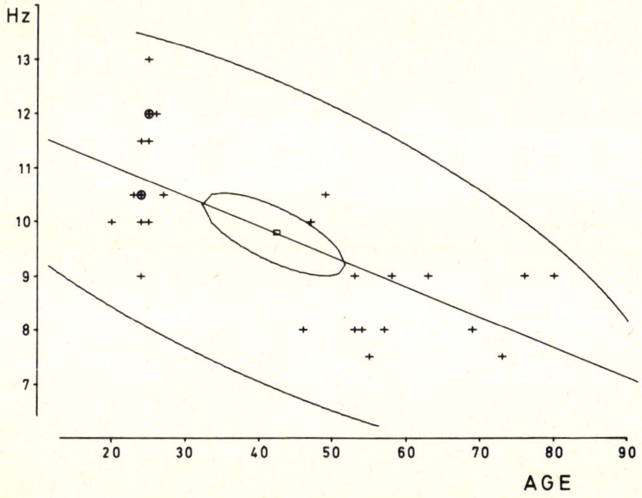

Fig. 5. Regression line of alpha frequency peak values (Hz) from right central-vertex region (C_z-C_4') as y, vs. age (years) of 28 normal subjects during control conditions. $y = 12.163\ (\pm 0.491) - 0.057\ (\pm 0.011)x$, correlation coefficient $r = -0.724$ ($p<0.001$). Confidence and tolerance ellipses as in Fig. 4. Two paired values (encircled crosses) are superimposed.

mentioned linear dependence. Instead, two statistical alpha peak groups emerged, with means (±S.D.) 10±1.1 Hz and 8.6±0.9 Hz, corresponding to the younger and older group, respectively. These two alpha peak groups were also significantly different from each other ($t = 6.150$, $p<0.001$).

TABLE 3

ALPHA POWER, ABSOLUTE

Mean reference alpha power values (±S.D.), in μV^2, in 33 normal subjects, preceding hand movement.

EEG Derivation	Central-Vertex Right	Central-Vertex Left	Midcentral Right	Midcentral Left	Central-temporal Right	Central-temporal Left
Control	3.1±4.1	4.0±4.9	3.5±4.2	3.8±3.9	5.2±6.1	5.0±4.8
Right Hand	3.2±4.6	3.4±4.5	2.9±2.8	3.6±3.7	4.8±5.4	4.9±5.0
Left Hand	3.4±4.7	4.0±4.8	3.5±4.1	4.3±4.6	4.9±5.4	6.0±5.8

Absolute reference alpha power. The mean reference alpha power values (±S.D.) for all 33 subjects, all scalp regions studied and experimental situations are listed in Table 3. Without reaching statistical significance, a tendency could be observed for the smaller values to concentrate on the upper central (central-vertex) areas and the largest values on the lower central (central-temporal) regions. The alpha power values were slightly smaller in the right hemisphere, but this was again nonsignificant. There were no remarkable differences between control and hand movement conditions.

TABLE 4

ALPHA POWER, PERCENTAGE

Mean reference alpha power values (±S.D.), in percentage of total EEG power, in 33 normal subjects, preceding hand movement.

EEG DERIVATION	Central-vertex Right	Central-vertex Left	Midcentral Right	Midcentral Left	Central-temporal Right	Central-temporal Left
CONTROL	20.9±11.5	23.1±13.3	21.0±11.3	21.9±11.9	22.1±11.0	23.0±12.2
RIGHT HAND	18.8±10.5	19.3±10.3	18.9±10.2	19.7±11.0	20.7±10.5	20.4±11.1
LEFT HAND	20.4±11.4	23.1±13.1	20.2±10.6	22.5±12.5	21.2±11.2	23.5±11.5

Relative reference alpha power. The mean reference alpha power values (±S.D.) expressed as percentage of total EEG power for all central areas analyzed, in all 33 subjects and during control or hand squeezing are presented in Table 4. Here all values distribute themselves uniformly across all areas and experimental conditions. Again, the values on the right hemisphere are somewhat smaller than those of the homologous left regions, but without reaching statistical significance.

The symmetry of these alpha percentage values, for homologous areas of both hemispheres was studied again by regression and correlation analysis, as summarized in Table 5. The mean slope values oscillated about slope 1.0 with small variations indicating an appreciable degree of symmetry, supported by highly significant correlation coefficients. The case for the central-vertex regions, in the control situation and preceding left hand squeezing is illustrated in Fig. 6.

TABLE 5

ALPHA POWER PERCENTAGE, INTERHEMISPHERIC CORRELATION

Mean slopes ($\pm$S.D.) of regression lines (S) of mean reference alpha percentage power values of left on right homologous derivations and correlation coefficients (C) from 33 normal subjects, preceding hand movement.

PAIRS OF EEG DERIVATIONS		Central-vertex	Midcentral	Central-temporal
	S	1.106$\pm$0.060	0.927$\pm$0.090	1.010$\pm$0.080
CONTROL	C	0.957	0.879	0.916
	p<	0.001	0.001	0.001
	S	0.895$\pm$0.073	0.947$\pm$0.090	0.971$\pm$0.077
RIGHT HAND	C	0.910	0.883	0.916
	p<	0.001	0.001	0.001
	S	1.073$\pm$0.077	1.053$\pm$0.093	0.873$\pm$0.096
LEFT HAND	C	0.928	0.897	0.853
	p<	0.001	0.001	0.001

The relationship between relative and absolute reference alpha powers was also studied by means of regression and correlation analysis. The set of correlation coefficients for a double logarithmic linearized regression had a mean ($\pm$S.D.) of 0.826$\pm$0.027, which was significantly larger ($t = 11.764$, $p<0.001$) than the mean of the correlation coefficients for a simple linear regression, (0.656$\pm$0.066). In Table 6, the linearized regression of ln (natural logarithm) of alpha percentage on ln of alpha power (μV^2) is summarized. The mean slope values show a uniform distribution, with minimal nonsignificant differences,

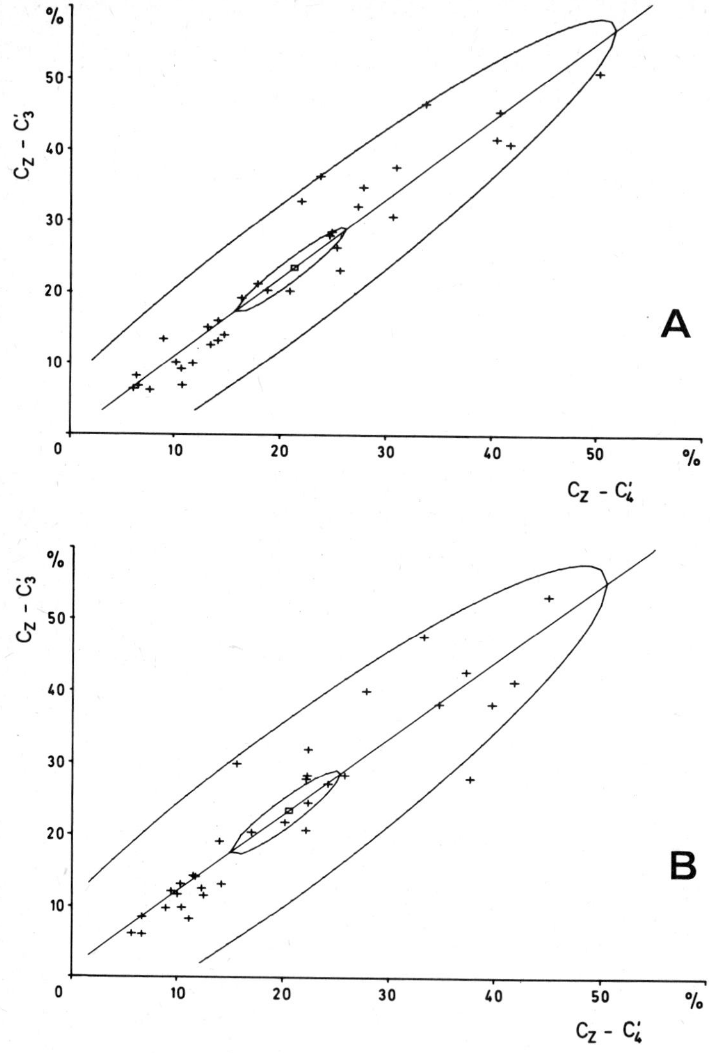

Fig. 6. Regression lines of alpha power percentage values for left central-vertex region (C_Z-C_3') as y, vs. corresponding values for right homologous area (C_Z-C_4') as x, from 33 normal subjects, under control conditions. (A) $y = 0.077\ (\pm 1.436)+1.106(\pm 0.060)x$, $r = 0.957(p<0.001)$, and left hand movement (B) $y = 1.218(\pm 1.795)+1.073(\pm 0.077)x$, $r = 0.928(p<0.001)$. Confidence and tolerance ellipses as in Fig. 4. No remarkable change in interhemispheric symmetry from A to B.

across all areas investigated and during all experimental situations. The corresponding correlation coefficients are highly significant and show also similar uniform distribution.

TABLE 6

PERCENTAGE-ABSOLUTE ALPHA POWER, CORRELATION

Mean slopes (±S.D.) of ln-ln regression lines (S) of alpha percentage values on alpha power (μV^2) in the reference interval, and correlation coefficients (C), in 33 normal subjects, preceding hand movement.

EEG DERIVATION		Central-vertex Right	Central-vertex Left	Midcentral Right	Midcentral Left	Central-temporal Right	Central-temporal Left
CONTROL	S	0.400± 0.050	0.421± 0.042	0.364± 0.042	0.407± 0.050	0.373± 0.041	0.399± 0.053
	C	0.822	0.874	0.842	0.827	0.853	0.805
	p<	0.001	0.001	0.001	0.001	0.001	0.001
RIGHT HAND	S	0.395± 0.042	0.375± 0.040	0.400± 0.052	0.377± 0.049	0.380± 0.044	0.415± 0.055
	C	0.860	0.858	0.812	0.810	0.840	0.803
	p<	0.001	0.001	0.001	0.001	0.001	0.001
LEFT HAND	S	0.370± 0.044	0.403± 0.047	0.389± 0.046	0.400± 0.051	0.382± 0.055	0.371± 0.054
	C	0.831	0.837	0.835	0.813	0.779	0.777
	p<	0.001	0.001	0.001	0.001	0.001	0.001

2. Alpha power changes related to movement or ERD

In previous studies in our laboratory, in relation to voluntary, repetitive, self-paced thumb movements, it has been shown that in most normal subjects a significant ERD is expected to occur at least during the 1 second interval preceding and the next 1 second interval at and following onset of movement [6].

Therefore, for the present investigation, the analysis of alpha power changes was concentrated on the two 1 second intervals preceding and following onset of hand squeezing, i.e. at time intervals t = 3-4 sec and t = 4-5 sec, with respect to the beginning of each 6-sec trial.

TABLE 7

ALPHA POWER CHANGE PRECEDING MOVEMENT

Mean percentage changes of alpha power ($\pm$S.D.) in the interval t = 3-4 sec, immediatly preceding hand movement, in 33 normal subjects.

EEG DERIVATION	Central-vertex Right	Central-vertex Left	Midcentral Right	Midcentral Left	Central-temporal Right	Central-temporal Left
CONTROL	-1.9$\pm$10.3	-3.9$\pm$13.3	1.6 $\pm$10.8	-1.4$\pm$12.5	-2.7$\pm$10.5	-3.3$\pm$11.1
RIGHT HAND	25.9$\pm$21.7	29.8$\pm$20.6	16.9$\pm$19.4	14.9$\pm$25.5	14.4$\pm$16.9	8.6$\pm$27.7
LEFT HAND	22.2$\pm$21.6	24.7$\pm$27.2	14.5$\pm$22.8	13.3$\pm$27.9	3.6$\pm$26.0	3.8$\pm$29.9

Table 7 shows the alpha power changes occurring in the 1 second interval immediately preceding movement onset. The values are expressed as mean percentage changes with respect to the reference alpha power, as defined in the foregoing sections. The results indicate a clear-cut significant alpha power decrease (ERD) preceding right or left hand squeezing, but not in the control condition. The difference between the series of values during control as opposed to either hand movement situation was of high statistical significance by nonparametric and t-test ($p<0.001$). Furthermore, the largest ERD values corresponded to central-vertex and midcentral regions, they were minimal in central-temporal areas.

TABLE 8

ALPHA POWER CHANGE FOLLOWING MOVEMENT

Mean percentage of alpha power ($\pm$S.D.) in the interval t = 4-5 sec, at and immediatly following onset of hand movement, in 33 normal subjects.

EEG DERIVATION	Central-vertex Right	Central-vertex Left	Midcentral Right	Midcentral Left	Central-temporal Right	Central-temporal Left
CONTROL	-3.3$\pm$13.6	-0.5$\pm$13.7	-2.0$\pm$14.4	-0.2$\pm$11.1	-4.3$\pm$12.8	-1.8$\pm$10.8
RIGHT HAND	42.2$\pm$27.7	40.4$\pm$23.9	32.1$\pm$26.8	27.4$\pm$23.4	23.6$\pm$26.8	18.0$\pm$24.8
LEFT HAND	37.1$\pm$23.2	37.0$\pm$28.9	28.8$\pm$23.0	22.9$\pm$30.3	15.3$\pm$26.2	16.8$\pm$27.8

Table 8 shows that during the 1 second interval at and following onset of movement, the ERD was quite clear during hand movement in contrast to the control condition. These mean ERD values were larger than those preceding movement onset (listed in Table 7), the difference showing high significance

($p<0.001$) with paired nonparametric and t-test.

In addition, as shown in Table 8, the mean ERD values during right hand squeezing were of greater relative magnitude than during left hand movement, this difference being significant ($p<0.001$) by nonparametric and t-test. It can also be noted that the mean ERD values for the right hemisphere had a tendency to be greater than those for the left hemisphere, this difference not reaching statistical significance.

TABLE 9

ALPHA POWER CHANGES, INTERHEMISPHERIC CORRELATION

Mean slopes (±S.D.) of regression lines (S) of percentage change of alpha power values of left hand on right homologous scalp EEG derivations and correlation coefficients (C), from 33 normal subjects, during hand movement.

PAIRS OF EEG DERIVATIONS		Central-vertex	Midcentral	Central-temporal
CONTROL	S	0.374±0.167	0.480±0.108	0.298±0.142
	C	0.373	0.625	0.353
	p<	0.05	0.001	0.05
RIGHT HAND	S	0.703±0.090	0.626±0.110	0.660±0.117
	C	0.814	0.716	0.712
	p<	0.001	0.001	0.001
LEFT HAND	S	1.103±0.105	0.968±0.159	0.702±0.143
	C	0.884	0.737	0.662
	p<	0.001	0.001	0.001

The symmetry of the ERD response in homologous areas of both hemispheres was studied by linear regression and correlation analysis. The mean slope values and correlation coefficients, in Table 9, indicate a low degree of symmetry in the control condition. However, during hand movement the mean slope values increased significantly (F-value confidence interval, $p<0.05$) indicating greater symmetry. ERD symmetry was greatest during left hand movement when the mean slope values approached slope 1.0.

Correspondingly the correlation coefficients (Table 9) were of greater statistical significance during hand movement as opposed to control conditions.

The mean slope values in Table 9 were larger for the left than for right hand movements, this being significant (F-value confidence interval, $p<0.05$) for upper and midcentral areas but not reaching significance for low central regions. In general, the mean slope values tended to approach but remained below 1.0, except for central-vertex regions during left hand squeezing. This was partly related to the fact that ERD values were slightly larger on the right hemisphere as already shown in Table 8.

An example of this remarkable change in symmetry of alpha power, from control to right hand squeezing, is illustrated in Fig. 7. Clearly, the slope of the regression line is significantly larger during right hand movement, since the line crosses above and away from the 5 % confidence elliptic border around the mean of the paired ERD values of central-vertex regions in the control condition. This figure also illustrates that the two statistical groups of paired ERD values of central-vertex regions, during control and right hand movement, display significantly different means (F-value confidence interval, $p<0.05$). Furthermore, the outer ellipses of each group, corresponding to the 5 % tolerance intervals, show that the paired ERD values contained within their borders can be reasonably considered to distribute themselves in a bivariate gaussian manner. Thus we are dealing, in control and hand movement conditions, with two well-defined, statistically different groups. This supports the notion that the ERD response during movement is a clear cut phenomenon which does not occur during control conditions.

Frequency of occurrence of voluntary movement ERD in individuals. As indicated in the section on methods, the ERD was evaluated for its significance by a nonparametric test, comparing the alpha power values during movement with those in the reference interval.

A subject was considered to have an ERD response, when the alpha power decrease during movement was significant at $p\leq0.001$ level and at least in one the 6 studied derivations. Following these criteria, only one among the 33 subjects (3 %) did not display a significant ERD during right hand squeezing, and only 3 (9 %) , during left hand movement.

Using other criteria, the grand mean of normal variation of alpha power calculated from the control values in Tables 7 and 8, in terms of percentages across all channels studied was -1.975 ± 1.696 % with a mean standard deviation of 12.075 ± 1.457 %. Ignoring topographical differences, we could consider that an alpha power decrease >24 % (1.98 S.D.) should be considered significant. A comparison of individual ERD measurements with this limit of 24 % of alpha power decrease, showed that only 3 of 33 subjects (9 %) did not show a clear ERD during right hand movement and 7 (21 %) during left hand squeezing.

Therefore, depending on the criteria used, in our series of 33 subjects 79 %

to 91 % had a clear ERD during left hand movement and 91 % to 97 % during right hand movement.

Dependence of ERD on reference alpha power. Because most of the subjects not presenting significant ERD were of the younger age group and showed small reference alpha power, the possible dependence of ERD on the reference alpha power was further investigated by linear regression and correlation analysis.

The mean slope values and correlation coefficients in Table 10 show that during control conditions the percentage changes in alpha power in the interval t = 4-5 sec did not clearly depend on their respective reference alpha power values. However, during right or left hand squeezing the ERD was clearly linearly dependent on its corresponding reference alpha power value. Judging from the correlation coefficients, this linear dependence was statistically significant for the upper and midcentral areas but not for the low central regions.

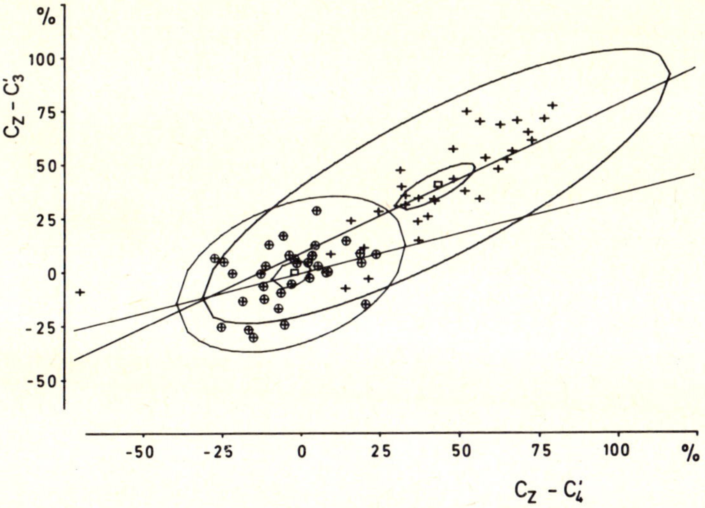

Fig. 7. Regression lines of alpha power percentage change values of left central-vertex region (C_Z-C_3') as y, vs. corresponding values of right homologous area as x, from 33 normal subjects, during control conditions (encircled crosses). $y = 0.726(\pm 2.307)+0.374(\pm 0.167)x$, $r = 0.373$ (non-significant) and during right hand squeezing (crosses) $y = 10.737(\pm 4.524)+0.703(\pm 0.090)x$, $r = 0.814$ ($p<0.001$). Confidence and tolerance borders as in Fig. 4. During hand movement, significant slope increase and elongation of elliptic contour indicate greater interhemispheric symmetry.

The above evaluation of the dependence of ERD on its correspnding reference alpha power should be considered to hold only within each studied scalp region.

That this cannot be generalized without regard to topography is well shown in Fig. 8, which also summarizes graphically some of the previous results.

TABLE 10

ALPHA POWER DECREASE - ALPHA POWER REFERENCE, CORRELATION

Mean slopes ($\pm$S.D.) of regression lines (S) of alpha power percentage decrease values on reference alpha power percentage (of total EEG power) values and correlation coefficients (C), in 33 normal subjects, during hand movement.

EEG DERIVATION		Central-vertex Right	Central-vertex Left	Midcentral Right	Midcentral Left	Central-temporal Right	Central-temporal Left
CONTROL	S	0.258$\pm$ 0.208	0.409$\pm$ 0.169	0.313$\pm$ 0.222	0.220$\pm$ 0.162	0.007$\pm$ 0.208	-0.046$\pm$ 0.159
	C	0.218	0.398	0.246	0.237	0.006	-0.052
	p<	n.s.	0.025	n.s.	n.s.	n.s.	n.s.
RIGHT HAND	S	1.371$\pm$ 0.406	1.217$\pm$ 0.355	1.516$\pm$ 0.384	0.853$\pm$ 0.352	0.538$\pm$ 0.450	0.356$\pm$ 0.397
	C	0.518	0.524	0.579	0.399	0.210	0.159
	p<	0.005	0.005	0.001	0.025	n.s.	n.s.
LEFT HAND	S	1.084$\pm$ 0.310	1.226$\pm$ 0.328	1.149$\pm$ 0.330	0.993$\pm$ 0.397	0.199$\pm$ 0.419	0.445$\pm$ 0.428
	C	0.532	0.557	0.530	0.410	0.085	0.184
	p<	0.005	0.001	0.001	0.025	n.s.	n.s.

In Fig. 8 no clear dependence of ERD on the reference absolute alpha power, in μV^2, across all regions can be noticed. In fact, the central-temporal areas having the largest reference alpha power showed the smallest ERD responses. Indeed in these areas, according to the data in Table 10, the linear dependence of ERD on reference alpha power did not hold true.

In the lower part of Fig. 8, similar graphical results are displayed for the two age subgroups during right hand squeezing; larger mean reference alpha power values and larger mean ERD (preceding and following movement onset) can be seen for the older group. In the older group, the differences between mean ERD values preceding and those following onset of hand movement were also clearly smaller than in the younger group.

DISCUSSION

The statistical results reported here are significant enough to define important features of the central activity within the alpha band, or mu activity, in resting condition as well as during voluntary, repetitive, self-paced movement.

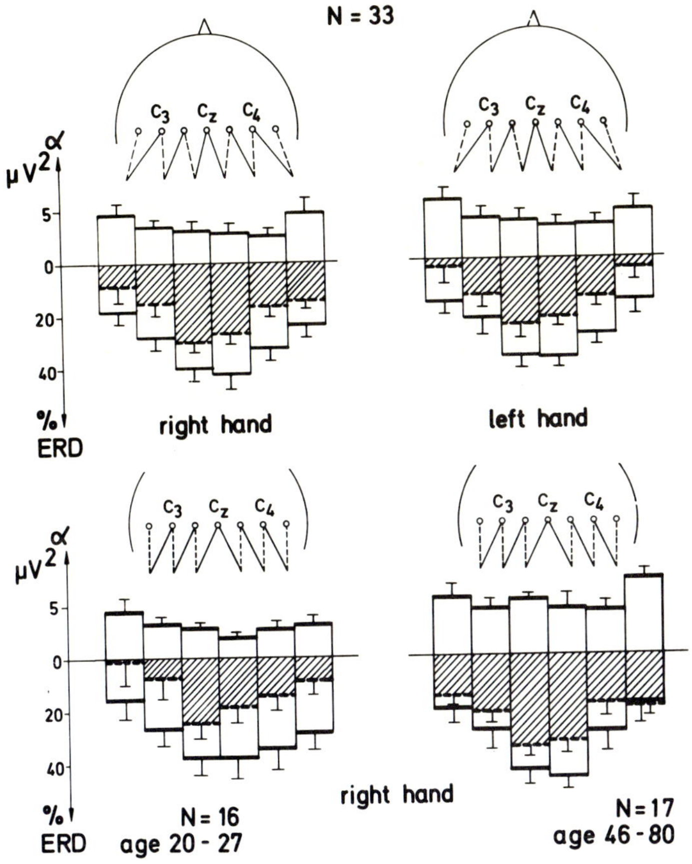

Fig. 8. Histogram displays of the transverse distribution, across central areas, of reference alpha power above zero line, in μV^2, and of ERD (below line) percentage changes in the 1-sec interval immediately before (broken line, hatched) and in the 1-sec interval during and following (full line) hand movements. Upper picture: results from the total group of 33 normal subjects, during right and left hand movements. Below: results for the two age subgroups: younger subjects (16) at left and older subjects (17) at right during right hand movement.

Reference alpha power. The first remarkable finding is that the reference alpha power was very stable in all central areas studied and for both experimental conditions, at rest or during hand movement (Tables 1,3 and 4).

In addition, this reference alpha power was rather symmetrical in homologous areas of both hemispheres.

Clearly definable alpha frequency peaks in the power spectra were only detectable in 28 subjects, or 84.4 % of our total series. An overall correlation of alpha peak with the subjects' age could not be established on a strong basis, because the ages did not make up a homogeneous group but concentrated into two statistically different groups. However, data such as those illustrated in Fig. 5 and the fact that the mean alpha frequency peak was significantly smaller in the older group, are indicative of a negative correlation between age and peak frequency of central mu activity. A similar relationship has already been documented for the occipital alpha activity [9].

The averaged peak frequencies (Table 1) fluctuated between 9.4 and 9.8 Hz for an age distribution of 42 ± 19 years (mean $\pm$S.D.). These are smaller values than those reported for the left hemisphere only (10.1 ± 1.9 Hz), in 7 subjects from electrode derivation C_3 to midway between C_3 and P_3, in which however the age distribution (25 ± 7) was that of a younger group than ours.[4]. Our values are also smaller than those reported for both hemispheres (10.5 ± 0.9 Hz)from derivations C_4-F_4 and C_3-F_3, in 16 subjects, whose age distribution (25 ± 9) was also younger [5] than in our group.

Symmetry between homologous areas of both hemispheres, was most notorious in the control condition, particularly for the alpha frequency peaks (Tables 1 and 2, Fig. 4) but also for the absolute (Table 3) and percent (Tables 4 and 5, Fig. 6) alpha power values. The argument that the common electrode C_Z would contribute to increase the correlations between the two central-vertex derivations is of little consequence, since equivalent correlation coefficients were obtained from the other mid and lower central regions (Tables 2 and 5).

In the past, authors have frequently reported, from conventional visual analysis,[10-13] voltage asymmetries of mu activity in central areas. Asymmetries to the extent of "25 % or more" were frequently observed [14].

More recently, an average peak power of 14.1 ± 3.6 dB (mean $\pm$S.D.) corresponding to 1.5 ± 0.1 μV^2/Hz measured at peak frequency was reported through automated analysis in 15 normal subjects and recording from derivations C_4-F_4 and C_3-F_3. Seven of these subjects had interhemispheric differences larger than 2 dB (0.1 μV^2/Hz) considered to indicate asymmetry. This statistical treatment, quite different from ours, considered the values of both hemispheres within a single normal monovariate distribution.

It appears noteworthy, from Table 3, that the average power values across all central areas and experimental conditions, varied between 2.9 and 6.0 μV^2.

which correspond to amplitudes between 1.7 and 2.4 μV. These amplitudes are quite small in comparison to those reported by conventional visual analysis [14].

That these mean amplitudes are small because in the bipolar montage used here the interelectrode distance was half that commonly used in the 10-20 system, is only a partial explanation. Our automated methods certainly allow detection of small mu amplitudes in many subjects that were formerly missed by simple visual inspection.

Small differences were observed between alpha power values in the reference interval expressed as absolute values and percentage of total EEG power, indicating more topographical variation of the absolute power (Table 3, Fig. 8) and more uniformity for the percentage values (Table 4).

In spite of these small topographic divergences between these central derivations, a fundamental relationship between percent and absolute values was defined as a double logarithmic linear dependence. The magnitude of this dependence was remarkably uniform throughout all central areas and for resting or movement conditions (Table 6). Thus, it could be considered to express or define some characteristic property of the central mu activity.

Voluntary movement ERD. The second most remarkable finding is the phasic decrease of alpha power immediately preceding, at and following onset of hand squeezing. This voluntary movement ERD was clearly of high statistical significance on two main comparisons: in each trial, that between the period during movement and the preceding reference one; and that between trials with movement and control trials without movement (Tables 7 and 8).

Perhaps the most intriguing finding is that the voluntary movement ERD was bilateral and symmetrical, as much during the one second immediately preceding movement onset (Table 7) as during the one-second interval at and following onset of hand squeezing (Tables 8 and 9).

The mean ERD values in this study (see Table 8) were between 22.9 and 42.4 % in upper and midcentral areas i.e. somewhat larger than the average, of about 25 %, measured in 5 subjects, from the left hemisphere (C_3 to midpoint toward P_3) while clenching the right fist repeatedly and using 4-sec analysis epochs [4]. Our values are however somewhat smaller than the mean ERDs measured in 15 subjects, from C_4-F_4 and C_3-F_3 derivations (61 ± 25 %) [5], while clenching both fists continuously for about 100 sec and using analysis epochs of 5.12 sec.

That the ERD response as it follows from the data on Tables 7 and 8, was predominant in upper and midcentral areas could be somewhat related to the fact that these regions correspond to the cortical sensorimotor representation of the hands. But if this were the only explanation, then the ERD response should be expected to be predominantly contralateral to unilateral hand movements.

The results of this study for hand squeezing are quite similar to previous results from our laboratory concerning thumb pressing [6], but otherwise apparently unexpected if one considers previous reports from other groups of investigators.

Previous authors have either assumed that the responsiveness of central mu activity during single hand movements should be mostly contralateral [4] and/or have made no effort to verify this assumption [4,5], or have based their impression of contralateral dominance of mu blocking on a few selected individual observations [14,15] without support of reliable group statistics.

As a matter of speculation, the bilaterality and symmetry of ERDs during voluntary movement could be related to their basic mechanism, as they may represent a general response to various types of stimuli, including photic stimulation [8]. Such thalamo-cortical mechanisms would supposedly involve bilateral, symmetrical desynchronizations of rhythmic alpha or mu activities.

In support of this speculation, the data in Table 9, most strikingly illustrated in the example of Fig. 7, reveal that during control conditions there was only a low degree of symmetry of the alpha power changes in homologous areas of both hemispheres during simultaneous intervals. On the other hand, a sudden increase in symmetry of the alpha power change, was observed during movement, as a common simultaneous decrease. This could be interpreted as a "bilaterally synchronized alpha (mu) desynchronization" or ERD, simultaneously occurring in homologous areas of both hemispheres during unilateral hand movement.

Interestingly enough, this dynamic symmetry was higher during left than during right hand movement in this all-right-handed group of normal subjects. In contrast, during right hand movement ERD responses were , on the average, slightly larger on the right hemisphere, i.e. right upper and midcentral areas (Tables 7 and 8).

The case for a better topographical sensorimotor specificity of the ERD response, in relation to voluntary movement, received further support from the finding that the ERD was linearly dependent on its reference alpha power in the upper and midcentral areas but not in the low central regions (Table 10). This can be correlated to previously mentioned results indicating that the low central areas had weaker ERD responses (Tables 7 and 8), in spite of showing essentially similar reference alpha power to the other central areas (Tables 3 and 4). Altogether this could be interpreted by assuming that in low central (central-temporal) regions a greater mixing occurs of mu activity responsive to movement and alpha activity propagated from the occipital areas and more reactive to photic stimulation, without necessarily showing different frequency peaks.

Thus, the voluntary movement ERD appears, from the present study, to be an expression of central mu activity showing topographical specificity in connection with cortical areas known to be related to sensorimotor functions.

This functional specificity of rolandic areas in humans has been documented since early publications based on corticograms [16,17] reporting blocking of rhythmic activity within the alpha and beta bands in time relation to hand movements.

Similar rhythmic activities responsive to movement were also reported in cats and baboons [18], known in the literature as sensorimotor rhythms [19] and mainly related to the high frequency alpha and low frequency beta bands.

The high frequency of occurrence of ERDs during hand squeezing among the individuals of our series (91 to 97 %) confirms our previous similar reports for thumb movement [6]. This degree of occurrence in humans is much higher than hitherto reported by others [4,5,20], using different methods of analysis, sometimes also of hand movements and/or electrode positions.

Voluntary movement ERD, an expression of central mu activity, is therefore a normal neurophysiological event in humans. It remains for future research to test the main statistical features defined by the present study, by checking them against results obtained from patients with disorders particularly affecting cortical or thalamo-cortical sensorimotor functions.

As an example of preliminary incursion into the clinical field, let us describe here the results from one patient. This was a 40-year-old male who had a left hemiplegic stroke due to occlusion of deep branches of the right middle cerebral artery. The lesion, according to the computer tomographic study made 64 days after onset ot the stroke (Fig. 9) was a small 1.5 cm x 1.0 cm ill-defined infarcted area on the right lenticular nucleus, particularly involving the globus pallidus and encroaching upon the posterior limb of the internal capsule. The routine EEG, read in the conventional visual manner, was reported as unremarkable.

Our special computer-assisted EEG analysis was performed 50 days after the onset of the stroke. The main results are summarized in the graphical output in Fig. 10. The most striking finding was a notorious ERD asymmetry in midcentral areas (C'_4-C_4 and C'_3-C_3) in this case illustrated during right hand squeezing (for comparison of results in a normal subject see Fig. 3). No asymmetry of reference alpha power values was found,being 1.9 μV^2 in both

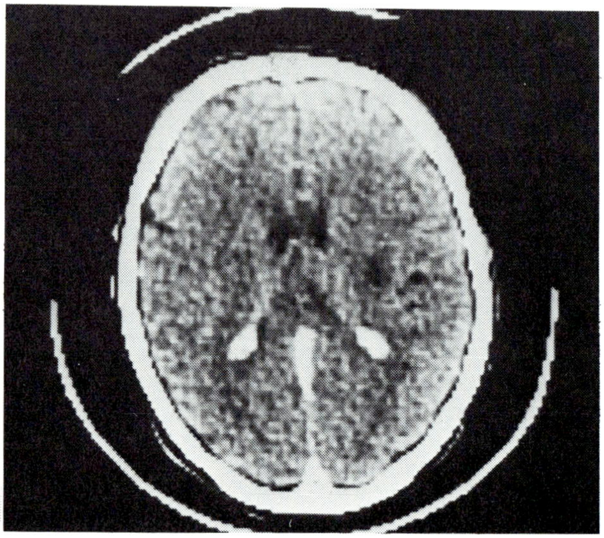

Fig. 9. Contrast enhanced computerized tomographic horizontal brain view, showing a hypodense lesion of ill-defined contours in the right lentiform nucleus impinging upon the posterior limb of the internal capsule. In this small area there was no uptake of contrast substance.

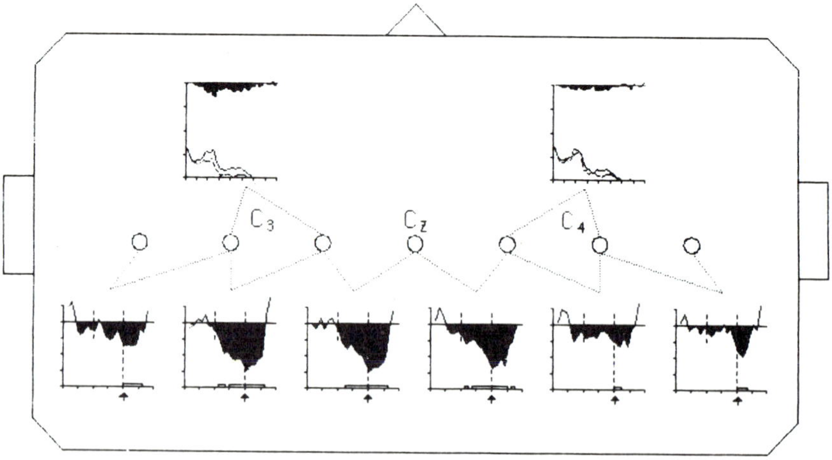

Fig. 10. Graphical computer output of alpha power changes in central scalp EEG derivations in a left hemiplegic patient, during right hand squeezing. In comparison to normal results, such as in Fig. 3, there is here a remarkable interhemispheric ERD asymmetry due to poor response on right midcentral region: Each of the upper curves represent two superimposed logarithmic power spectra from the midcentral areas.

right and left midcentral areas, or in terms of alpha percentage of total EEG power: 22.6 % on the right and 24.4 % on the left. The alpha frequency peaks, which appear in the logarithmic power spectra in the upper row, were asymmetric, 8.0 Hz on the right and 10.0 Hz on the left. The second logarithmic spectra superimposed in each graph, showed more decrease of alpha power on the left and only minimal on the right.

The maximum ERD values in midcentral regions, obtained for the 1 second interval at and following onset of right hand movement, were 66.4 % decrease on the left and only 30.8 % on the right midcentral areas. This represents considerable ERD asymmetry because of insufficient mu rhythm responsiveness to movement on the right midcentral area.

Although at this point further speculation is not warranted, it seems promising to continue investigating in the clinical field, where new sources of anatomo-physiopathological relationships may help to clarify the underlying mechanisms of voluntary movement ERD.

SUMMARY

The characteristic features of the central mu activity, during repetitive, voluntary, self-paced hand movements were statistically analyzed in 33 right-handed, neurologically normal, human subjects. After band-pass filtering, the scalp EEG retained mainly alpha band frequencies. Three types of situations were used: control trials, trials with right and trials with left hand squeezing.

The baseline alpha power was roughly symmetrical across hemispheres under control and movement conditions. Its absolute magnitude increased from the vertex towards central-temporal regions, but the relative percent of total EEG power remained essentially the same. However, significant correlation coefficients and double-logarithmic linear regression indicated dependence of alpha percent on its absolute magnitude.

In 28 subjects, clear reference alpha peaks, symmetrical across hemispheres, were detected. These alpha peaks showed linear dependence and inverse correlation with age.

An alpha attenuation or event-related desynchronization (ERD) absent during control situation was present in both hemispheres and essentially symmetrical during single hand movements, being of larger absolute power during right hand movement. The relative magnitude of the ERD was linearly dependent on the reference alpha power, with parallel significant correlation coefficients, which were highest on central-vertex areas and decreased towards lower central-temporal regions.

These findings are interpreted in relation to: 1) degree of mixing of alpha (mu) rhythm reactive during movement with alpha rhythms more responsive to light stimulation and 2) somatotopic, sensorimotor cortical representation.

ACKNOWLEDGEMENT

The authors wish to thank Dr. H. Maresch for helping with the computer programming, Mrs. E. Mandl, Dipl.-Ing. W. Pribyl and Mr. J. Resch for the EEG recording and data processing, Mr. P. Winkler and Mr. P. Lipp for the statistical analysis. Prof. Dr. N. Geyer and Dr. W. Wege, Stroke Unit of the Hospital of Neurology and Psychiatry (LSKH) Graz, kindly provided control and hemiplegic patients and clinical data and Doz. Dr. W. D. Sager, Department of Radiology, University of Graz, the computer tomography reports. Mrs. M. Kainz typed the original manuscript.
This investigation was supported by the "Fonds zur Foerderung der wissenschaftlichen Forschung", Project No. 3535.

REFERENCES

1. Kornhuber, H.H. and Deecke, L. (1964) Hirnpotentialänderungen beim Menschen vor und nach Willkürbewegungen, dargestellt mit Magnetband-Speicherung und Rückwärtsanalyse. Pflügers Arch. ges. Physiol. 281, 52.
2. Vaughan, H. G. (1975) The motor potentials. In A. Remond (Ed.) Handbook of EEG and clin. Neurophysiol. Elsevier, Amsterdam 8A, 86-91.
3. Deecke, L., Grözinger, B. and Kornhuber, H.H. (1976) Voluntary finger movement in man: cerebral potentials and theory. Biol. Cybernetics 23, 99-119.
4. Kuhlman, W.N. (1978) Functional Topography of the Human mu rhythm. Electroenceph. clin. Neurophysiol. 44, 83-93.
5. Storm van Leeuwen, W., Wienecke, G., Spoelstra, P. and Versteeg, H. (1978) Lack of bilateral coherence of mu rhythm. Electroenceph. clin. Neurophysiol. 44, 140-146.
6. Pfurtscheller, G. and Aranibar, A. (1979) Evaluation of event-related desynchronization (ERD) preceding and following voluntary self-paced movement. Electroenceph. clin. Neurophysiol. 46, 138-146.
7. Pfurtscheller, G. (1977) Graphical Display and Statistical Evaluation of event-related desynchronizsation (ERD). Electroenceph. clin. Neurophysiol. 43, 486-487.
8. Pfurtscheller, G. and Aranibar, A. (1977) Event-related cortical desynchronization detected by Power Measurements of Scalp EEG. Electroenceph. clin. Neurophysiol. 42, 817-826.
9. Obrist, W.D. (1963) The electroencephalogram of healthy aged males. In J.E. Birren et al. (Eds.) Human aging: A biological and behavioral study. PHS Publication No. 986 U.S. Govt. Printing Office, Washington, D.C. pp. 79-93.

10.Gastaut, H., Terzian, H. and Gastaut, Y. (1952) Etude d'une activité électro-encephalographique méconnue: "le rhythme rolandique en arceau" Marseille-méd. 89, 296-310.

11.Magnus, O. (1954) The central alpha rhythm ("rhythme en arceau"). Electroenceph. clin. Neurophysiol. 6, 349-350.

12.Canali, G. and Carpitella, A. (1956) Rilievi statistici sul ritmo "en arceau". Sist. nerv. 8, 8-12.

13.Roger, A. and Bert, J. (1959) Etude des corrélations entre les différentes variables. EEG Rev. neurol. 101, 334-360.

14.Chatrian, G.E. (1976) The mu rhythm. In A. Rémond (Ed.) Handbook of EEG and clinical Neurophysiology, Elsevier, Amsterdam 6A, 46-69.

15.Chatrian, G.E., Petersen, M.C. and Lazarte, J. (1959) The blocking of the rolandic wicket rhythm and some central changes related to movement. Electroenceph. clin. Neurophysiol. 11, 497-510.

16.Jasper, H. H. and Penfield, W. (1949) Electrocorticograms in man: Effect of voluntary movement upon the electrical activity of the precentral gyrus. Arch. Psychiat. Nervenkr. 183, 163-174.

17.Gastaut, H. (1952) Etude électrocorticographique de la réactivité des rhythmes rolandiques. Rev. neurol. 87, 176-182.

18.Rougeul-Buser, A., Boyer, J.J. and Buser, P.A. (1978) Transitional states of awareness and short-term fluctuations of selective attention: Neurophysiological correlates and hypotheses. In P.A. Buser and A. Rougeul-Buser (Eds.) Cerebral correlates of concious experience. North-Holland, Amsterdam - New York - Oxford pp. 215-232.

19.Roth, S. R., Sterman, M.B. and Clemente, C.D. (1967) Comparison of EEG correlates of reinforcement, internal inhibition and sleep. Electroenceph. clin. Neurophysiol. 23, 509-520.

20.Schoppenhorst, M., Brauer, F. and Freund, G. (1977) Mu rhythm and its detection by coherence functions. IX Int. Congress Amsterdam. Electroenceph. clin. Neurophysiol. 43, 505

Rhythmic EEG Activities and Cortical Functioning
G. Pfurtscheller et al., editors

SPATIOTEMPORAL CHARACTERISTICS OF INTRINSIC EVOKED AND EVENT RELATED POTENTIALS IN THE HUMAN CORTEX

D.PAPAKOSTOPOULOS, H.J.CROW and P.NEWTON
Burden Neurological Institute, Bristol BS16 1QT (England)

INTRODUCTION

From the beginning of electroencephalography the clinical and scientific value of electrocorticography (ECoG) has been well appreciated, and a considerable volume of data about the rhythmic activities of the brain has been collected by direct cortical recordings[1-10]. Systematic studies of average evoked potentials from the human scalp[11,12] have been clarified by some direct cortical recordings of these activities[13-22].

The discovery of brain potentials which develop under conditions of active interaction of the subject with his environment imposed restrictions on the population of patients from whom such corticographic recordings can be obtained. Obviously only alert and co-operative individuals can be considered for these recordings. In addition, the study of the Contingent Negative Variation (CNV), the P300, and the movement related brain macropotentials[23] due to the relatively small size of the potentials, can only be carried out in laboratories with sufficient computing facilities. Under these circumstances it is not surprising that only a few groups have been able to record directly from the brain potentials like the CNV[24-28], the P300[26], Bereitschaftspotential[21,29], and the other movement related brain macropotentials[30]. Even those papers which deal with corticographically recorded event related macropotentials do not contain data about the spatiotemporal characteristics of these potentials, or the relationships between the waveform and reactivity of the evoked potentials and the intrinsic activities of the brain locations from where the event related activities had been obtained. The aim of the present investigation is to bridge this gap by studying the spatiotemporal characteristics and interactions of corticographic data from various brain areas with the subject either relaxed or engaged in certain behavioural sequences known to elicit event related changes in brain macropotentials.

METHOD

Our 8 subjects were psychiatric patients (3 male and 5 female) selected for treatment by multiple focal leucocoagulation[31,32]. All subjects except one were right handed and their ages ranged from 27 to 54 years. Subdural electrodes were inserted to try and improve the monitoring of the treatment[1]. The gold wires of 150 um diameter with an uninsulated tip of 4 mm were twisted in 2 sheafs of 6 electrodes, with 4 mm between electrodes, and were placed subdurally during the operation for the insertion of intra-cerebral electrodes. One of these sheafs was directed either towards the tip of the frontal cortex or towards the temporal cortex and the other towards the sensorimotor cortex about 3-5 cm lateral to the midline. All subjects but one had the electrodes on the right hemisphere. Recordings from these 12 electrodes were obtained with an Elema-Schonander EEG machine fitted with universal amplifiers, each with an input impedance of 100 megohms so that the effective time constant of the intracerebral electrodes was sufficiently long for recording the slow potentials. In all studies, an average common reference electrode was used, the 64 electrodes implanted in frontal and anterior cingulate white matter being combined together to provide the reference electrode.

Almost all the data to be presented and discussed were recorded from the subdural electrodes. A few references are made to data recorded from intracerebral electrodes.

A PDP 12 Computer was employed for data collection and for on or off line processing. It was programmed to sample up to 16 channels simultaneously.

Stimuli used were a 70 dB auditory click presented through lightweight earphones, single flashes delivered by a stroboscope, 30 cm in front of the patient's face, and electrical stimuli of 0.1 msec duration applied to the median nerve at the wrist with a DISA-ministim stimulator. The intensity was enough to produce a small muscle twitch.

From all subjects ECoG records were obtained during periods of relaxation with the eyes open or closed, and during alternating, 16 sec periods, of relaxation with eyes open and left or right fist clenching.

Somatosensory visual and auditory evoked cortical potentials to the appropriate stimuli, regularly or irregularly presented, were subsequently studied. For eliciting the CNV and P300 potentials the experimental design consisted of the presentation of two stimuli, a click and a flash 1 or 2 sec apart, firstly without instruction and later with instruction to press a switch on receiving the second stimulus. The intertrial interval was variable and

---

[1] The technique of multifocal leucocoagulation which has been in use for 21 years consists of intracerebral implantation of fine gold wire electrodes (4000 x 150 microns) for periods of 3 to 6 months. This permits progressive intermittent coagulation over time thus minimising the lesions and adjusting them to the needs of each individual patient.

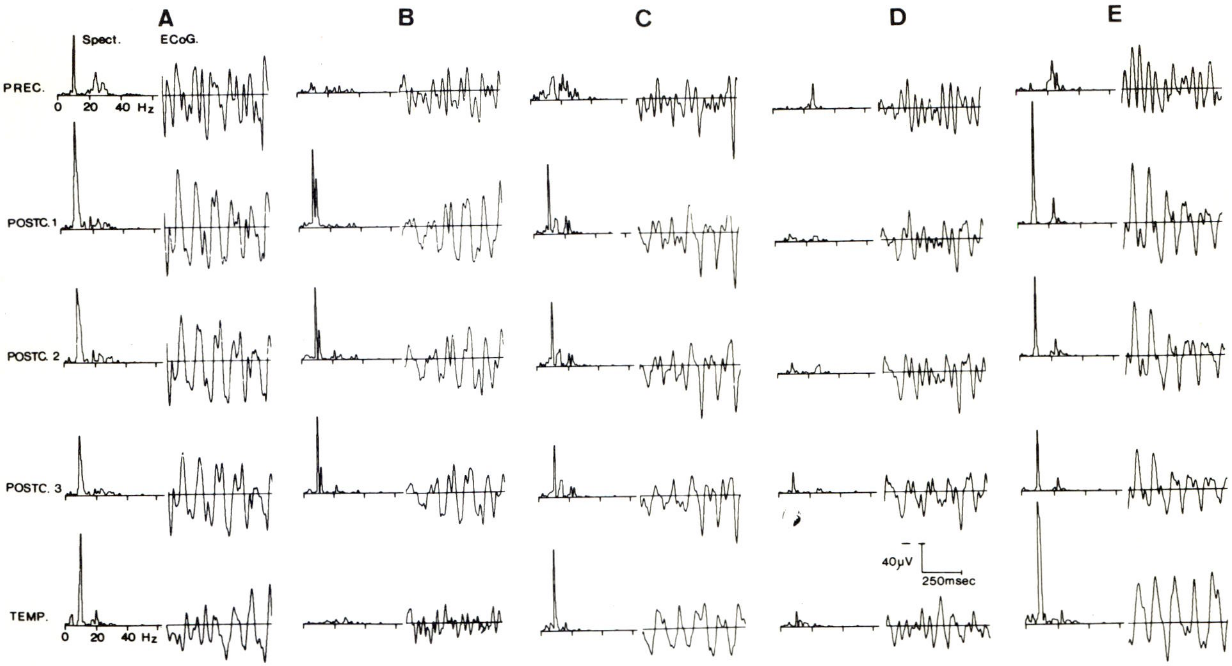

Fig. 1. Samples of corticogram (ECoG) and their spectra (spect.) from 5 cortical areas simultaneously registered at five different instances in time, A, B, C, D und E, during a 2 min recording session with the subject relaxed with eyes open. Each spectrum derived from 1 sec ECoG activity of which 5oo msec is shown in the figure. PREC is the precentral, POSTC 1 is the first postcentral electrode, POSTC 2 the electrode 4 mm posterior of POSTC 1, POSTC 3 4 mm posterior of POSTC 2. Electrodes length 4 mm. All electrodes were referred to a common average reference electrode.

ranged from 6 to 20 sec.

RESULTS

The pre- and postcentral areas of all subjects tested were identified by the polarity and latency of the first cortical component of the somatosensory evoked potential which followed contralateral electrical stimulation at the wrist[33]. The polarity of this component was negative for the postcentral area and positive for the precentral area. The take off and peak of the precentral response was 1-2 msec later than the postcentral peak. The latency of the postcentral peak ranged between 18-21 msec.

The intrinsic rhythms of the prefrontal, precentral, postcentral, parietal and temporal areas, when studied with the subject in a relaxed state, varied from area to area, and also varied within the same area across time. Yet the variability within and between areas was not such as to preclude identification of the area. A typical example of this combined versatility and predictability of cortical rhythms can be seen in Fig.1 (A-E).

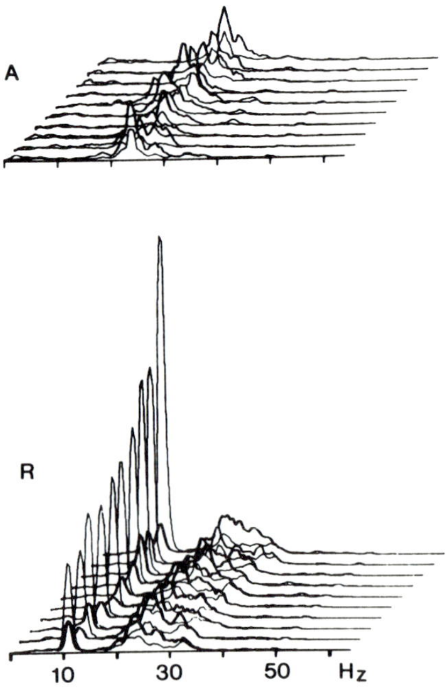

Fig. 2. Simultaneously recorded and superimposed averaged spectra from pre-central areas (thick traces) and postcentral areas (thin traces) of the right hemisphere with the subject relaxed (R) and clenching his left fist (A). Each spectrum derived from 16 sec. of ECoG activity.

In Fig.1-A an 11 Hz rhythm was recorded from all electrodes; at 1-B it was clear only at the postcentral locations; at 1-C a temporal rhythm was added to the 1-B pattern; at 1-D there was no clear rhythmicity at 11 Hz anywhere; and at 1-E the rhythm was most prominent at the temporal and postcentral locations. As well as the 11 Hz rhythm, a 25 Hz rhythm was recorded. The power spectra of the latter rhythm, at times A, B, C, D and E show different spatial patterns. Yet it was a consistent observation that despite the spatiotemporal variability of both the 11 and the 25 Hz rhythms, the precentral area displayed the faster rhythm more prominently than the slower. The trend was in the opposite direction in the postcentral area.

In Fig.2-R, ten samples of power spectra are superimposed, each having been derived from 16 secs of ECoG activity simultaneously recorded from the pre- and postcentral areas. The spectra have been ranked from front to back in ascending power of the 11 Hz rhythm in order to clarify the presentation. The predominance of the 11 Hz rhythm in the postcentral area is obvious. It is also clear that the precentral 11 Hz rhythm does not follow the postcentral amplitude pattern of that rhythm.

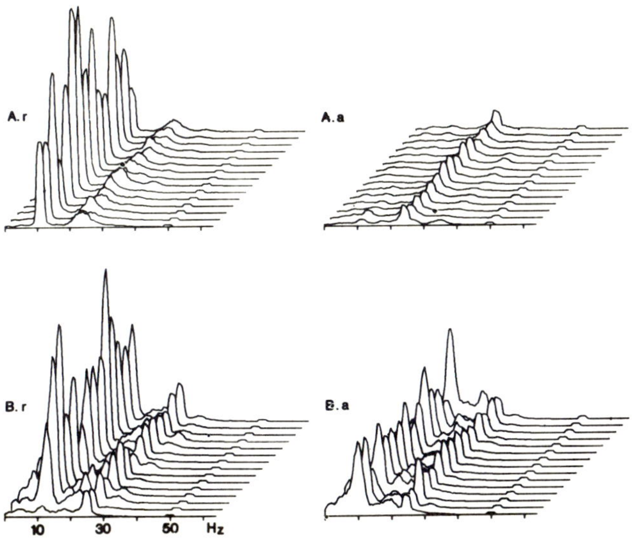

Fig. 3. Simultaneously recorded average spectra from a right precentral (top traces A.r. and A.a.) and a right temporal (lower traces B.r. and B. a) electrodes with the subject relaxed (A.r. and B.r.) and clenching his left fist (A.a. and B.a.). Each spectrum derived from 16 sec of ECoG activity.

In spite of the independence of these features of the 11 Hz rhythm in the pre- and postcentral areas, it was completely abolished in both areas by fist clenching. This is illustrated in Fig. 2-A.

The temporal 11 Hz rhythm behaved differently to fist clenching, which only attenuated it, but consistently failed to abolish it. This is illustrated in Fig.3.

The pre- and postcentral 11 Hz rhythms could also be shown to react differentially in other circumstances. This occurred during passive movements of the fingers. The precentral 11 Hz rhythm was abolished by this manoeuvre, but the postcentral rhythm persisted throughout. This is illustrated in Fig.4.

Independence of similar rhythms recorded from various areas could also be shown by employing the averaging procedure. Fig.5-1 illustrates results obtained from two subjects, where in one the precentral (A) alpha band activity was used to trigger simultaneous recording from the pre- and postcentral areas, and in the

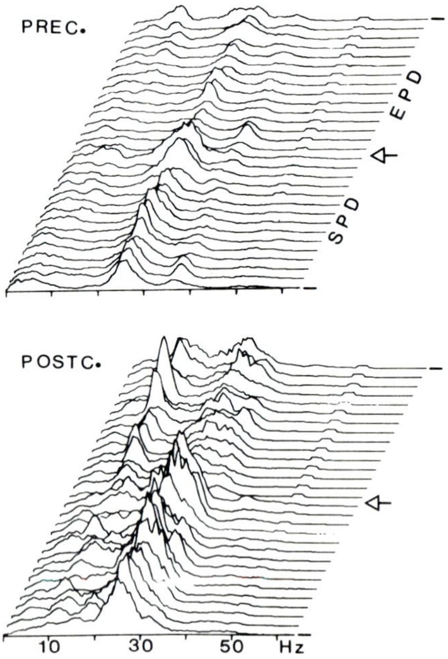

Fig.4. Simultaneously recorded average spectra from right precentral (PREC) and right postcentral (POSTC) electrodes during self-paced displacement (SPD) and externally paced finger displacement (EPD). Open arrow indicates the transition from self paced to externally paced movements. The subject had her eyes open on both occasions. Each spectrum derived from 16 sec of ECoG activity.

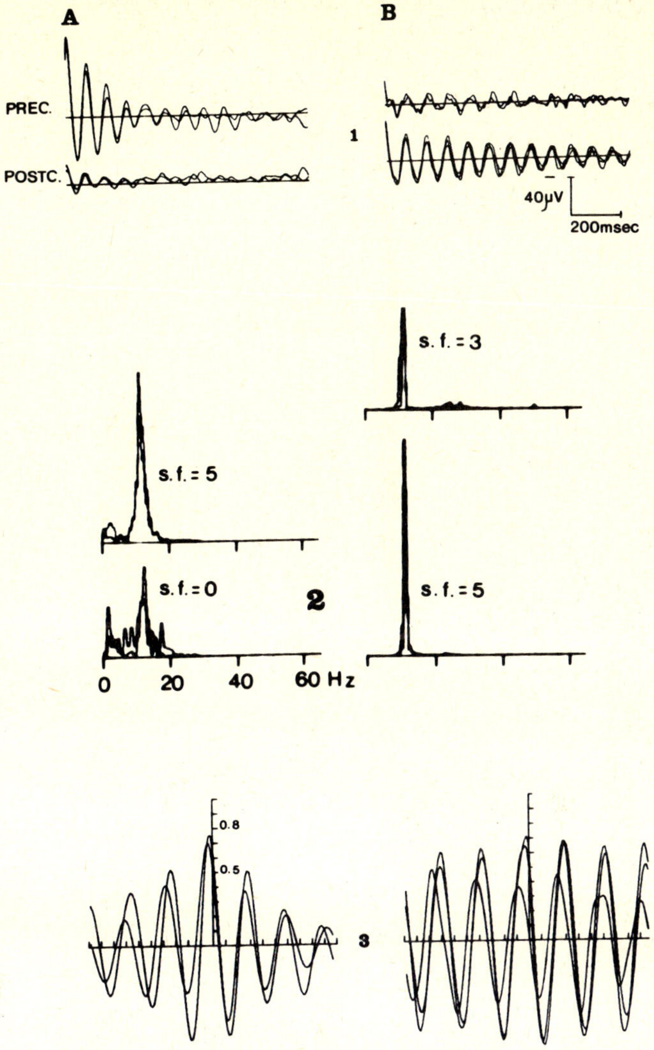

Fig.5. Superimposed averaged (N=30) intrinsic activities (1) of two subjects, A and B, from precentral (PREC) and postcentral (POSTC) electrodes. The corresponding spectra (2) and cross-correlograms between the two areas (3). For subject A the precentral and for subject B the postcentral "a" activity triggered the averager. Note the difference in scale factors of the spectra between subjects and locations.

other the postcentral (B) rhythm was used to trigger the simultaneous recording from both areas. A prominent averaged rhythm is seen at the location providing the trigger pulse, while the rhythm at the other area is much diminished. Quantification of the differences is seen in the power spectra of these rhythms, in Fig. 5-2.

The intracerebral electrodes which were to be used in treatment were introduced perpendicularly from the frontal convexity to the orbital cortex. Their position was identified by radiography. In one subject simultaneous recordings were obtained from two electrodes (4.0 mm apart) traversing the orbital cortex. Rhythms were seen to be shifting in phase from moment to moment. This is illustrated in Fig.6. The rhythms are out of phase at time A; slightly shifted at time B; in phase at time C; and present at only one electrode at time D.

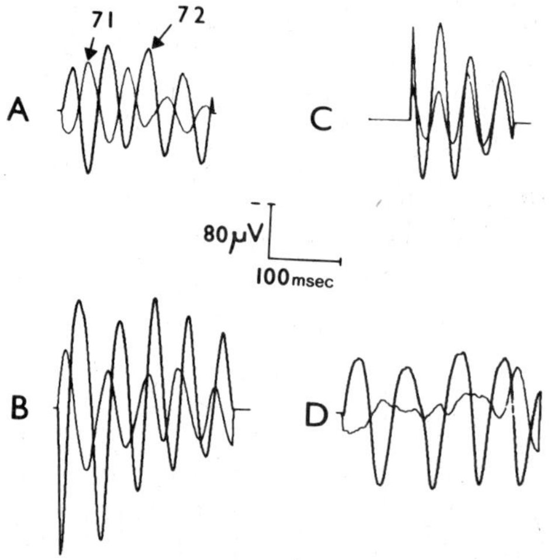

Fig.6. Superimposed ECoG activities simultaneously recorded across the orbital cortex from electrodes 71 (thin trace) and 72 (thick trace) during four different instances in time, A, B, C and D, during a short recording session with the subject relaxed with eyes open.

In view of this demonstration of variability in transcortical phase relationships, we tried repeatedly to isolate instances of phase reversal of the central rhythms. On no occasion did we find phase reversal of the central rhythm recording from pre- and postcentral areas, although a phase shift equivalent to 8-12 msec time interval was usually present (Fig.5-3), with the postcentral leading. On some occasions, however, the opposite lead was observed, and in a few instances the time lag could be 16 or 20 msec. It should be emphasised that in these pre- and postcentral areas, there was always characteristic reversal of the early somatosensory response.

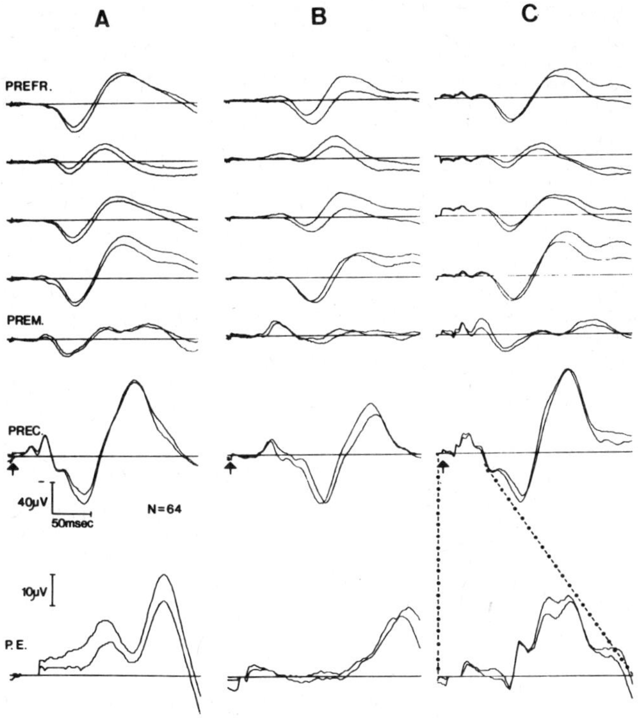

Fig. 7. Superimposed averaged evoked potentials from one subject to auditory (A), visual (B) and left median nerve stimulation at the wrist (C) recorded from 4 prefrontal (PREFR) one premotor (PREM) and one precentral (PREC) electrodes. The lower traces (P.E.) represent the expanded and amplified early part of all precentral potentials as indicated (-o-) in C.

Area specificity was also observed when evoked potentials were elicited from the various locations. This is illustrated in Fig.7 where can be seen the spatial distribution of the auditory, visual and somatosensory evoked potentials in the prefrontal, premotor and precentral cortical areas. An obvious difference in waveform can be noticed between the responses of the three areas.

The size of the area from which similar waveforms of evoked potentials were recorded was different for different parts of the hemisphere. The somatosensory EPs recorded from the primary receiving areas were complex in waveform and considerably different from adjacent electrodes. The EPs recorded from prefrontal areas were much more simple in waveform, and their similarity across areas very obvious. In addition, comparison between evoked responses to stimuli of different modalities recorded at a distance from the primary receiving areas revealed obvious similarities in waveform, as can be seen by comparing the auditory and visual evoked potentials of Fig.7.

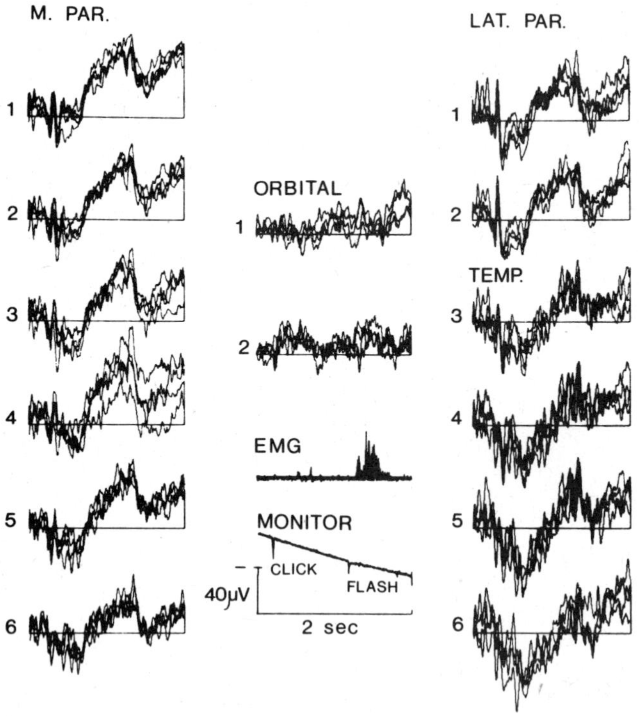

Fig.8. Superimposed CNV averages simultaneously recorded from six paramedian parietal electrodes (M.PAR.1-6), two lateral parietal electrodes (LAT.PAR.1-2), four temporal electrodes (TEMP.3-6) and two orbital electrodes (ORBITAL 1-2) in the right hemisphere. EMG: The corresponding rectified electromyogram of the left arm muscles. Each trace is the average of 16 trials.

This more widespread development of similar EPs in the non primary receiving areas, and the across modality similarity, were also characteristics of the stimulus or response time locked brain electrical activities referred to as event related potentials, such as the CNV and the P300 potentials. This is illustrated in Fig.8. Yet in spite of the widespread development of the event related potentials, closer inspection reveals that the area specificity is still present. The CNV amplitude and waveform, for example, for some of the 14 locations registered simultaneously in Fig.8, show considerable differences. The orbital cortical areas do not develop a CNV at all. The antero-posterior line of paramedian parietal electrodes (1 to 6) shows CNVs obviously different in amplitude; and the waveform of the CNV of the temporal areas is different from all other locations. The positivity following the warning click, or S1 signal, is prominent in the more posterior of the temporal areas, while any positivity following the imperative flash or S2 stimulus is very small. In contrast the more anterior temporal electrodes consistently record a post-S2 positivity with 300 msec latency.

The prominent post-S1 positivity of the posterior temporal locations is so prolonged that the CNV develops only during the last third of the foreperiod. Yet the differences in CNV amplitude and form, in different areas, which have been illustrated in Fig.8, are not necessarily associated with differences in the waveform of the evoked potentials following the S1 stimulus. From certain brain areas very similar potentials were evoked by the S1 and S2 stimuli, presented to the subject either without instruction or within the context of a CNV paradigm. But as can be seen from Fig.9-A the CNV amplitude of electrode 10 was considerably larger compared to that of electrode 11. In contrast electrode 11 compared with electrode 12 was recording similar potential in both conditions. Electrodes 10, 11 and 12 were separated from each other by 4.0 mm, in an antero-posterior line.

The similarity in evoked potentials and CNV recorded from electrodes 11 and 12 in the case illustrated in Fig.8 was not accompanied by similarity of rhythmicity in those two areas. In contrast, the power spectra of spontaneous rhythms recorded at electrodes 10 and 11 (with similar CNVs) were virtually identical.

The relationship between CNV, evoked potentials and the type of rhythmical activity recorded from a particular area was a complex one. The complexity is seen to be increased even more when one takes into account the fact that apparently similar rhythmic activity, recorded from various areas, with the subject in a relaxed state, reacted in different ways at the various locations, when the subject was engaged with some task. This is illustrated in Fig. 10 where the power spectra from the most anterior and the most posterior electrodes of a paramedian parietal line of electrodes are shown together with those of the most anterior of the lateral parietal and the most posterior of the temporal electrodes. During both the eyes open and eyes shut conditions, apparently similar alpha band activity was recorded from the four areas. The frequency of this activity and its lack of reactivity to eye closure was common in all locations. Yet its amplitude and reactivity to contralateral fist clenching was different from area to area, and in the T.6 location almost absent. The CNV and P300 activities in these locations have been already presented for this subject in Fig. 8 and their dissimilarity has been pointed out.

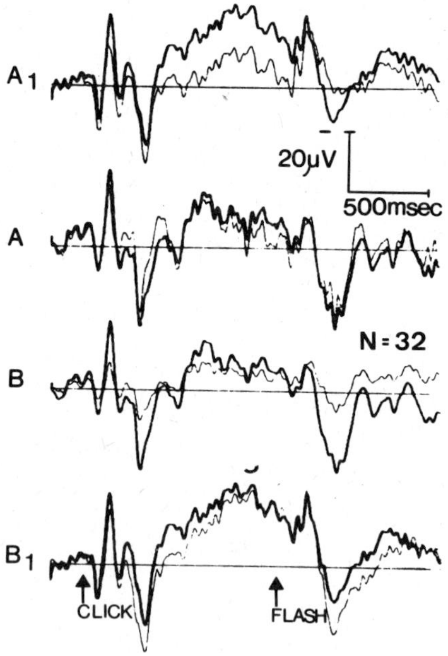

Fig.9. Superimposed averaged corticographic activities to a pair of auditory (click) and visual (flash) stimuli. The interstimulus interval was 1 sec. At A and B the stimuli were presented to the subject without interruption, at A1 and B1 with instruction to press a button with his left thumb immediately after the flash. Thick line: activity from a middle parietal electrode, electrode 11. Thin line at A1 and A: activity from an electrode, electrode 10, 4 mm anterior to electrode 11. Thin line at B1 and B: activity from electrode 12, 4 mm posterior to electrode 11.

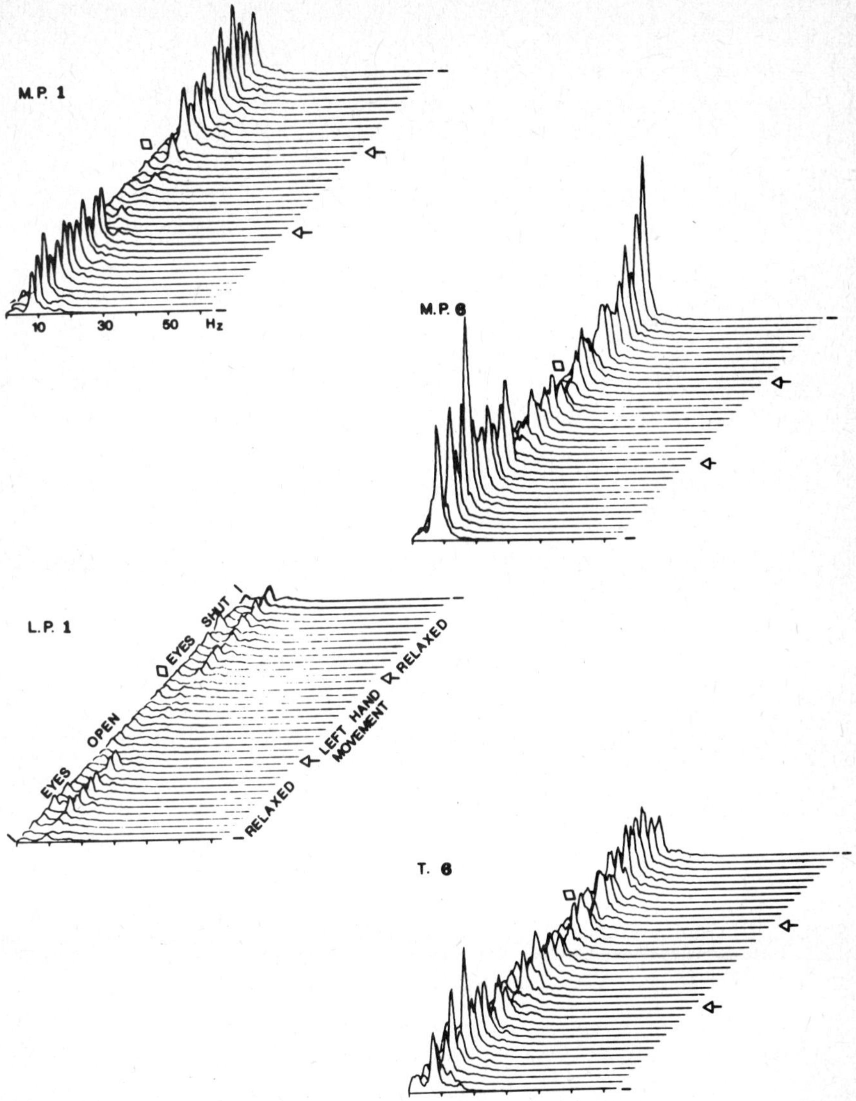

Fig. 10. M.P.1 and M.P.6 the sequentially recorded average spectra for the most anterior middle parietal electrodes. L.P.1 the first lateral parietal, T.6 the last temporal electrode. With the subject relaxed, clenching his left fist and relaxed again (indicated by open arrows). During the last relaxed sequence the subject had his eyes shut (indicated by open diamonds). Each spectrum derived from 16 sec of ECoG activity.

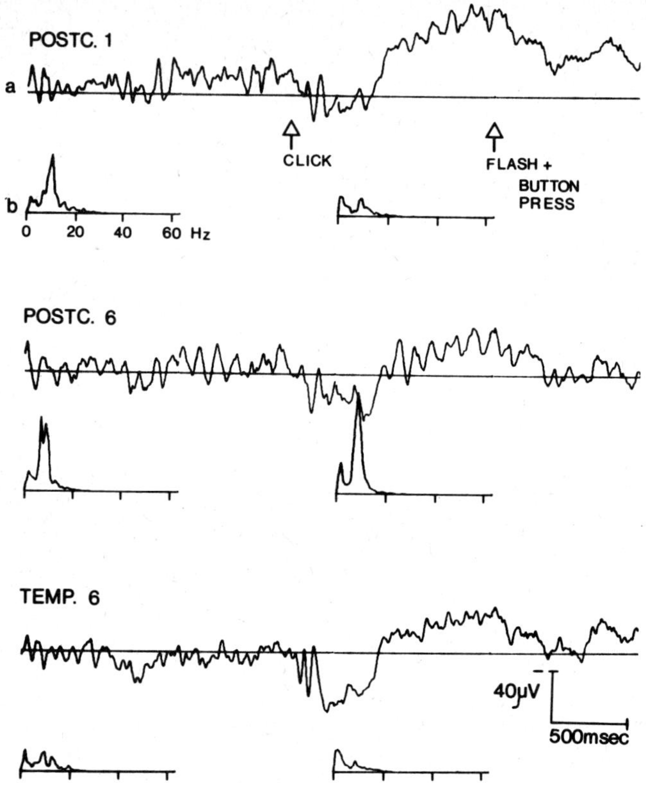

Fig. 11. CNV averages (a) from 9 single trials simultaneously recorded from the first postcentral electrode (POSTC.1) an electrode 4 cm behind (POSTC.6) and another one at the temporal area (TEMP.6). At b the average spectrum of the same single trial data (recorded with short time constant) for the corresponding 1 sec. part of trace a.

The rhythmicity of a particular location at rest, and the reactivity of this rhythmicity during the CNV development were also area specific. This is illustrated in Fig.11 where the activities before and during CNV are shown either as straightforward averages (a) or as averaged power spectra (b). It can be seen that from the first post-central electrode a large CNV is recorded, during which the alpha band activity abundantly present 1 sec before S1 is almost abolished. The last postcentral (POSTC.6) and the temporal (TEMP.6) electrodes register a rather similar CNV, but the power of the alpha band activity of the last postcentral electrode increased in comparison with its pre-S1 level, and the temporal electrode had a very small rhythmicity in the alpha band either before or during CNV.

DISCUSSION

The present corticographic studies of various rhythmic macropotentials, evoked and event related, are mostly in agreement with, or extend previous observations. The data also contain additional elements which occasionally appear to contradict certain current notions, and suggest the need for alternative approaches when formulating models of electrocortical organization underlying behaviour.

The observations of Scarff and Rahm[2], Jasper[3] and Last[9] about the predominance of alpha band activity postcentrally is in agreement with our data. But in contrast with Scarff and Rahm, who in only one of their 17 cases detected alpha band activity precentrally, in all of our cases alpha band activity was intermittently present in the precentral area. It seems that our data are in some agreement with Brazier[34] who detected a persistent rhythmic activity of 13-14 Hz in the motor cortex of the unanaesthetized cat, but in disagreement with Howe and Sterman[35] who localised similar activity only in the postcruciate somatosensory cortex of the unanaesthetized alert cat.

To what extent the pre- and postcentral motorsensory alpha band activity in man is controlled by different thalamic nuclei cannot be answered by our data. It should yet be emphasized that the precentral alpha band activity and its fluctuations in amplitude and duration presented a pattern which was most of the time different from that of the postcentral alpha band rhythm. There were, however, instances when the two activities were correlated, so that using the one as trigger the other could be averaged. This procedure further illustrated the observation made already from single trial data that there is no phase reversal of the pre- and postcentral alpha band activity. In some of the subjects there have been observed phase differences between the two activities, equivalent to 8-12 msec time interval, with the postcentral activity leading.

Thus in terms of duration of each episode of alpha band activity in the two areas, in terms of different fluctuations in amplitude within the same episode, and in terms of phase differences between areas, it may be argued that different neuronal systems underly the generation of the pre- and postcentral alpha band activities.

Both pre- and postcentral alpha band activities were abolished by self paced contralateral hand movements. To what extent this common influence of movement upon the two rhythms indicates some similarity or sharing of function of the alpha band activity of the two areas is open to speculation. However, the dissociation of reactivity of the alpha band activity of the two areas with externally paced (passive) finger movement is of considerable interest and importance. The precentral rhythm was mostly affected by contralateral movements externally paced. This is in agreement with an independent sensory input to the precentral cortex proposed on the basis of evoked potential data[20,21]. It should be noted that the simultaneously recorded temporal alpha band activity was almost unaffected by the finger movements. This difference in reactivity of similar rhythmicity recorded from different areas, indicates that one must qualify any inference about function drawn only from the waveform characteristics of any macropotential.

During the relaxed state rhythms of the beta band between 16-32 Hz were recorded from pre-, postcentral and temporal areas. These rhythms were more variable and independent in the pre- and postcentral areas than the alpha band activity. The reputed predominance of these rhythms precentrally was not evident in our data. On the contrary, the power spectra within the 16-32 Hz band were postcentrally of equal amplitude to those of the precentral area. Probably due to the diminished amplitude and duration of the alpha band activity, the beta rhythms can be easily recognised precentrally in the raw record.

During self-paced movements some of these beta rhythms decreased, some of them were not affected or even became more prominent. Thus, with regard to the reactivity of the beta rhythms, our data are in partial disagreement with those of previous authors who only mentioned decrease of those rhythms with self-paced movement[3]. This apparent discrepancy could be the result of applying more powerful procedures than those previously available to quantify the data.

It was also a consistent observation across all of our subjects that the alpha and beta band rhythms were dissociated both in spatial distribution and in reactivity. In that respect our data are more in agreement with Chatrian[36] who suggested dissociation between mu rhythms and beta activity, and with Waller and Feldman's[37] data in cat which showed independence of their A (10-15 Hz) and B (16-30 Hz) activity in somatosensory thalamic area and the sensorimotor cortex.

The absence of consistent relationship of the alpha and beta band activities between and within areas was intermixed with instances when larger or smaller cortical areas were highly correlated. As has been mentioned previously, during those instances the phase of the postcentral alpha band rhythmicity was usually leading, so that a latency of 8-12 msec between the two rhythms could be measured. This observation is in agreement with previous results by Pfurtscheller and Cooper[38] who noticed such difference, but did not discuss the location of their electrodes. The proposal (Kulman, this volume) that the pre- and postcentral alpha band rhythms are phase reversed is not upheld by our data, although the first component of the somatosensory evoked potential is of different polarity in the two areas.

In general, recording either epicortically or transcortically there were no indications of volume conduction in our recordings. This is in agreement with Petsche et al.[39] and Speckmann et al.[40]. Our 4 mm electrodes, 4 mm apart, are too large to permit calculations of the size of the individual cortical domain, but it can be said that the size of this domain is smaller than our electrodes. It also can be inferred that the organization of this domain is far from being rigid. Which factors shape the temporary size of a domain, and what their value is set to at a given point in time will depend on past, present, or anticipated future input-output of the domain. Particular domains may have many, few, or no factors in common with other domains. If there are common factors, they may vary in unison or with varying degrees of independence in the different domains. Mountcastle et al.[41] summarized the problem as follows: "Some of our observations suggest that another mode of organization is from time to time superimposed on the basically columnar pattern and that this additional set formation is dynamic and conditional in nature." Creutzfeldt[42] pointed out that "if the columnar hypothesis means that a discrete group of cells within a cylinder or strip of cortex 0.5 mm wide should be considered a separate functional unit and that the cortex is thus divided into separate functionally discrete columns,each assigned to a separate purpose, the hypothesis becomes untenable".

Evoked and event related potentials recorded from the same subjects and at the same electrodes from which the intrinsic rhythms were recorded, were in agreement with the inferences based upon rhythmicity as far as the dynamics of electrocortical organization is concerned. Waveforms were recorded from each area studied which were in size, or form, or both size and form, different from the simultaneously recorded activities of other areas. Indeed the presence of a particular activity at a particular location was dependent on the task which the subject was engaged with. The distribution of the P300 potential in parietal areas in a "go" type experiment, and the sweep of the potential in prefrontal areas in "no go" type experiments, is a typical example[26]. The distribution of the P300 following the S1 and S2 stimuli, illustrated in Fig.8, is another example. The CNV at first glance in Fig. 8 appears to have some spatial similarity, but closer inspection reveals that this is not the case. Both its amplitude or its take off time vary across areas 4 mm apart.

It should yet be appreciated that in spite of the area specificity it is also true that some areas could develop very similar potentials. This similarity of waveform across areas was mostly seen on the evoked potentials recorded from electrodes located at a distance from the primary receiving areas. In such conditions it could be argued that the potentials are somehow either volume conducted or picked up by the reference electrode. This is contradicted by the preserved difference of waveform between primary and non-primary areas when simultaneously recorded, and it is also in contradiction with data showing that a potential with similar form can develop independently in one or another area according to the task the subject is engaged with. It seems more probable that the neuronal system underlying the evoked responses from non primary areas are different from and more widespread than those generating the primary area responses.

Responses of non primary areas to visual, auditory and somatosensory stimuli were described in chloralose anaesthetized animals for the first time by Amassian[43] and verified by other authors from both anaesthetized and alert preparations[44-47]. To what extent these non-primary area responses are the result of a common input[44,45] and thus are modality unspecific, or are modality specific[48,49,47], or a combination of both dependent on the areas studied[46], cannot be resolved by our data. What can be inferred from the across modality similarity of the EPs is that the termination of the input and the organization of the cortical substratum activated by this input are probably similar across modalities for those cortical locations we have been able to explore in man.

Such modality independent widespread systems should not be assumed as unspecified "in reality"; as Jung[50] put it, "the so-called non-specific, reticular system has rather specific function". Probably there are many such systems which interact in different ways in different behavioural conditions, one of them being more predominant in one condition and another one in a different condition. It is in the understanding of this dynamic interplay that the organization of behaviour should be looked at; and if this is the case we should not ignore information about the various systems revealed to us by changes in other macropotentials which simultaneously develop in space, in time, or both space and time. The waveform, spatiotemporal characteristics and reactivity of the various macropotentials should not be studied in isolation from each other, because it could be that their interaction is the most important feature in understanding the organization of behaviour by the brain.

SUMMARY

Multichannel corticographic studies of intrinsic, evoked, sustained and the so-called P300 macropotentials were obtained from 8 patients carrying chronically implanted gold electrodes for treatment purposes. The recorded data indicate that:

1. Each one of the rhythmic, evoked and event related brain macropotentials studied displays its own spatial distribution.
2. The amplitude and spatiotemporal characteristics of each of the recorded activities are stable provided the experimental conditions remain unchanged.
3. Some of these macropotentials are well localised, and some are more widespread on the cortex.
4. The amplitude and extension in time of a particular activity varies, and depends on the cortical area from which it is recorded.
5. Change of experimental conditions results in change of amplitude and spatiotemporal characteristics of many macropotentials.
6. The spatial expansion of any particular macropotential is not invariable, but changes from a localised to a widespread distribution according to the conditions of the performance the subject is engaged with.
7. During different experimental conditions there are different location-specific interactions between activities.

By accepting that those amplitude-space-time characteristics and their interactions do have relevance with the subject's behaviour one is also forced to conclude that the various rhythmic, evoked, and event related macropotentials should not be studied in isolation from each other as though one man's signal could be considered the other man's noise because the brain behaviour problem, probably, will not be answered with only one technique's voice.

ACKNOWLEDGEMENTS

The authors wish gratefully to thank Dr.R.Cooper for comments and advice during the data acquisition period and on the manuscript, Dr.P.Pocock and Mr.W.J.Warren for assistance in computer programming, and Mrs.Freda Darby for typing and secretarial assistance.

REFERENCES

1. Schwartz,H.G. and Kerr,A.S. (1940). Electrical activity of the exposed human brain: description of technique and report of cases. Arch. Neurol. Psychiat. (Chic.), 43, 457-559.
2. Scarff,J.E. and Rahm,W.E. (1941). The human electrocorticogram. A report of spontaneous electrical potentials obtained from the exposed human brain. J.Neurophysiol., 4, 418-426.

3. Jasper,H. (1949). Electrocorticograms in man. Electroenceph. clin. Neurophysiol. Suppl.No.2, pp16-29.

4. Delgado,J.M.R. and Hamlin,H. (1956). Surface and depth electrography of the frontal lobes in conscious patients. Electroenceph. clin. Neurophysiol., 8, 371-384.

5. Sem-Jacobsen,C.W., Petersen,M.C., Dodge,H.W.Jr., Lazarte,J.A. and Holman,C.B. (1956). Electroencephalographic rhythms from the depths of the parietal occipital and temporal lobes in man. Electroenceph. clin. Neurophysiol., 8, 263-278.

6. Walter,W.G. (1962). Oscillatory activity in the nervous system. In: R.G. Grenell (Ed.). Neural Physiopathology, pp222-257. Harper and Rowe, Publ.

7. Bates,J.A.V. (1963). Special investigation techniques. Indwelling electrodes and electrocorticography. In: D.Hill and G.Parr (Eds.). Electro-encephalography. Macmillan, New York, pp429-479.

8. Cooper,R., Winter,A.L., Crow,H.J. and Walter,W.G. (1965). Comparison of sub-cortical, cortical and scalp activity using chronically indwelling electrodes in man. Electroenceph. clin. Neurophysiol., 18, 217-228.

9. Last,S.L. (1967). Electrocorticography in Man. J. Roy. Coll. Phycns., Lond., Vol.1, No.2, 109-117.

10. Aimone-Marsan,C.A. (Ed.). (1973). Direct, Cortical and Depth Evaluation of the Brain. Parts B and C. In: A. Remond (Ed.). Handbook of electroence-phalography and clinical Neurophysiology, Vol. 10.

11. Katzman,R. (Ed.). (1964). Sensory Evoked Responses in man. Ann. N. Y. Acad. Sci., 112.

12. Donchin,E. and Lindsley,D.B. (Eds.). (1969). Average evoked potentials. Methods, results and evaluations. NASA SP-191, Washington,D.C., 79-84.

13. Hirsch,J.F., Pertuiset,B., Calvet,J., Buisson-Ferey,J., Fishgold,H. et Scherrer,J. (1961). Etude des reponses electrocorticales obtenues chez l'homme par des stimulations somesthesiques et visuelles. Electroenceph. clin. Neurophysiol., 13, 411-424.

14. Giblin,D.R. (1964). Somatosensory evoked potentials in healthy subjects and in patients with lesions of the nervous system. Ann.N.Y.Acad.Sci.,112,93-142

15. Kelly,Jr.,D.L., Goldring,S. and O'Leary,J.L. (1965). Averaged evoked somato-sensory responses from exposed cortex of man. Arch.Neurol.(Chic.), 13, 1-9.

16. Domino,E.F., Matsuoka,S., Waltz,J. and Cooper,I.S. (1965). Effects of cryogenic thalamic lesions on the somesthetic evoked response in man. Electroenceph. clin. Neurophysiol., 19, 127-138.

17. Broughton,R.J. (1969). In: E.Donchin and D.B.Lindsley (Eds.). Average Evoked Potentials. pp79-84. NASA SP-191.

18. Goldring,S., Erturan,A. and Weber,D.C. (1970). Comparative study of sensory input to motor cortex in animals and man. Electroenceph. clin. Neurophysiol., 29, 537-550.

19. Goldring,S. and Ratcheson,R. (1972). Human motor cortex: Sensory input data from single neuron recordings. Science, 175, 1493-1495.

20. Papakostopoulos,D., Cooper,R. and Crow,H.J. (1974). Cortical potentials evoked by finger displacement in man. Nature, 252, 582-584.

21. Papakostopoulos,D., Cooper,R. and Crow,H.J. (1975). Inhibition of cortical evoked potentials and sensation by self-initiated movement in man. Nature, 258, 321-324.

22. Velasco,M. and Velasco,F. (1975). Differential effect of task relevance on early and late components of cortical and subcortical somatic evoked potentials in man. Electroenceph. clin. Neurophysiol., 39, 353-364.

23. Otto,D.A. (Ed.). (1978). Multidisciplinary perspectives in Event-Related Brain Potential Research. EPA-600/9-77-043.

24. Haider,M., Ganglberger,J.A. and Groll-Knapp,E. (1972). Computer analysis of subcortical and cortical evoked potentials of slow potential phenomena in humans. Confinia Neurol., 34, 224-229.

25. McCallum,W.C., Papakostopoulos,D., Gombi,R., Winter,A.L., Cooper,R.and Griffith,H.B. (1973). Event related slow potential changes in human brain stem. Nature (Lond.), 242, 466-467.

26. Papakostopoulos,D. and Crow,H.J. (1973). Electrocorticographic studies of the Contingent Negative Variation and P300 in Man. In: W.C.McCallum and J.R.Knott (Eds.). The Responsive Brain, pp201-204. John Wright and Sons Ltd., Bristol.

27. Tsubokawa,T. and Moriyasu,N. (1978). Motivational slow negative potential shift (CNV) related to thalamotomy. Appl. Neurophysiol., 41, 202-208.

28. McCallum,W.C. (1978). Relationships between Bereitschaftspotential and Contingent Negative Variation. In: D.A.Otto (Ed.). Multidisciplinary perspectives in event-related brain potential research. EPA-600/9-77-043, pp124-130.

29. Groll-Knapp,E., Ganglberger,J.A. and Haider,M. (1977). Voluntary movement-related slow potentials in cortex and thalamus in man. In: J.E.Desmedt (Ed.). Attention, voluntary contraction and event-related cerebral potentials. Prog. clin. Neurophysiol., Vol.1, pp164-173 (Karger, Basel).

30. Papakostopoulos,D. (1978). The serial order of self-paced movement in terms of brain macropotentials in man. J. Physiol. (Lond.), 280, 70p-71p.

31. Crow,H.J., Cooper,R. and Phillips,D.G. (1961). Controlled multifocal leucotomy for psychiatric illness. J.Neurol.Neurosurg.Psychiat., 24,353-360

32. Crow,H.J. (1973). Intracerebral polarization and multifocal leucocoagulation in some psychiatric illnesses. Psychiat.Neurol.Neurochir.(Amst.),76,365-381.

33. Papakostopoulos,D. and Crow,H.J. (1979). Direct recordings of the somato-sensory evoked potentials from the cerebral cortex of man and the difference between precentral and postcentral potentials. In: J.E.Desmedt (Ed.). Prog. clin. Neurophysiol. Vol.7, Karger, Basel (in press).

34. Brazier,M.A.B. (1963). The problem of periodicity in the electroencephalogram: Studies in the cat. Electroenceph. clin. Neurophysiol., 15, 287-298.

35. Howe,R.C. and Sterman,M.B. (1972). Cortical-subcortical EEG correlates of suppressed motor behavior during sleep and waking in the cat. Electroenceph. clin. Neurophysiol., 32, 681-695.

36. Chatrian,G.E., Petersen,M.C. and Lazarte,A. (1959). The blocking of the rolandic wicket rhythm and some central changes related to movement. Electroenceph. clin. Neurophysiol., 11, 497-510.

37. Waller,H.J. and Feldman,M. (1973). Correlations between somatosensory thalamic activity and cortical rhythms. Brain Research, 57, 417-441.

38. Pfurtscheller,G. and Cooper,R. (1975). Frequency dependence of the transmission of the EEG from cortex to scalp. Electroenceph. clin. Neurophysiol., 38, 93-96.

39. Petsche,H., Muller-Paschinger,I.B., Pockberger,H., Prohaska,O, Rappelsberger,P. and Vollmer,R. (1978). Depth profiles of electrocortical activities and cortical architectonics. In: M.A.B.Brazier and H.Petsche (Eds.). Architectonics of the Cerebral Cortex, pp257-280.

40. Speckmann,E.J., Caspers,H. and Janzen,R.W.C. (1978). Laminar distribution of cortical field potentials in relation to neuronal activities during seizure discharges. In: M.A.B.Brazier and H.Petsche (Eds.). Architectonics of the Cerebral Cortex, pp191-209.

41. Mountcastle,V.B., Lynch,J.C., Georgopoulos,A., Sakata,H. and Acuna,C.(1975). Posterior parietal association cortex of the monkey: Command functions for operations within extrapersonal space. J. Neurophysiol., 38, 871-908.

42. Creutzfeldt,O.D. (1978). The Neocortical Link: Thoughts on the generality of structure and function of the Neocortex. In: M.A.B.Brazier and H.Petsche (Eds.). Architectonics of the Cerebral Cortex, pp357-383.

43. Amassian,V.E. (1954). Studies on organization of a somesthetic association area including a single unit analysis. J. Neurophysiol., 17, 39-58.

44. Thompson,R.F., Johnson,R.H. and Hooper,J.J. (1963a). Organization of auditory somatic sensory and visual projections to association fields of cerebral cortex in the cat. J. Neurophysiol., 26, 343-364.

45. Thompson,R.F., Smith,H.E. and Bliss,D. (1963b). Auditory, somatic sensory and visual response interactions and interrelations in association and primary cortical fields of the cat. J. Neurophysiol., 26, 365-388.

46. Buser,P. and Bignall,K.E. (1967). Nonprimary sensory projections on the cat neocortex. Intern. Rev. Neurobiol., 10, 111-165.

47. Robertson,R.T., Mayers,K.S., Teyler,T.J., Bettinger,L.A., Birch,H., Davis,J.L., Phillips,D.S. and Thompson,R.F. (1975). Unit activity in posterior association cortex of cat. J. Neurol., Vol.38, 780-794.

48. Buser,P. and Borenstein,P. (1959). Responses somesthesique, visuelles et auditives, recueilles au niveau du cortex "associatif" suprasylvian chez le chat curaris non anesthesié. Electroenceph. clin. Neurophysiol, 11, 285-304.

49. Buser,P., Borenstein,P. and Bruner,J. (1959). Etude des systemes "associatifs" visuels et auditifs chez le chat anestésié sous chloralos. Electroenceph. clin. Neurophysiol., 11, 305-324.

50. Jung,R. (1974). Concluding session: Panel discussion. Brain Research, Vol. 71, 552-555.

CHANGES OF EEG RHYTHMS PARALLEL TO EVENT-RELATED POTENTIALS

© 1980 Elsevier/North-Holland Biomedical Press
Rhythmic EEG Activities and Cortical Functioning
G. Pfurtscheller et al., editors

SYNCHRONIZATION OF SINGLE EVOKED RESPONSE COMPONENTS: ESTIMATION AND INTERPRETATION OF REPRODUCIBILITY MEASURES

C. PARVIN, F. TORRES and E.A. JOHNSON
Division of Health Computer Sciences, Department of Laboratory Medicine and Pathology, Box 511 Mayo Memorial Building, 420 Delaware Street SE, University of Minnesota, Minneapolis, Minnesota 55455 (U.S.A.)

INTRODUCTION

In visual evoked response research, a number of assumptions about the evoked response and its relation to brain electrical activity were made. Many of these, such as assumptions about a constant evoked response, independent random spontaneous activity, and time-separated peak events, make possible convenient extraction techniques and data reduction methods. However, while these assumptions are physiologically plausible, they are not based on any current physiological knowledge. Indeed, some of the assumptions are suspect. It has been recognized by many that the characteristics of the after discharge are very similar to alpha activity. Cohn[1], Peacock[2], Barlow[3] and Levonian[4] have published papers that claim some degree of correlation between spontaneous alpha activity and the evoked response; however they disagree on what the relationsship is and whether the after discharge is truly alpha activity. Kooi and Bagchi[5] have noted relationships between early components of the evoked response and occipital rhythms. Rodin, et al.[6] reported correlation between the evoked response and other background rhythms of the EEG.

To collect the information necessary to investigate the assumptions made about the relationship between the visual evoked response and cortical rhythms some method of analysis at the single epoch level is generally required. Attempts at single epoch analysis have become more numerous, but still little has been done, This is partially due to the additional computing power needed for such work and the difficulty in finding good methods of analysis.

We have been working to develop methodologies that allow for a more comprehensive set of summary data to be collected from an evoked response session. The hope is to characterize better the evoked response, to improve upon methods of quantifying the differences between evoked response sessions, and to define parameters that enable more detailed investigations of assumptions and alternative hypotheses of generation. Efforts were directed at analysis of the single epoch and its relationship to the average evoked response (AER).

DATA ACQUISITION

The visual evoked responses (VERs) were recorded in a clinical research laboratory divided into two areas. The subject recording area was electrically shielded and sound-attenuated. It contained a lounging chair and a lamp enclosed in a soundproof box with a white opaque screen. The recording and stimulating equipment was located in the other area. There was a 16-channel Model 78 Grass polygraph and a 7-channel FM magnetic tape recorder connected to record in parallel with the polygraph. A Grass Model PS-2 photic stimulator was used to produce the light flashes. The bandwidth of the system was set at 0.1 Hz to 90 Hz.

Subjects sat in a slightly reclined position in the darkened recording room. Intense (setting 8 on the Grass photic stimulator) flashes of white light were presented at one per second. Duration of the flashes was 10 μsec. The light source was place approximately 30 cm from the subject's eyes, diffused by a screen placed between the lamp and the subject. The subject was instructed to relax, stay attentive and look at a point marked on the center of the screen One session was comprised of 100-200 flashes.

Seven channels of data were recorded: six channels of EEG data and one channel recording the pulses synchronous with the light flashes. The six scalp locations varied from subject to subject, but always conformed to the 10-20 international lead system and always included the occipital leads. Referential recordings were made with the ipsilateral ear serving as reference electrode. Only data from the left occipital lead was used in this study. Tin electrodes, 6 mm in diameter, were used with a salt gel and applied with collodion and compressed air. Electrode impedance was always kept below 5,000 Ω and was generally kept at approximately 3,000 Ω.

The FM analog tape was digitized on a PDP 12 computer. A sampling interval of 3.3 μsec (303 samples per second) was used with 10 bits of resolution for the range -50 μV to 50 μV. The digitized tape was run on a CDC 3300 computer, to determine where the stimuli occurred and to accumulate the appropriate epoch data into a single record. An epoch was defined as 90 data values (297 msec) prior to the stimulus and 166 data values (548 msec) after the stimulus. Thus, one epoch was comprised of 256 data values spanning approximately 845 msec.

EEG artifact was defined to be extreme amplitude changes relative to the background activity. This definition allows detection and elimination of those epochs that have the most severe effect on the subsequent analysis, while keeping the artifact detection algorithm simple and straightforward. The general algorithmic approach was to define a moving window four epochs long. The mean and standard deviation within the window were calculated. Threshold values were obtained in terms of standard deviation units above and below the mean. The number of standard deviation units was calibrated to agree with an EEG technologist on a random subset of data. Each amplitude value of the next epoch was compared with these thresholds. If two or more successive amplitude values exceeded the thresholds, the epoch was considered artifact and tagged as such. All left occipital data were checked for artifact in this way. In the worst session approximately 7.5 % of the epochs were eliminated. Most sessions had substantially less than 5 % of their epochs eliminated. The artifact appeared to be randomly distributed throughout the sessions. After the artifact was eliminated, only the first 90 remaining epochs of a session were used for further analysis. Repeated sessions for several different subjects were analyzed. Subjects included epileptics and normals.

FILTERING THE EVOKED RESPONSE

In an effort to understand better the average evoked response, it was decided to filter it into frequency bands and determine to what extent the conventional EEG frequencies contributed to the AER. The original intention was to see if greater consistency could be found in the AERs when they were broken into frequency bands than was apparent from looking at the unfiltered AERs. Frequency bands were chosen to correspond roughly to conventional EEG frequency bands: 1-7 cps (delta and theta), 7-14 cps (alpha), and 14-25 cps (beta). The 0-1 cps band and frequencies above 25 cps were considered to be noise.

Digital filtering was performed using a split cosine bell as the filter function. The filters were designed in such a way that 100 % of the contribution of each frequency is retained in the filtering process; that is, no part of the power of any frequency band was lost due to the filtering. The sum of the three pass band filters substracted from the unfiltered data defined the residual. Thus, the three frequency bands plus the residual exactly equalled the unfiltered data. Figure 1 shows examples of three AERs, their amplitude spectra, the three filtered bands and the residual, with their amplitude spectra.

The following characteristics have been observed in our data. The amplitude spectra of the AERs often contain major peaks in the 1-7 cps and 7-14 cps frequency bands. If a peak occurs in the 14-25 cps band, it is usually much smaller. The residual contains relatively little energy in the spectrum. The location and width of the peaks are such that they are generally located within a single frequency band. No information is lost if the peaks do not fit nicely

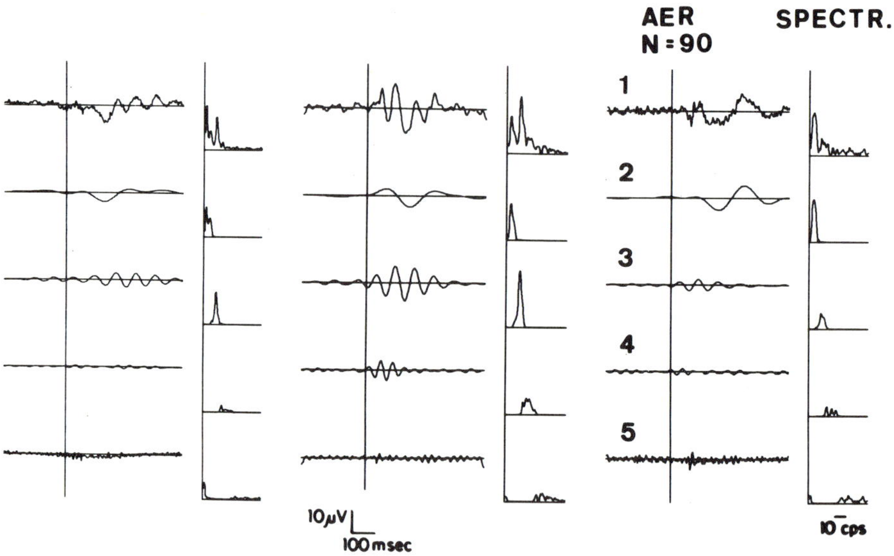

Fig. 1. Filtered average evoked responses and their amplitude spectra:1) the unfiltered AER, 2) 1-7 cps filtered band, 3) 7-14 cps filtered band, 4) 14-25 cps filtered band, 5) residual. The AERs are plotted such that positivity at the occipital electrode relative to the reference electrode results in a downward deflection.

into the filtered bands, but if they do, it makes interpretation of the filtered components easier.

The 1-7 cps band always had a positive wave (plotted downward in the figures) occurring after the stimulus. The wave varied in its location, with its peak usually occurring 150-300 msec after the stimulus. The amplitude of the peak varied from approximately 1 to 10 µV. In some cases the positive wave was either preceded or followed by a negative wave of approximately the same amplitude. In the majority of cases, though, the positive wave was the only one of singificant amplitude.

The 7-14 cps band was more variable than the 1-7 cps band; however, variability between the sessions for a single individual was still much less than between individuals. The predominant frequency was similar in character to alpha activity, usually ranging between 9-11 cps. The time at which the maximum activity occurred varied from shortly after the stimulus to late in the epoch. Most cases had some significant activity in this band; significant in the sense that there was an increase in activity after the stimulus.

There appeared to be little important activity in the 14-25 cps band, in most cases. In those situations where there was a significant increase in activity, it was usually of short duration and occurred soon after the stimulus (within 100 msec).

Rather than defining evoked response differences in terms of changes in significant peak amplitudes and latencies of the unfiltered AER, it proved advantageous to describe differences in terms of changes in filtered components. There were a number of reasons to be optimistic about this approach. First, the narrow frequency range of the filtered components produced simpler waves. Second, the filtered waves always added up to give the unfiltered AER. Thus, a complex-looking unfiltered AER might be much more easily interpreted in terms of simpler waves adding up in a certain way. Third, the difficulties relating to the problem of peak separability in time were lessened. Two peaks that occurred at approximately the same time could be separated as long as their "shapes" were sufficiently different, for instance, if one peak had s sufficiently longer duration than the other. Fourth, the filtered components of the AER could have a significantly higher signal-to-noise ratio than the unfiltered AER in the following sense. Whatever portion of the true ER that fell in a particular filtered band was only obscured by noise that fell in the same band. In those narrow frequency ranges where there were significant amounts of ER activity, the amount of signal relative to the amount of noise should have been improved. Finally, the filtered bands had the potential to partition the variability between evoked responses in an advantageous manner. The filtered bands offered the possibility that certain bands would contain most of the variability of the ER while other bands would be more consistent. This had already been observed in the 1-7 cps band which seemed to have a consistent pattern that was much less variable than the other bands.

Figure 2 shows some specific examples of the way in which different AERs can be interpreted using the filtered components as opposed to the unfiltered AER (1-25 cps). The numbered peaks on the unfiltered AERs were chosen by a computer algorithm and then reviewed and corrected as necessary by one of us, taking into account the general configuration and number of peaks of most AERs from a given subject.* In Figures 2A and 2B the major difference in the AERs can be explained by the middle frequency band, where, in one case, the activity occurs earlier in the epoch than it does in the other. The difference in Figures 2C and 2D can be explained primarily by the presence of activity in the middle frequency band in Figure 2D that is absent in Figure 2C.

*Revision of the numbered peaks chosen by the computer algorithm never involved inclusion of additional peaks. Revisions were limited to removing peaks or renumbering peaks to improve consistency between AERs.

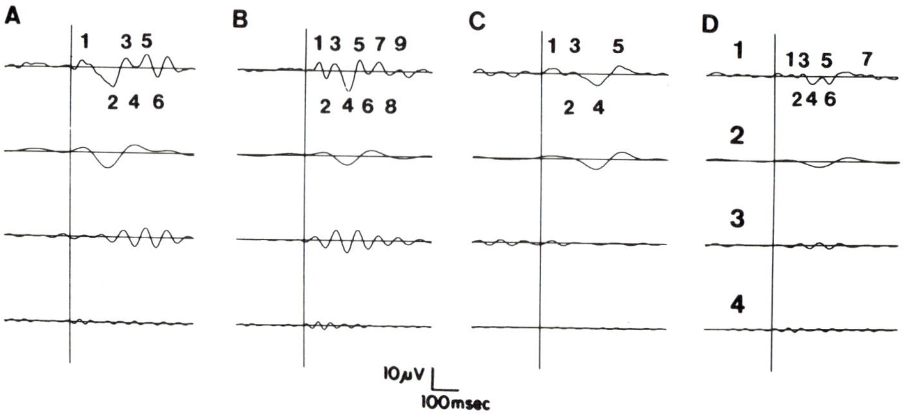

Fig. 2. Filtered average evoked responses to 90 stimuli. Comparison of A with B and C with D suggests that interpretations of differences in the unfiltered AERs (1-25 cps) are improved when looking at the filtered components.

SUPERIMPOSITION OF FILTERED BANDS

Superimposition of waveforms has the advantage of supplying qualitative information about epoch to epoch variability that an average does not. However, the ability to give meaningful interpretation to the way in which unfiltered epochs add together to give the AER is limited due to the complexity of the waveforms. The simpler waveforms resulting from band pass filtering improve the chances of detecting consistencies between epochs.

Figure 3 shows examples of overplotted epochs of the unfiltered data and the three filtered bands. Only the first 6 of the 90 epochs are overplotted,but the AERs shown are when all 90 epochs are averaged. Figure 3A represents a session in which no stimulus was presented; that is, all procedures were identical to a normal evoked response session except that the light source was blocked from the subject. The random appearance of the overplotted epochs is consistent with the resulting AER which shows no significant activity. Figure 3B is a true ER session for the same subject. Figures 3C and 3D are for two different subjects. Note the varying degree of epoch reproducibility that is present in the low- and middle-frequency bands. Also note that where a high degree of reproducibility occurs, significant activity in the AER results.

It can be seen that the superimposed epochs of the filtered components contain information about the evoked response session that is not conveyed in the AER. This additional information relates to 1) the way individual epochs add to give the AER (variability, signal to noise ratio), 2) the characteristics of the spontaneous EEG before and after the stimulus, and 3) the relationship of the ER and spontaneous EEG. How this information is interpreted and used depends on the underlying assumptions made about the ER and EEG.

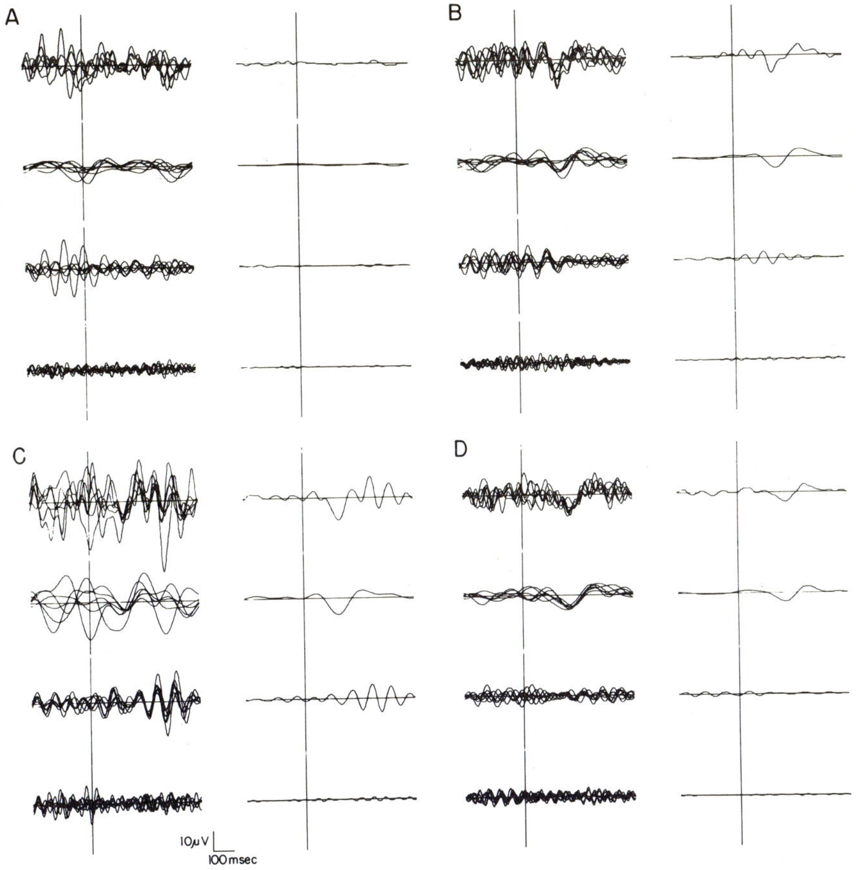

Fig. 3. Examples of differing degrees of epoch reproducibility. Only six epochs from a session of 90 epochs are superimposed. In session A, no stimulus was presented.

The standard additive model assumes that a constant evoked response adds to spontaneous EEG which is independent of the ER and unaffected by the stimulus. This model is based on the supposition that spontaneous EEG is much larger in amplitude than the ER, effectively burying it. Thus, averaging a large number of epochs reduces the amplitude of the random EEG allowing the ER to emerge. Since this model does not expect a large signal (ER) to noise (EEG) ratio for the individual epoch, no attempt is generally made to collect or interpret data at this level. However, comparing the AERs and the superimposed epochs of the filtered frequency bands, it becomes clear that in many cases the evoked response is quite visible. It has also been pointed out previously that a number of the assumptions made in the standard model are questionable. These include the assumptions of constant ER, random spontaneous EEG, EEG unaffected by the stimulus, and EEG independent of the ER. Trying to incorporate the relaxation of these assumptions into the additive model becomes difficult, especially since, once spontaneous EEG is allowed to be affected by the stimulus, the concept of an evoked response becomes less clear. Thus, in order to quantify the information contained in the superimposed epochs,it proved to be more convenient to approach the evoked response generative model in a different manner.

Rather than assuming a constant evoked response added to spontaneous EEG, consider interpreting the evoked response as spontaneous EEG that responds to the stimulus. The response can take any form, but can be thought of in terms of changes in amplitude, frequency, or phase of the EEG waveform. Since it is still assumed that the ER is transient, the changes are expected to be of finite duration. This is a more general model than the standard additive approach. To give an example of this interpretive approach, refer to Figure 3C. Here, the 7-14 cps frequency band can be explained as spontaneous alpha activity that has a random phase prior to the stimulus. After the stimulus there is a phase change that gradually converges to an almost constant value and then diverges again. This is accompanied by an increase and subsequent decrease in amplitude. Likewise, the other examples can be interpreted in terms of phase and amplitude changes.

The conceptual difference in the two models relates to their interpretation of the generative processes involved in the evoked response. In the standard model, the neurons or neuronal aggregates which are responsible for generating the evoked response are assumed to be different from those responsible for generating the spontaneous EEG. This follows from the assumption of independence. Also, the "evoked response neuronal complexes" are assumed to be idle during the unstimulated EEG. This follows from the additive assumption and the finite nature of the response. The new model assumes that the neurons that generate the spontaneous EEG are also responsible for generating the evoked response (or that they at least influence it). A flash of light causes the appropriate neurons to alter their activity in some consistent way. Thus, the shape of an evoked response to a specific light flash is related to the state of the neurons at the time of the flash.

While, conceptually, the difference is important as far as hypotheses of generation are concerned, mathematically, however, the two approaches can be shown to be very similar when restricted to a narrow band of frequencies. In fact, the standard additive model turns out to be a special case of the spontaneous EEG response model, where the change in amplitude and phase of the spontaneous EEG are restricted to a particular family of curves. Thus this new model can be thought of as nothing more than an alternative way of looking at the evoked response session that aids in the interpretation of the observed phenomena in the superimposed epochs.

EPOCH SUMMARY PLOTS

In an effort to represent what is seen in the superimposed epochs in a more convenient and measurable fashion, measures of phase and power as functions of time were developed. It was desired to keep the calculations for these measures simple and inexpensive. The strategy used to calculate phase as a function of time was to determine the time at which the EEG signal changed polarity and the direction of the polarity change. These times of zero crossing give a reference by which the phases of the different epochs relative to one another can be measured. For this approach to yield good results, however, the signal must be contained in a narrow band of frequencies to avoid the problem of a high-frequency wave riding on a lower-frequency wave. This criterion is fulfilled in the middle- and high-frequency bands, but in the low-frequency band baseline drift can occur to a certain extent. However, over all the epochs in a session, the picture given of the phase relationships should be an accurate one.

Using this approach, an easily interpretable picture of the presence or absence of phase-locking in an evoked response session can be obtained. This is done by accumulating separately the number of zero crossings that occur in the positive and in the negative direction over a number of successive time intervals for all epochs in the session. Two histograms result, one that can be plotted upward and the other downward. If the epochs are completely random, then the histograms should have approximately equal numbers of zero crossings in each time interval. On the other hand, if significant phase-locking occurs, there should be certain time intervals where most of the zero crossings occur and others where very few occur.

The approach used to estimate power as a function of time is similar to that used by Pfurtscheller[7]. A time interval is defined which approximately equals the median period length of one of the frequency bands. For a particular epoch, the amplitude values for each time point in an interval are squared and summed over the interval. The power values are then averaged over all epochs in the session to give a measure of the average power over time for the session. Similarly, the average evoked response is squared and summed over the same intervals to give the power as a function of time in the AER. The two functions are then plotted on the same axis. If only completely random epochs are present, the power in the AER should only be $(\frac{1}{N})$th the average power in an epoch (where N is the total number of epochs making up the average). However, if a significant response occurs, the ratio of the powers should be larger than $\frac{1}{N}$. Note that the AER power can never be greater than the average epoch power.

Figure 4 shows six superimposed epochs and the AER (for all 90 epochs), the zero crossing histogram and power plots for a particular evoked response session. Time intervals for the zero crossing histograms were chosen so that the

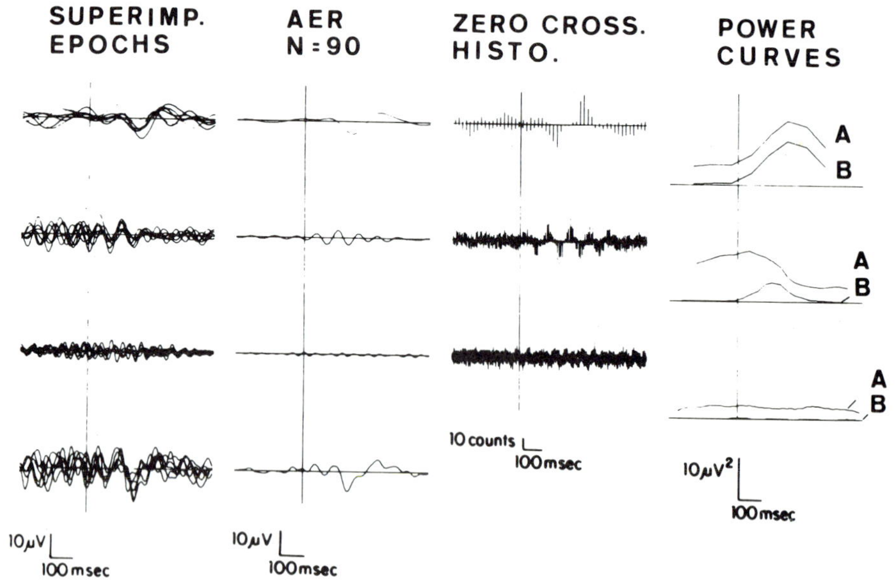

Fig. 4. Example of the zero crossing histogram and power plots for the AER shown. In the power plots, A represents the average epoch power and B represents the AER power.

expected number of zero crossings per time interval in the completely random case was approximately constant. Thus the time interval used in the high-frequency band, where more zero crossings will occur, is narrower than in the low-frequency band. The histogram plotted upward corresponds to zero crossings plotted upward in the AER. Power values are plotted at the midpoints of the time intervals used for their calculations. Thus, the distance from the edge of the axis to the first value plotted is one-half of this interval. A step size of one-third this interval length is used to plot successive points.

Prior to the stimulus, the zero crossing histograms are fairly uniform, but after the stimulus, in the low- and middle-frequency band, an alternating sequence of peaked distributions develops. The distributions occur at half-period intervals and alternate in direction. The "width" of the distributions indicates the degree of reproducibility in the superimposed epochs. The upper curve in the power plots (labelled with an A) is the average epoch power curve. Note that in this particular example, it shows that in the low-frequency band the power increases after the stimulus, in the middle-frequency band it decreases after the stimulus, and in the upper-frequency band it remains almost the same.It also reflects the relative power in each frequency band before and after the stimulus. Prior to the stimulus, the majority of the power is in the middle-frequency band. The smaller curve (labelled with a B) is the power in the AER. The size of this curve relative to the average epoch power curve reflects the degree of reproducibility in the epochs.

EPOCH SUMMARY MEASURES

While important and useful information can be obtained from these plots by simple visual inspection, more quantitative measures can be obtained. One possible set of parameters to summarize the important information in the zero crossing histograms and power plots is as follows. For each frequency band:

1) the minimum variance in the zero crossing histogram and the time point at which it occurs
2) the average epoch power prior to the stimulus and after the stimulus
3) the AER power prior to and after the stimulus
4) the maximum ratio of AER power to average epoch power and the time-point where it occurs.

Other parameters can be derived from these, such as 1) the ratio of AER power to average epoch power prior to and after the stimulus, 2) the difference in average epoch power before and after the stimulus, and 3) the percentage of the epoch power occurring in each frequency band.

Estimating the variance of the successive peaked distributions in the zero crossing histograms will measure the spread in the distributions. The minimum variance will therefore indicate the best level of reproducibility and where it occurred. The distribution of interest will contain 90 observations, one zero crossing from each epoch. If this distribution is approximately symmetrical, then the average curve should cross zero near the mean. However, if the tails overlap, it may be difficult to tell where one distribution ends and the next begins. The amount of overlap in the tails will depend on the degree of reproducibility in the epochs and the frequency variability within the filtered band. There is no way of correctly separating the tails when they overlap other than by re-examining each individual epoch. Therefore, the following approach was used to arrive at an estimate for the distribution and its variance. The time at which a zero crossing occurred in the AER served as the starting point. Moving out in both directions from this point, the total number of zero crossings was accumulated, until adding a further single time interval in each direction caused the total to exceed 90. The appropriate percentage of zero crossings in these last two intervals was retained so that the total equaled 90. With the endpoints of the distribution defined in this way, the usual variance estimate was calculated. This procedure can be repeated for all other zero crossings. There will be a tendency for this procedure to underestimate the true variance. This is primarily due to the artificial way in which the endpoints of the distribution are determined, which will result in endpoints that are equal to or narrower than the true endpoints. However, the values obtained will still accurately reflect the relative amount of reproducibility present.

Using one power value to represent the entire post-stimulus interval is approximately equivalent to replacing the post-stimulus power curve with a constant curve equal to its mean. This has the disadvantage of ignoring the increases and decreases in power that occur within the post-stimulus interval. However, the alternatives would either require 1) objectively determining an appropriate subinterval of the post-stimulus curve, 2) subjectively choosing an appropriate subinterval, or 3) retaining more than one value to represent the post-stimulus curve. Retaining more than one value might be appropriate in certain situations, but in general it defeats the goal of data reduction. Subjective choice of subintervals would probably decrease the variance of the parameter, but introduces the danger of statistical bias. Employing an objective method for obtaining a significant subinterval of the post-stimulus interval was attempted. The method was based on the ratio of the AER power to the average epoch power. It used the model

$$y_i(t) = A_i \sin(2\pi ft + \phi_i) \qquad i = 1,\ldots,N$$

where N equals the number of epochs in the session and the frequency, f, is assumed to be constant and approximately equal to the midpoint of the filtered band. A critical value was determined such that when it was exceeded, the activity in the AER was considered significantly different from completely random activity. The points at which the ratio went above the critical value and when it again fell below the value determined the appropriate subinterval. If more than one significant subinterval occurred, the one that contained the maximum power ratio was used. It was found that when calculating post-stimulus power in this way, as opposed to using the entire post-stimulus interval,the two values appeared to change in the same way from session to session.

RESULTS

Preliminary investigation of this set of parameters has revealed that they accurately reflect the information contained in the epoch summary plots and, in most cases, they tend to be less variable within an individual than between individuals. An exception to this was the time of minimum variance in the zero crossing histograms. This parameter suffered for two reasons. First, in the case where there was no appreciable phase-locking, the time of occurrence of the minimum variance was, in essence, random. Second, in many cases where significant phase-locking occurred, the time of minimum variance had a tendency to jump by half-periods from session to session, which made the estimated variability of the parameter far too large. Methods of improving upon this parameter were tried, but none were successful.

The primary goal of this study was the development of methodologies for obtaining more useful information from an evoked response session. Given the small data set analyzed, no statistically significant trends or differences (between epileptics and normals, for example) were expected. However, certain general observations were made. One of the trends that was suggested, but not fully substantiated by the data, was that of a characteristic phase-locking ability for an individual subject. While it is true that some individuals had consistently better epoch reproducibility than others, as measured by the minimum variance of their zero crossing histogram or their post-stimulus power ratio, it could not be determined whether an individual tended to have a particular ability to phase-lock regardless of average epoch power, because there was not enough range in the epoch power values.

Another question of interest concerns the alternative hypothesis of generation of the ER: phase-locked spontaneous activity versus an additive constant ER. We should remember that these are not mutually exclusive, but rather that the additive model is a special case of the phase-locking model. Thus, for the additive model to be appropriate, certain phenomena should always be observed. For example, there should always be an increase in the average epoch power after the stimulus at least as great as the power increase observed in the AER. If this is not the case, then either the evoked response and spontaneous EEG are not additive, or the spontaneous EEG is not stationary. Comparing the AER power increase to the difference in average epoch power before and after the stimulus suggests that for the low-frequency band the relationship seems to hold, while for the middle-frequency band there are cases where it does not. This raises the interesting possibility that the low-frequency band obeys an additive model, while the middle-frequency band responds in a different way.

CONCLUSION

The ability to address questions such as these is important for increased understanding of the evoked response. The ease with which the necessary parameters are obtained using the methods described make the investigation of such questions practical for routine studies. As more studies are carried out with attention focused on the relationship between the cortical rhythms and the visual evoked response, as well as the changes that occur in the AER, many questions will be answered that are usually never asked because of the restrictive assumptions made. This additional information should increase understanding of the generation of the visual evoked response and enhance its usefulness as a diagnostic tool.

ACKNOWLEDGEMENTS

This work was supported by NIH Contract NO-INS-52327 and by a grant from the Minnesota Medical Foundation

REFERENCES

1. Cohn, R. (1964) Ann. N.Y. Acad. Sci., 112, 281-291.
2. Peacock, S.M. (1970) "Average 'After Activity' and the Alpha Regeneration Cycle," Electroenceph. clin. Neurophysiol., 28, 287-295.
3. Barlow, J.S. (1960) "Rhythmic Activity Induced by Photic Stimulation in Relation to Intrinsic Alpha Activity of the Brain in Man," Electroenceph. clin. Neurophysiol., 12, 317-326.

4. Levonian, E. (1966) "Evoked Potential in Relation to Subsequent Alpha Frequency," Science, 152, 1280-1282.

5. Kooi, K.A., Bagchi, B.K. (1964) "Observations on Early Components of the Visual Evoked Response and Occipital Rhythms," Electroenceph. clin. Neurophysiol., 17, 638-645.

6. Rodin, E.A., Grisell, J.L., Gudobba, R.D. and Zachary, G. (1965) "Relationship of EEG Background Rhythms to Photic Evoked Responses", Electroenceph. clin. Neurophysiol., 19, 301-304.

7. Pfurtscheller, G. (1977) "Graphical Display and Statistical Evaluation of Event-Related Desynchronization (ERD)", Electroenceph. clin. Neurophysiol., 43, 757-760.

© 1980 Elsevier/North-Holland Biomedical Press
Rhythmic EEG Activities and Cortical Functioning
G. Pfurtscheller et al., editors

EVENT RELATED CHANGES OF RHYTHMICAL CORTICAL POTENTIALS

W. STORM VAN LEEUWEN[1,2)] and A. KAMP[2)]

1) Department of Clinical Neurophysiology, University Hospital Utrecht, Catharijnesingel 101, Utrecht (The Netherlands)

2) Brain Research Department, Institute of Medical Physics TNO, Da Costakade 45, Utrecht (The Netherlands)

INTRODUCTION

Alpha rhythms can be regarded as the first detected example of event-related changes in rhythmical cortical potentials. The alpha rhythms recorded from the scalp, represent electrical potential variations occurring synchronously or near-synchronously in comparatively large populations of neurons. For this reason the alpha rhythm , like most other EEG rhythms, can be related only to rather gross changes of physiological state. For the detection of subtle changes in brain activity the electrical potential variations of much smaller populations have to be studied. This can be done in animals by bringing small electrodes into direct contact with brain structures. For obvious reasons, such studies cannot be carried out in normal man. In patients suffering from brain diseases which necessitate the introduction of electrodes into the brain for diagnostic and/or therapeutic reasons, relations between smaller neuronal populations and psychological variables can be investigated without harm to the patient. In fact, such investigations are of potential value to the patient since they may help to localize the brain sites where the electrodes have been placed and/or to assess the effect of therapy.

We have been able to carry out such investigations in a small number of psychiatric patients in whom electrodes had been implanted for therapeutical reasons[1]. The motivations and the methods for doing this have been described and discussed extensively by Grey Walter, Crow and Cooper and their collaborators[2] and need not be repeated here. Moreover, we have carried out similar studies in a few epileptic patients in whom electrodes were implanted subdurally or within the brain to detect epileptogenic trigger foci which might be surgically removed. These studies have revealed the existence of various event related cortical rhythmical potentials which were not obtainable from scalp electrodes.

METHODS

Introduction of electrodes

In the psychiatric patients the electrodes were introduced by means of a method closely resembling that of Crow et al.[2]. Bundles of gold wire electrodes were placed stereotaxically in various cortical structures. They remained in place over a considerable period of time (up to 8 months). The present study mainly concerns event related activities encountered in the orbito-frontal cortex.

In the epileptic patients part of the electrodes were introduced within brain structures - mainly the hippocampus and amygdala - following the above described stereotaxic method. In addition,sheets of electrodes were pushed into the space between cortex and dura. This method allowed recording of comparatively large numbers of electrodes (up to approximately 60) in direct contact with cortical structures with little or no damage to neurons. In these patients the electrodes remained in place over periods of several weeks (up to 8 weeks).

Recording

The recording from intracerebral electrodes was carried out against a common reference electrode in the white matter of the orbito-frontal areas, using a time constant of 1 sec and a high frequency cut-off at 70 Hz. For CNV recording the time constant was 10 sec and linked mastoid electrodes were used as a common reference.

The recording was usually carried out using a conventional EEG recording system by means of an Elema Schönander 16 channel electroencephalograph. In some instances the electrical brain signals were transmitted by a 16 channel EEG radio telemetry system[3] with which the patient was allowed to move about in a room adapted for this purpose.

In all cases the electrical signals were stored on tape (EMI 24 track or Ampex 14 track) and subsequently analysed.

Analysis methods

In the course of time, various forms of analysis have been applied, most recently a hybrid analysis system has been employed[4]. The EEG signals were played back from tape and fed to a number of analogue bandpass filters with resonance frequencies between 2 Hz and 32 Hz.

The outputs of these filters were digitized, stored and subsequently statistically analysed. Apart from the brain signals a number of pulses were also stored on tape, marking the time at which certain events occurred, such as the presentation of stimuli (S1 and S2) and/or the giving of the subject's responses (R). These pulses were used for computing relations between EEG signals and certain events forming part of a number of stimulus response paradigms. The computer was programmed to average the brain signals in relation to a certain event marker used as trigger point over a period of 8 sec[5]. The trial period was divided into a period of 4 sec (indicated as -4) preceding the trigger point (indicated as 0) and 4 sec following it (indicated as +4). The -4 sec period was subdivided into a period of 3 sec (-4 to -1) and a period of 1 sec immediately preceding the trigger point (-1 to 0). The average of the EEG spectral activity during the 3 sec period -4 to -1 sec was used as a reference period for the subsequent periods.

In the final stage of statistical analysis the spectral data in the reference period and the post event data for each trial were compared using a non-parametric statistical test (Mann-Whitney). The computer was programmed to determine whether the spectral data in the samples over the -1 to +4 periods were larger or smaller than the reference values (the mean computed over the -4 to -1 period) for each separate trial. From these data graphs were drawn (see Fig. 1) indicating the number of trials in which the spectral data in the various frequency bands were significantly larger or smaller than reference (statistics according to Mann-Whitney at $p < 0.05$) for each separate time unit. Thus it was possible to determine significant changes notwithstanding the considerable variance from presentation to presentation. In order to reduce the effect of spontaneous fluctuations in EEG signals the computer was programmed to consider a change of activity as relevant only if a significant ($p < 0.05$) deviation from reference value occurred over at least 3 consecutive samples (approx. 1 sec). Moreover, a presentation was omitted if significant deviations from reference occurred in the -1 to 0 sec period.

Paradigms

Over the years various stimulus response paradigms have been used:

1) A conventional CNV paradigm in which S1 was a click sound and S2 a series of bright flashes which the patient was instructed to suppress by pressing a button.
2) A time estimation paradigm in which S was a click sound. The patient was asked to press a button 1 second after the sound was heard.
3) The patient, sitting on a chair, was presented with a brief projection of a simple picture. The patient was asked to get up, walk to a blackboard and draw the projected picture on it, walk back to the chair and sit down again.

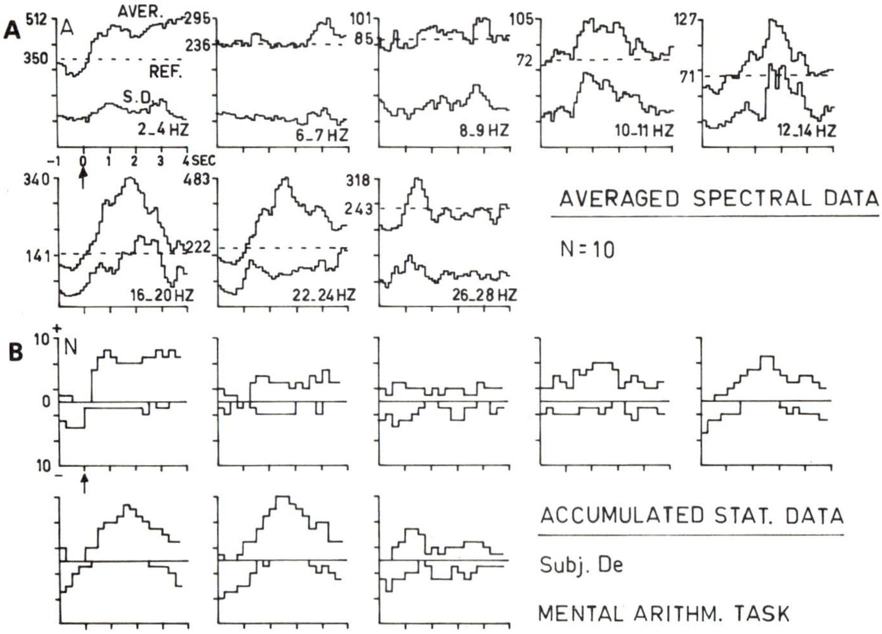

Fig. 1. Results of averaged sequential frequency analysis and statistical analysis in eight frequency bands obtained during performance of a mental arithmetic task. Trigger point at T = 0 sec corresponds with the verbal presentation of a number. Number of trials is 10. A: frequency analysis: AVER = averaged sequential spectral data determined, between T = -1 sec and T = 4 sec, at 40 sample points; REF = averaged baseline data obtained between T = -4 sec and T = -1 sec; SD = standard deviation of averaged spectral data. B: statistical analysis: accumulated sequential statistical data presenting, at 19 sample locations, the number of trials in which a significant increase (+) or decrease (-) took place ($p < 0.05$) with regard to the 24 baseline samples between T = -4 sec and T = -1 sec.

4) Arithmetic paradigm (MAT). The patient was asked to keep in mind a certain number. The patient was then given another number and was instructed to add this to the first number and tell the outcome of the addition as soon as possible.

5) Word association test (WTL and WTLN). The patient was given a word and was instructed to verbalize an association as soon as possible.

   a) Part of the words given to the patients were psychologically "loaded" for the patient (WTL).

   b) The other words had no specific psychological "load" (WTNL).

   The words termed "loaded" and "non-loaded" had been previously determined by the psychiatrist in charge of the patient. The words were presented at intervals varying randomly between 5 and 15 sec.

In all paradigms involving speech, the sound produced by patient and/or investigator was recorded. The first sound of a spoken word was used as one of the trigger pulses.

RESULTS

One of the first event-related rhythmic activities observed by us, happened in an epileptic patient in whom electrodes had been implanted in various temporal and frontal cortical areas. In this patient 6-9 c/sec activity was encountered in derivations from electrodes lying on the medial temporal cortex and 10-18 c/sec activity in derivations from electrodes on the medial frontal lobe. In this case the patient was presented at irregular intervals with a click sound and was instructed to press a button 1 sec later (paradigm 2).

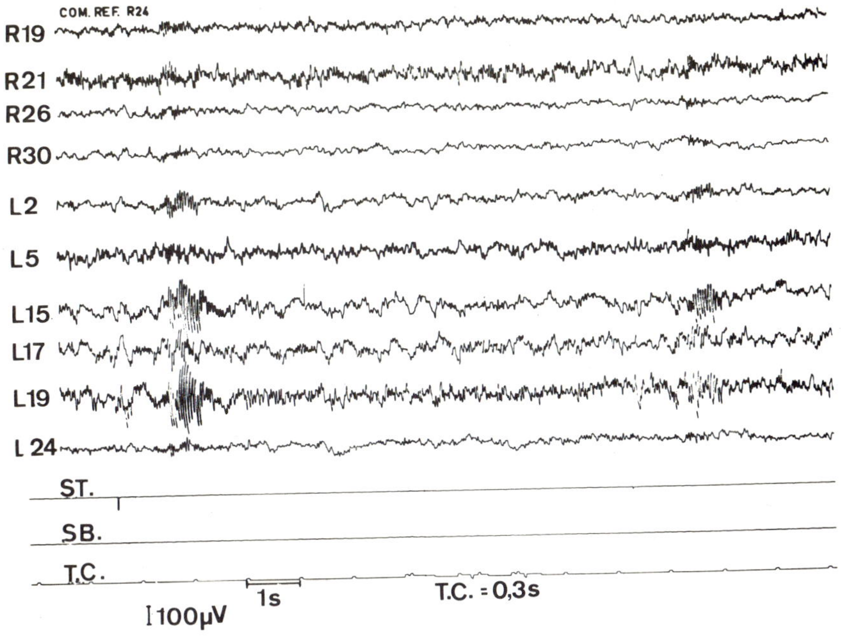

Fig. 2. Occurrence of beta burst during CNV paradigm (equivocation). The vertical deflection in the trace indicated by ST marks S1 (a click sound). R19-R30: electrodes in the right orbito frontal cortex; L2-L24: electrodes in the left orbito frontal cortex. Trace SB indicates if subject presses a button (not performed in this trial).

Immediately preceding and following the pressing, there was an increase of 10-18 c/sec activity in the frontal derivations and a decrease of 6-9 c/sec in the temporal derivations. In this same patient an increase of 10-18 c/sec activity in the frontal derivations was observed also during the paradigm 3.

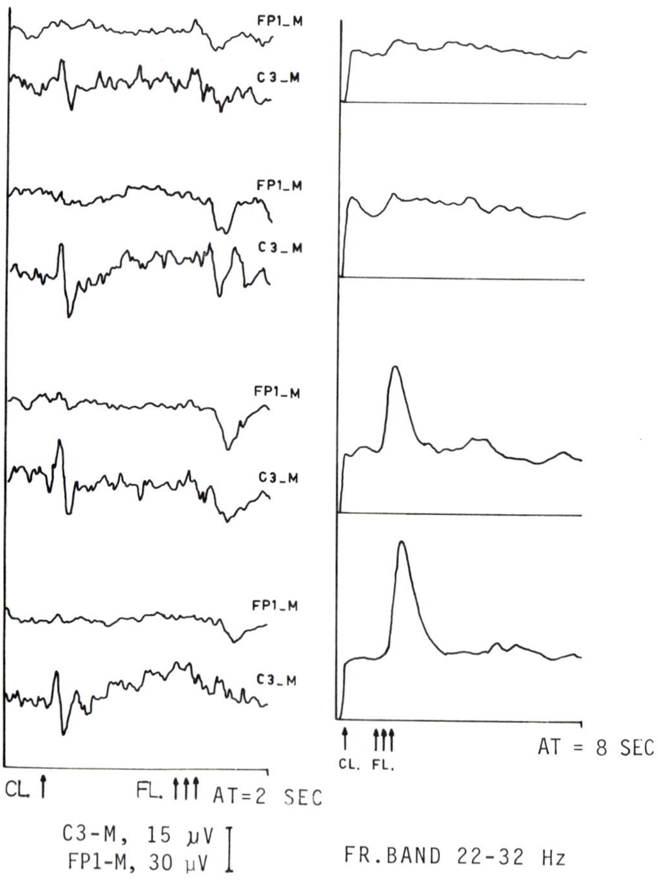

Fig. 3. The curves at the left side represent CNVs during 4 series of CNV paradigms (psychiatric patient). The 2nd and the 4th show clear CNV excursions at 2 scalp derivations, the 1st and 3rd show no marked CNVs. The 4 curves on the right-hand side show the simultaneous reactivities in the 22-32 Hz band obtained from a medio-frontal intra cerebral electrode (note different time scale). Comparison of CNV and 22-32 Hz reactivities indicate that there is no unambiguous relation between the occurrence of a CNV and an increase of beta activity.

A more extensively studied event related activity was observed in a psychiatric patient. In this case it concerned bursts of rhythmic activity at approximately 18-28 c/sec occurring during several stimulus response paradigms, one of which was the conventional CNV paradigm. The activity was called "beta burst" and has been described in some detail elsewhere[6]. For this reason it may suffice here to refer to a figure of this communication showing a beta burst in a CNV experiment during equivocation (Fig. 2), and to a figure showing that there is no consistent relation between the size of the beta burst and of the simultaneously recorded CNV (Fig. 3). In this communication a slight but significant relation between beta burst and decrease of heart frequency is also shown.

Subsequently beta bursts during CNV have been observed in various other psychiatric patients, though not in all. It was observed in some of these patients that the time at which beta burst occurred varied from one electrode to another, even in neighbouring electrodes at a distance of only 4 mm apart (Fig. 4). In this figure beta bursts are represented by an increase of 22-24 Hz activity in the electrodes L21 (left lateral orbito-frontal cortex). In both patients there is a considerable increase of beta activity at L21 approximately 1/2 sec after S1, while a significant increase at electrode L20 occurs approximately 1 sec later than the increase at electrode L21. In both patients therefore the 22-24 Hz increase in L20 and L21 follows different patterns.

Technical improvements and the use of computer analysis have made it feasible recently to study event related activities at various frequencies occurring simultaneously in 2 electrodes on the right and 2 electrodes in homologous areas on the left of the orbito-frontal cortex (lateral part) in 3 different patients. In the Figures 5, 6, 7 and 8 histograms are presented, obtained during the paradigms 4, 5a and 5b (arithmetic : MAT, word association test with "loaded" words: WTL and with "non-loaded" words: WTNL) in the frequency bands 2-4 Hz, 16-20 Hz, 22-24 Hz and 26-28 Hz. From Fig. 5 it appears that consistent change occurred in the 2-4 Hz band, where this activity increased in all subjects during most of the paradigms.

In subject V there was an initial increase of this activity in the right side during WTL but it was followed by a marked decrease. In subject F a similar though less pronounced phenomenon occurred in the left side during WTL.

In contrast, the changes of activity in the 16-20, 22-24 and 26-28 Hz bands (Figs. 6, 7 and 8) vary from one subject to another, from one electrode to another, from one frequency to another and from one paradigm to another. For example, in subject F there was a decrease in these frequency bands of both

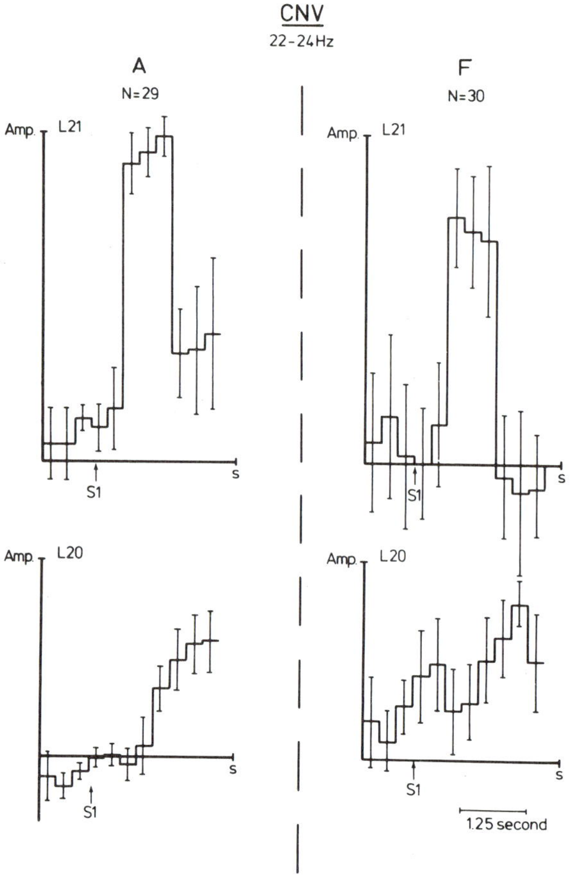

Fig. 4. Reactivity of 22-24 Hz during CNV paradigm in the left orbito-frontal cortex in 2 patients (A and F) in 2 adjacent electrodes (L21 and L20), 4 mm apart). Note sudden increase in L21 immediately after S1 and less marked increase approximately 1 sec later in L20.

sides during the 3 paradigms with the exception of an increase of 16-20 Hz, WTNL on the left and a decrease followed by an increase of 26-28 Hz, WTNL on the right. In subject V there was a marked decrease of the 3 bands on the left during the 3 paradigms while on the right there was a decrease of 26-28 Hz during WTL but a marked increase of 16-20 Hz and 22-24 Hz during MAT and a less pronounced increase during WTNL and no consistent increase or decrease during WTL.

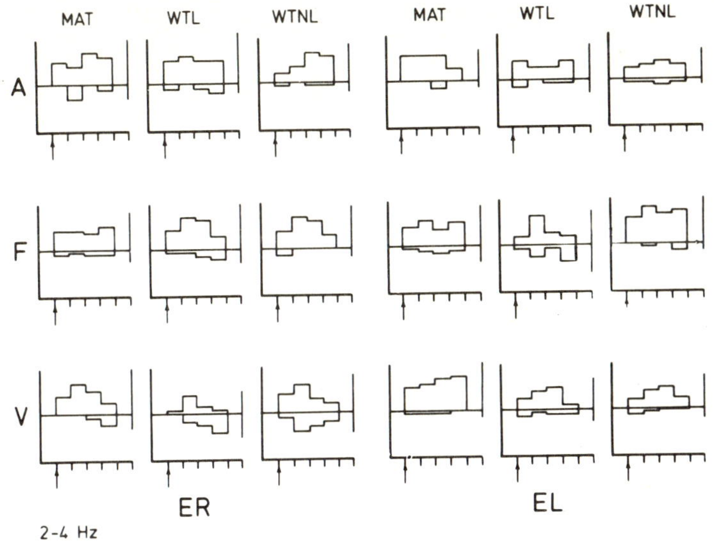

Fig. 5. Reactivity of 2-4 Hz in an orbito-frontal electrode on the right (ER) and the left (EL) in 3 psychiatric patients (A, F and V) during 3 paradigms (MAT is arithmetic; WTL is word association test with "loaded" words; WNTL is word association test with non-loaded words). Vertical scale: percentage of trials in which a significant increase or decrease during at least 1 sec occurred ($p<0.05$), full scale is 100 %. Horizontal scale: 1 division 1 sec, arrow indicates moment of stimulus presentation.

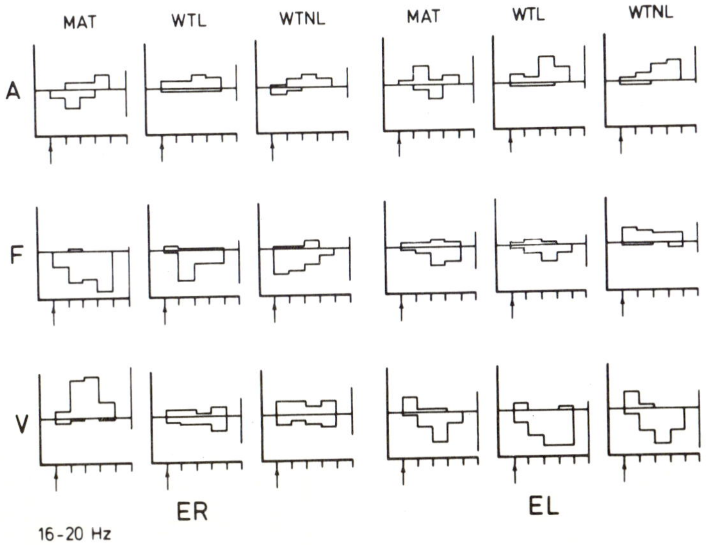

Fig. 6. Reactivity of 16-20 Hz, otherwise same as Fig. 5.

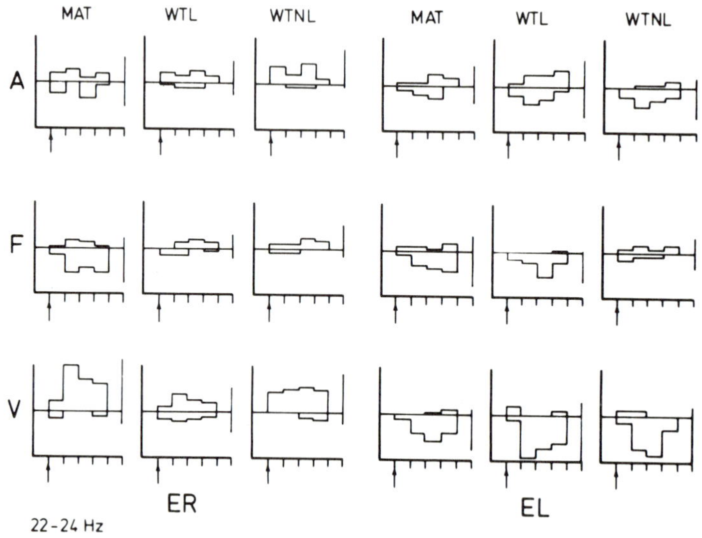

Fig. 7. Reactivity of 22-24 Hz, otherwise same as Fig. 5.

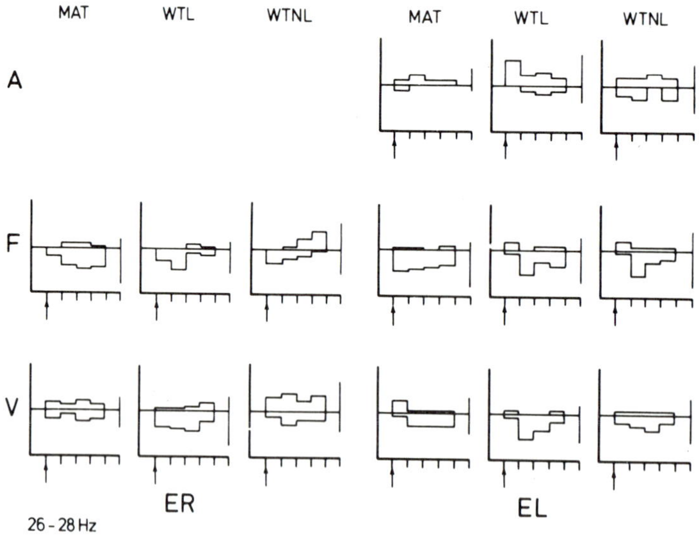

Fig. 8. Reactivity of 26-28 Hz, otherwise same as Fig. 5.

To illustrate further the consistency of the reactivities in the lowest frequency band and the inconsistency in the higher frequency bands, further data reduction was carried out for all frequency bands, in 3 subjects during the 3 paradigms and the CNV paradigm. The results are shown in Fig. 9. In this

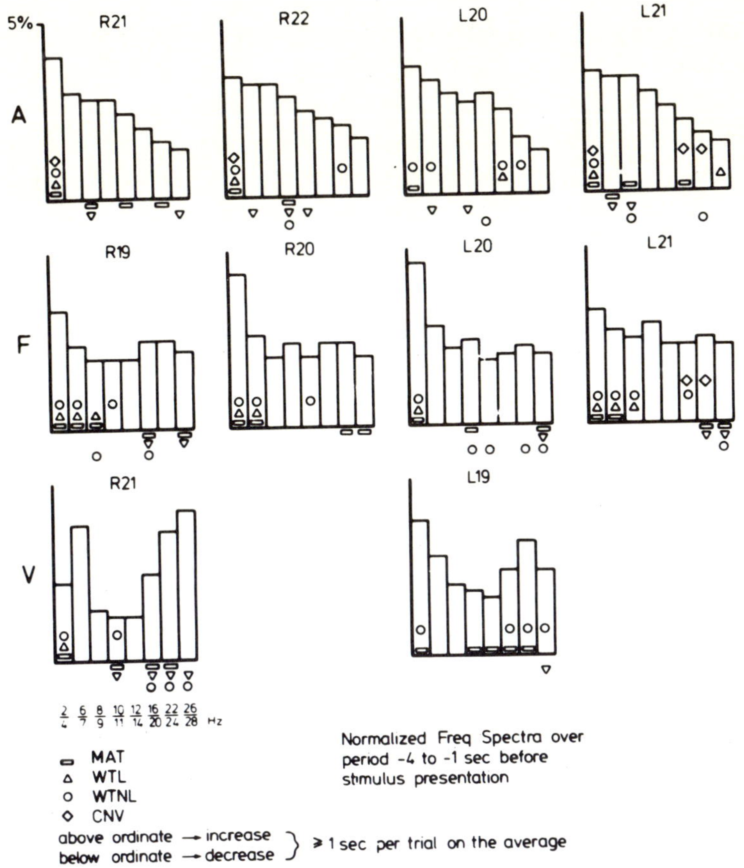

Fig. 9. Comparison of reactivities during 4 paradigms in 8 frequency bands at 2 electrodes in the orbito-frontal cortex on the right side (R numbers) and 2 electrodes on the left (L numbers) in patient A and F and 1 electrode on the right and the left in patient V. A reactivity is indicated if a significant increase or decrease of an activity occurred ($p<0.05$) in an interval of 4 sec following the stimulus presentation on the average per trial at least during 1 sec.

figure the only consistent reactivity in the higher frequency bands is the increase of 16-20 and 22-24 Hz in an electrode on the left side (L21) in the subjects A and F during CNV paradigm. This increase represents the beta burst during CNV, it was not encountered in subject V.

DISCUSSION

As has been shown in the section "results" change of activity in various frequency bands related to a variety of stimulus response paradigms can occur in many cortical areas. Considering the fact that the response of the subject forms part of behaviour these phenomena come under the heading of "event related changes in cortical rhythmic activities". Some of these changes are reasonably consistent, as, for example, the increase of 2-4 Hz changes during the arithmetic and word association tests and the beta bursts during the CNV paradigm. Other changes can be highly significant for one particular electrode during one particular paradigm as far as an increase or decrease is concerned, but it may be impossible to detect an overall pattern of activity. This is exemplified clearly by the changes in the 16-20, 22-24 and 26-28 Hz bands during the arithmetic and word association tests (see Figs. 6-9). During one paradigm there may be a highly significant increase in one of the frequency bands at one electrode while there is a decrease or no obvious change in another frequency band and/or at another electrode. During another paradigm the pattern may be different. Thus, it has not been possible to detect a consistent generalized pattern in these frequency bands.

The following questions arise: 1) what the physiological meaning of the event-related rhythmic potential changes may be, 2) whether in patients they can be used as an aid for the localization of the electrodes and 3) whether they can be used as an aid for assessing the effects of therapy. To this end it is necessary,however,to discuss first whether these phenomena which have been observed in patients can be regarded as a form of "normal activity". We are of the opinion that this indeed is the case. The reasons for this opinion are 1) similar forms of EEG activity, though over much larger neuronal populations, can be recorded from the scalp of normal subjects as,for example,the reactivities of the alpha and mu rhythms, 2) similar forms of reactivity can be encountered in animals as for example in dogs[7,8] and it has been proposed[9] that,as far as certain EEG phenomena are concerned,the dog may be regarded as a model for man, 3) the patients in whom we have encountered these event-related changes suffer quite different forms of brain affliction, for example obsessional neurosis and focal epilepsy.

Physiological meaning

In a previous communication we described a slight decrease of heart frequency during the occurrence of beta bursts[6]. Tentatively,it was proposed that this might represent a relation with a physiological reaction medicated by the autonomic nervous system.

The present series of experiments do not contradict this proposal as far as the reactivity of high frequency components during the CNV paradigm is concerned, since,if during this paradigm any reactivity occurred in the beta band, it always consisted of an increase, even if the increase at one electrode did not always precisely coincide in time with an increase at another electrode.

The reactivity in the beta band during the other paradigms, however, differed considerably from the CNV reactivity: at one electrode there might be an increase at one paradigm, a decrease at another,while at another electrode a quite different pattern of reactivity was present. It should be noted that each separate stimulus of the other paradigms (MAT, WTL and WTNL) differed from those of the CNV in that they contained new information. Therefore, the other paradigms were much less uniform and more complex to the subject than the CNV paradigm. This difference might be reflected by the different reactivities to these paradigms in the beta frequency bands. In contrast to the CNV they cannot be related to a general factor but appear to be specific to circumscribed brain areas. In this connection it is relevant to mention that the increase of beta activity in response to the warning stimulus S1 in the CNV paradigm may be observed in one electrode,but not in another electrode at 4 mm distance, which may, however,show an increase at the imperative stimulus S2 (see Fig. 4). In contrast to the activities in the high frequency bands the reactivities in the low frequency bands are much less variable. They consist almost always of an increase during all paradigms at all electrodes. This reactivity pattern,therefore,may well be related to a general factor. In view of the discriminative function ascribed to the frontal cortical areas and the fact that in all paradigms the low frequency reactivities were related to the presentation of a meaningful stimulus, the question arises whether a general factor underlying these reactivities could be a change in the functional state of the orbito-frontal cortical areas in question. Although on the basis of the present data such an assumption can only be hypothetical, it is relevant to note that the reactivities related to the unequivocal stimulus S1 of the CNV paradigm are less pronounced and occur at fewer electrode sites than the reactivities related to the complex stimuli of the other paradigms. In connection with this, it may be of importance that both, beta- and low frequency reactivity occur simultaneously at the same electrode. The low frequency reactivity might indeed be related to a change in the general functional state of relatively large cortical areas, whereas the beta reactivity might be related to neuronal activity in circumscribed cortical areas. In this respect the relations which have been found between regional patterns of EEG activity and increased regional cerebral metabolism and bloodflow during sensory stimulation and task performance[10] are of interest.

Aid for localization

As far as the aid the recording of these phenomena might give to the localization of site of the electrodes is concerned, the beta bursts during the CNV paradigm have so far been encountered only in the lateral part of the orbito-frontal cortex, but it has not been possible to improve on the x-ray localization by means of the occurrence of this phenomena.

Aid for assessing effects of therapy

Finally, until now it has not been possible to detect relations between these phenomena and therapeutical effects. In this respect it should be stressed however, that in order to study such relations systematically one has to have a much larger body of data than that at present available.

In conclusion it can be stated, therefore, that the nature of these event-related rhythmic brain potential still remains in the dark. We are of the opinion that further elaborate study is necessary and may answer at least some of these important questions.

SUMMARY

In some epileptic and psychiatric patients in whom electrodes have been implanted in direct contact with cortical structures for diagnostic therapeutical reasons an investigation has been carried out into relations between rhythmical cortical electrical potentials and events in various stimulus response paradigms.

One of the relations described earlier[6] was the occurrence of rhythmic activity at 22-28 Hz, "beta burst", during the CNV paradigm.

Various other paradigms have been investigated. It was found that reactivity took place in temporal and frontal cortical structures in various frequency bands during one or more of these paradigms. The most consistent of these were increase of 2-4 Hz activity in the lateral part of the orbito-frontal cortex during arithmetics and word association tests. Reactivities in higher frequency bands, though in some cases highly significant, did not exhibit a specific pattern.

It was submitted that these phenomena are not inherent in brain affliction but most probably occur also in normal subjects.

It is discussed what the physiological meaning of these phenomena might be and whether they can be used as an aid for the localization of the electrodes or for giving aid in assessing the effect of therapy.

REFERENCES

1. Storm van Leeuwen, W. (1973) Intracerebral interventions in patients with behavioural disorders. Psychiat. Neurol. Neurochir. (Amsterdam), 76, 345-351.

2. Crow, H.J., Cooper, R. and Phillips, D.G. (1961) Controlled multifocal frontal leucotomy for psychiatric illness. J. Neurol. Neurosurg. Psychiat., 24, 353-360.

3. Storm van Leeuwen, W. and Kamp, A. (1969) Radiotelemetry of EEG and other biological variables in man and dog. Proc. Roy. Soc. Med., 62, 451-453.

4. Kamp, A. and Vliegenthart, W. (1977) Sequential frequency analysis: a method to quantify event related EEG changes. Electroenceph. clin. Neurophysiol., 42, 843-846.

5. Storm van Leeuwen, W. and Kamp, A. (1979) Proc. of Symposium in psychophysiology of today and tomorrow. Leningrad, April 1979 (in press).

6. Kamp, A., Schrijer, C.F.M. and Storm van Leeuwen, W. (1972) Occurrence of beta bursts in human frontal cortex related to psychological parameters. Electroenceph. clin. Neurophysiol., 33, 257-267.

7. Storm van Leeuwen, W., Kamp, A., Kok, M.L. and Tielen, A.M. (1963) Relation entre l'activité électrique de certaines structures cérébrales et le comportement du chien. Rev. Neurol., 109, 258-259.

8. Storm van Leeuwen, W., Kamp, A., Kok, M.L. and Tielen, A.M. (1968) Relations between behaviour in dogs and electrical activities in various parts of the brain. Progr. Brain Res., 22, 181-194.

9. Lopes da Silva, F.H. and Storm van Leeuwen, W. (1978) The cortical alpha rhythm in dog: the depth and surface profile of phase. In: Architectonics of the cerebral cortex (Eds. M.A.B. Brazier and H. Petsche), Raven Press, New York, pp. 319-333.

10. Ingvar, D.H., Rosén, I. and Johannesson, G. (1979) EEG related to cerebral metabolism and bloodflow. Pharmakopsychiat., 12, 200-209.

© 1980 Elsevier/North-Holland Biomedical Press
Rhythmic EEG Activities and Cortical Functioning
G. Pfurtscheller et al., editors

EVENT-RELATED CHANGES OF EEG ALPHA ACTIVITY IN RELATION TO SLOW POTENTIAL SHIFTS

GERHARD GRÜNEWALD, ERIKA GRÜNEWALD-ZUBERBIER and JOHANNES NETZ
Institut für Hirnforschung der Universität, D-4000 Düsseldorf (Federal Republic of Germany)

INTRODUCTION

Numerous investigations have reported 'conditioned' or 'contingent'[1] EEG alpha[a] attenuation responses (AARs) in the interstimulus interval (ISI) of a sensory-sensory (S1-S2) conditioning paradigm or, more reliably, in the ISI of a warned reaction (S1-S2-respond) experiment. Reviews are given by Morrell[2], Grossman[3], and Nebylitsyn[4]. Some of these investigations[5,6,7,8] have controlled non-associative effects (sensitization, pseudoconditioning), and shown that a substantial part of the AAR in the ISI is attributable to the paired presentation of the two stimuli. An example is demonstrated in fig.1, where the AAR to S1 in a S1-S2-respond paradigm is compared with the AAR to S1 in a 'random control procedure'.

The aim of the present study was to investigate the relationships between AAR and the slow potential shift that develops during the ISI, i.e. the contingent negative variation (CNV). Observations on the subject were previously reported[9,10,11,12,13,14,15,16]. Obviously, there is no close direct and dependent relation between these two event-related EEG changes. The time course and behaviour of alpha amplitudes and of the simultaneously recorded slow potential shift in a S1-S2-respond paradigm will be compared below (a) under different task conditions, and (b) for their dependence upon a subject-related variable (ability to concentrate). Considering our earlier finding[17] of similar task-dependent variations in the 'late CNV' and the Bereitschaftspotential (BP), we have also studied the alpha activity in the period prior to non-cued responses, comparing the time courses of alpha amplitudes and those of the BP.

METHODS

<u>Experiment I.</u> (See Grünewald-Zuberbier et al.[14].)

3 parallel groups, each with 7 children, with reliable high, medium and low abilities to concentrate, served as subjects (Ss); ages were 11-12 years.

---

[a] The term 'alpha' is used here for EEG waves of 8-13 Hz, without consideration of location and functional properties.

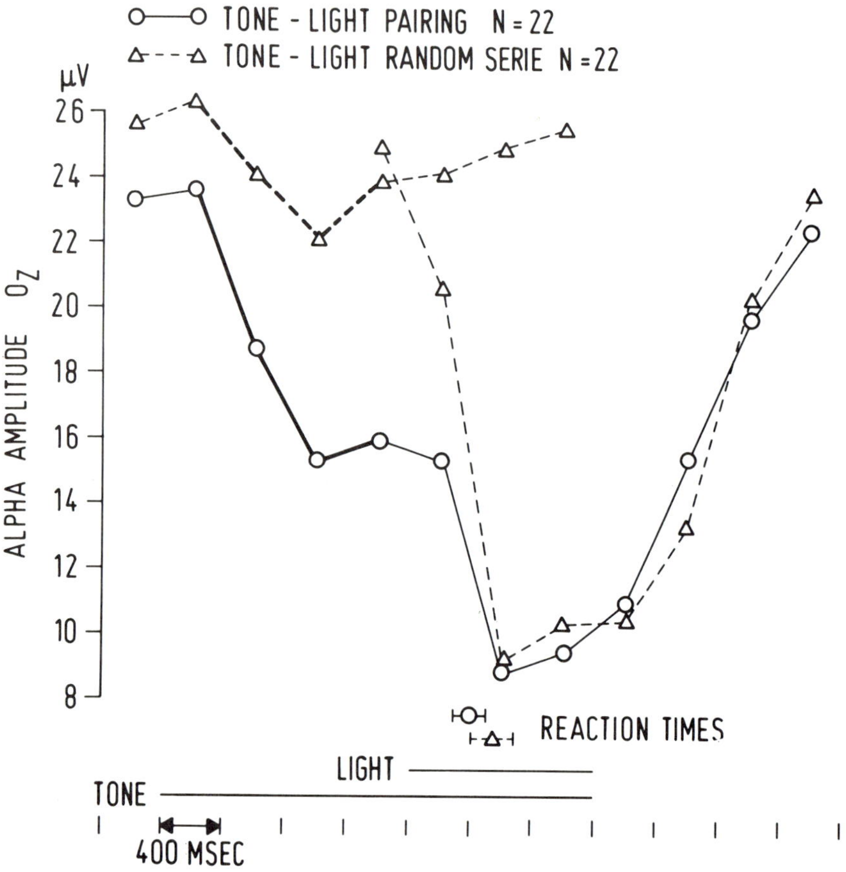

Fig.1. Changes in EEG alpha (8-13 Hz) amplitude (a) during the trial epoch of a tone-light-respond paradigm, and (b) elicited by the same stimuli presented unpaired. 22 children (age: 11-13 years) served as subjects. S1 was a tone of 250 Hz and 55 dB, delivered through headphones. S2 was a tachistoscopically presented white field of 17 mL and 1.2 sec duration. Eyes were kept open during the experiments. Conditions: (a) With random intertrial intervals ranging from 5 to 20 sec, application of 20 overlapping S1-S2 pairings with an ISI of 1.6 sec, both stimuli ending at the same time. Subjects had to press a response key as quickly as possible at the onset of S2. (b) S1 and S2 were presented unpaired, with the same intertrial intervals as in the pairing condition, according to independently established random series for both stimuli. Also here, subjects had to respond at the onset of S2.
Group means over averaged consecutive 400 msec intervals, based on the individual alpha amplitude averages (N between 10 and 17). Trials with low voltage fast activity in the pre-S1 interval were eliminated automatically. The pre-S1 amplitude levels do not differentiate statistically between the conditions. The enhanced amplitude reduction within the ISI of tone-light pairing in contrast to random presentation (thick lines) is highly significant ($p < 0.001$).(Modified from Grünewald et al. (7), and Grünewald-Zuberbier et al. (8).)

Ss were seated in a darkened room facing a dimly and homogeneously lit projection screen. They were asked to continuously focus on a lighted circle (0.5^0; 0.33 mL) projected on the screen during the experiments, and to refrain from blinking except between trials. In a S1-S2-respond experiment, S1 consisted in a frequency change of a constant masking tone of 160 Hz, 59 $\pm$3 dB to 250 Hz at the same intensity. The sound was delivered through headphones. S1 was followed after 2 sec by S2, a patterned light stimulus of 200 msec duration, projected on the fixation point, both stimuli ending simultaneously. In 50% of the trials S2 was a square and in 50% a triangle of equal luminance (0.9 mL) and area (0.7^0 and 1^0 side length, respectively). Both types of S2 were given in a random order. Ss had to respond (a) to each type of S2, and (b) selectively to one type of S2 by pressing a key as quickly as possible. Each condition consisted of 60 trials. Before (a) and (b), Ss were given series of 20 trials with no instruction to react. The intertrial interval (ITI) varied randomly from 4 to 8 sec. The sequential order of the reaction conditions was counterbalanced within the groups.

Experiment II. (See Grünewald et al.[17].)

Ss were 8 male right-handed adults. The general experimental conditions were as in experiment I. (a) In a S1-S2-respond paradigm, S1 consisted of a frequency change of a constant masking tone at 160 Hz, 59 $\pm$3 dB, to 250 Hz at the same intensity during 2 sec and then followed by S2, a tone at 500 Hz, 59 $\pm$3 dB and 150 msec duration, to which Ss had to respond as quickly as possible. The response was a right-handed horizontal displacement of a freely movable rod from its fixed starting position to the left towards a contact plate at a distance of 3.4 cm. The response was performed rapidly ('ballistic') in one session, and slowly ('ramplike') in another session. The ITI varied randomly from 5 to 10 sec. (b) In a non-cued response condition, Ss heard only the masking tone. They paced their movements themselves; the minimum interval was suggested to be about 5 sec. The 4 experimental conditions: cued and non-cued respond paradigms, with either rapid or slow movements, each condition consisting of 60 trials, were performed in a balanced order within the group.

EEG analysis.

Electrodes. Scalp Ag/AgCl electrodes, especially prepared for DC recording, were located in frontal, central, parietal and occipital positions according to the international 10-20 system, all referred to linked mastoids. Eye movements were monitored by recording from right supraorbital and canthal electrodes. Time constants for all derivations were 5 sec, and the upper cutoff was 70 Hz.

In experiment II, the electromyogram (EMG) was also recorded from a bipolar pair of electrodes located on the moving arm over the m. flexor carpi radialis. All potentials, stimuli and responses were recorded on paper and on FM tape for off-line computer analysis.

Slow potential shifts. In the CNV condition, the trial epochs beginning 500 msec before S1 and ending 1 sec after S2 offset, and in the BP condition, the trial intervals commencing 2.5 sec before and continuing for 1 sec after response onset were digitized at a resolution of 100 points/sec. Artifact-free trials were averaged for each S and each condition. Amplitude measurements were made with reference to a baseline averaged in the first 500 msec of the sections.

Alpha amplitude changes. Earlier quantitative studies have almost exclusively defined an AAR in a binary form, i.e. an alpha attenuation or blocking response is either present or not. In addition, the latency and duration of the response have been measured[4,18,19], although reliable determinations of the onset and termination are critical issues. These procedures hardly take either the magnitude of the AAR or its gradual changes over time into consideration. Recent investigations[7,8,9,11,15,16,20,21,22,23] have quantified alpha and other rhythmical brain activities along the time scale with the aid of computer techniques.

In our experiments, a period analysis[24,25] of the following type was used. The EEG was first filtered with a bandpass filter (8-13 Hz; 48 dB/octave), and digitized (100 points/sec). For each trial epoch, the sequence of wave periods was then determined by using EEG zero crossings, the mean of all digital values of the trial epoch serving as the zero line. Zero crossings were calculated by linear interpolation between adjacent digital values with different signs relative to the zero line. A wave period was defined as the interval between two successive EEG zero crossings in the same direction. The amplitude of the wave was measured as half the difference of the maximal and minimal values in this interval. Then, the amplitudes of waves were stored in all addresses corresponding to the temporal extension of the waves in an array of the same number of addresses as the digitized trial epoch. In alpha-free intervals, a zero amplitude was noted. In other words, the time course of alpha amplitudes was approximated by a step function, with step width determined by the duration and magnitude by the amplitude of alpha waves. Finally, the same trials as were used in the slow potential analysis were averaged.

RESULTS

Experiment I

Fig.2 shows CNV and AAR group averages of children with high (+group) and with low (-group) abilities to concentrate in a tone-light-respond paradigm whose reactions to S2 were either absent, simple or discriminative. Changes in alpha amplitude are expressed as percentages of the pre-S1 level. No statistical[a] differences in this reference level were found between groups or conditions. At C_z, in the no-reaction condition, both CNV and AAR are poorly developed and there is no differentiation between the groups. In the simple and in the discriminative reaction tasks, however, both 'waveforms' exhibit significant group differences. While the CNV in the first second of the ISI does not distinguish between the groups, the AAR to S1 during this interval is considerably enhanced in the +group. In the second half of the ISI, the AAR has already vanished, while the CNV in the same period is comparatively larger in the +group.

At O_z, no substantial CNV can be observed. The AAR to S1, on the other hand, shows magnitudes and differentiations similar to those found at the central derivation. Moreover, in the discriminative reaction task a second long-latency AAR -obviously related to S2- develops in the +group. Fig.3 shows that this 'anticipatory' desynchronization also occurs in children with medium ability to concentrate. In fig.3 alpha amplitude changes in the post-S2 period are also presented to demonstrate the comparatively larger light-evoked AAR, which is dominant in the occipital derivation (see also fig.1). No group effects could be found with respect to this component. However, the occipitally stronger post-S2 reduction of amplitudes in the discriminative reaction task is highly significant. A general tendency for alpha augmentation in the recovery period of the arousal reaction to light can be observed, but this does not become significant.

Experiment II

Fig.4 illustrates the CNV and AAR grand averages in a tone-tone-respond paradigm, with rapid and slow movements (RMs and SMs) in reaction to S2. There are some parallels with the findings of experiment I. At central electrode positions the CNV does not differentiate between the movement conditions in the first second of the ISI, while the AAR does. The alpha reduction within this

---

[a] Here, and in experiment II, differences in amplitudes between the experimental conditions were analysed statistically using non-parametric exact probability tests according to Fisher and Pitman (26). Analyses of the mean amplitudes were performed for 200 msec intervals overlapping at successive 100 msec segments, and beginning 400 msec after S1.

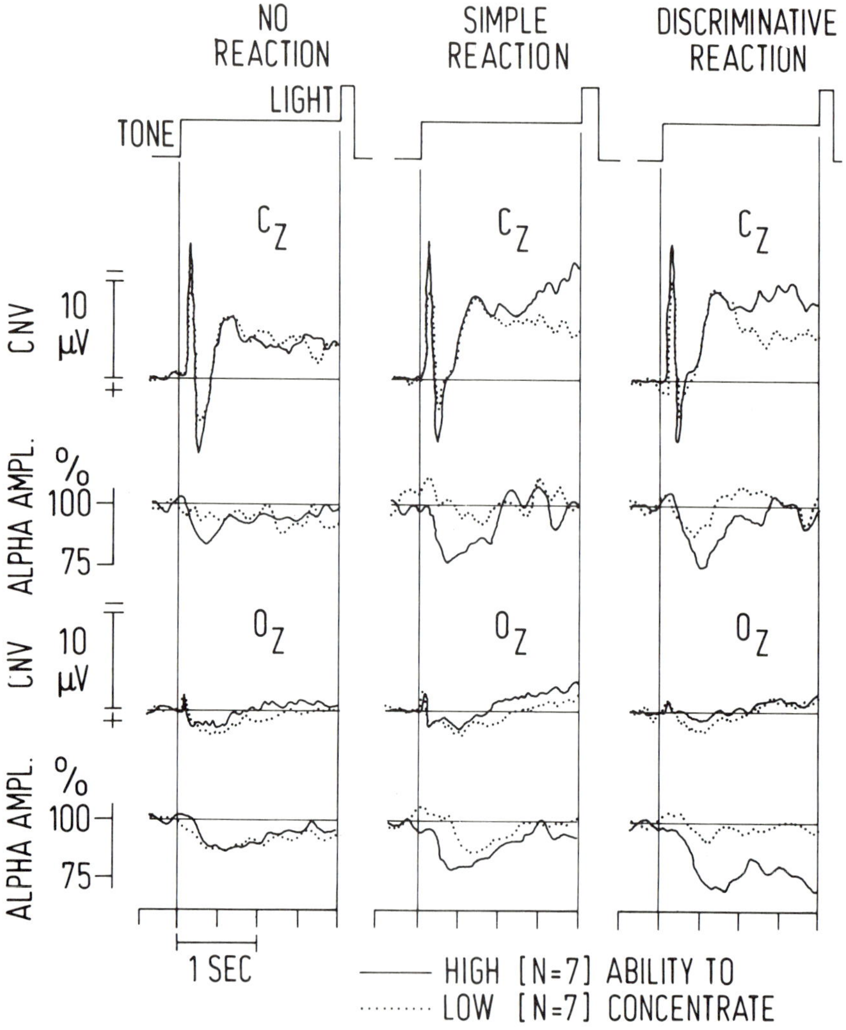

Fig.2. Contingent negative variations (CNVs) and corresponding changes in EEG alpha (8-13 Hz) amplitude during the ISI of tone-light pairings. Alpha amplitudes expressed as percentages of the amplitude level, averaged in the 500 msec interval prior to S1. S2 was either a triangle or a square, both types given in random order. In 3 experiments, Ss had (a) not to respond to S2, (b) to respond to every S2 by pressing a key with the right index finger as quickly as possible and (c) to respond selectively to one type of S2. Two parallel groups, each containing 7 children, with high and low abilities to concentrate. Group means over averaged consecutive 50 msec intervals, based on the individual CNV averages and on the individual percentage of alpha amplitude averages of the same trial blocks (N between 30 and 54). (Modified from Grünewald-Zuberbier et al. (14), and completed by unpublished results.)

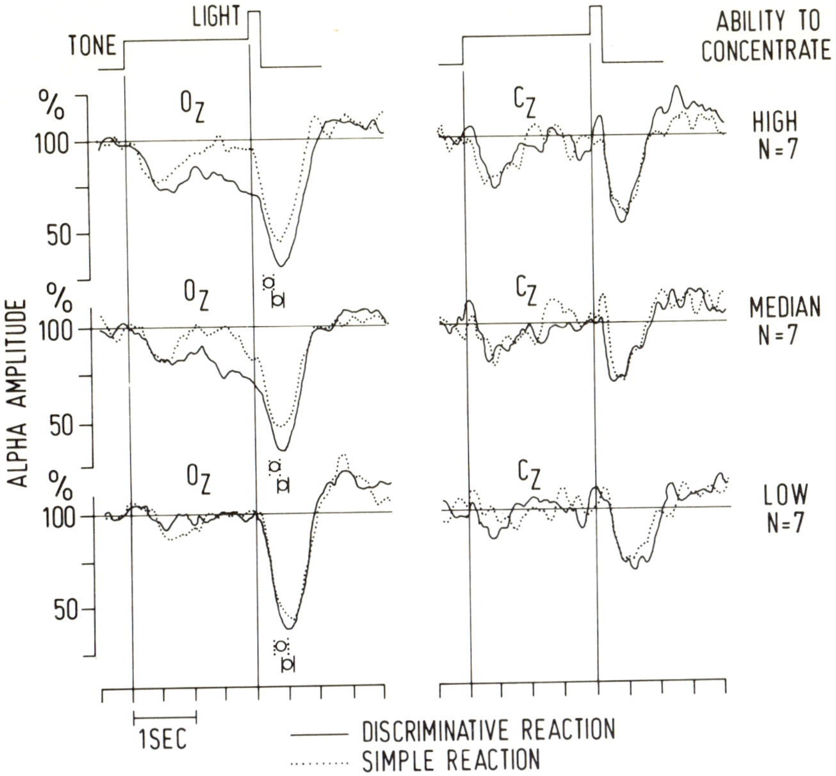

Fig.3. Changes in EEG alpha (8-13 Hz) amplitude during the trial epoch of a tone-light-respond paradigm. Alpha amplitudes expressed as percentages of the pre-S1 amplitude level. In 2 experiments, Ss had to respond (a) to every S2 (simple reaction), and (b) selectively to one of two types of S2 (discriminative reaction), given in random order. 3 parallel groups of 7 children each, with different abilities to concentrate. Group means over averaged consecutive 50 msec intervals, based on the individual percentage of alpha amplitude averages (N between 29 and 54). Mean reaction times are also indicated. (Modified from Grünewald-Zuberbier et al. (14), and Grünewald et al. (15), and completed by unpublished results.)

interval is significantly larger in the RM than in the SM condition, with maximal differentiation at the vertex. (In the SM condition the AAR only reaches a significant extentin comparison with the pre-S1 level at C_3'.) In the second half of the ISI, especially in the interval prior to S2, the AAR has ended. On the contrary, during the same period, the CNV develops a significantly steeper rise in negativity in the RM than in the SM condition, here also with maximal differentiation at the vertex.

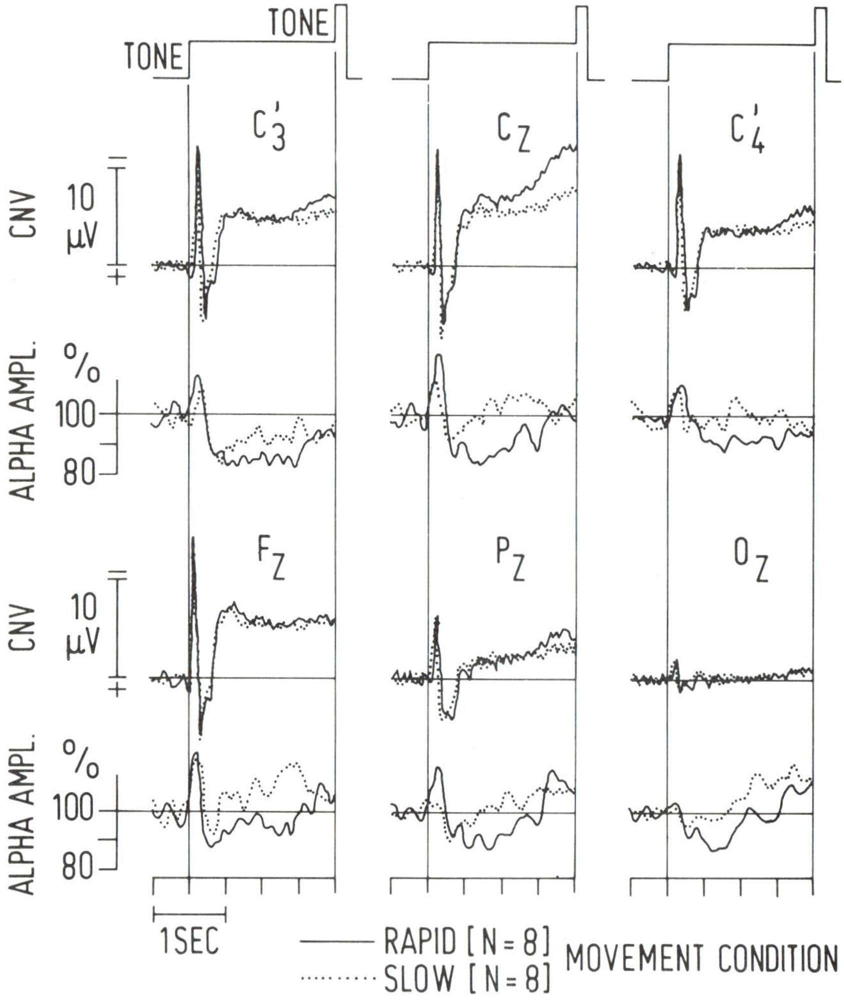

Fig.4. Contingent negative variations (CNVs) and corresponding changes in EEG alpha (8-13 Hz) amplitude during the ISI of a tone(S1)-tone(S2)-respond paradigm. Alpha amplitudes expressed as percentages of the pre-S1 amplitude level. C_3' and C_4' electrode positions, 1 cm anterior to C_3 and C_4, respectively. In 2 experiments, Ss had to respond to S2 as quickly as possible (a) with a rapid ('ballistic'), and (b) with a slow ('ramplike') movement of the right hand. Group means of 8 right-handed adults over averaged consecutive 30 msec intervals, based on the individual CNV averages and on the individual percentage of alpha amplitude averages of the same trial blocks (N between 36 and 56). (Modified from Grünewald et al. (17), and completed by unpublished results.)

At the other electrode positions (lower part of fig.4), no effects of the experimental conditions -with the exception of P_z- are demonstrable in the CNV. On the other hand, the AAR to S1 shows differentiations like those of the central leads. The alpha augmentations in the second half of the ISI are significant at F_z and O_z in the SM condition.

Fig.5 represents grand averages of slow potential shifts before self-paced RM and SM. Corresponding to the late CNV, the slopes and final amplitudes of the central BPs are larger before RM than when preceding SM. Moreover, as in

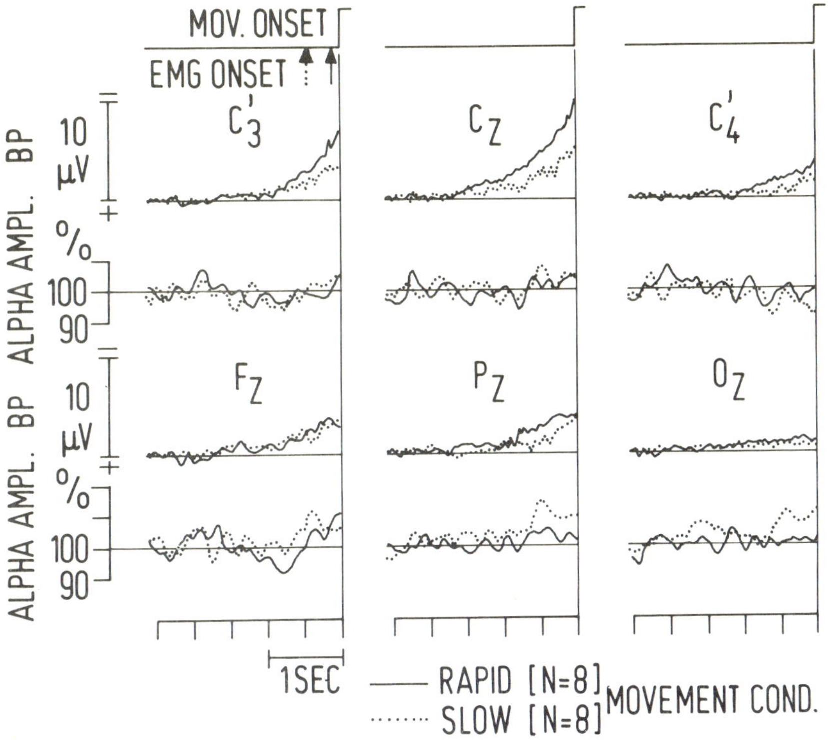

Fig.5. Bereitschaftspotentiale (BPs) and corresponding changes in EEG alpha (8-13 Hz) amplitude in the period before self-paced rapid and slow movements with the right hand. Alpha amplitudes expressed as percentages of the amplitude level, averaged in the 2.5-2.0 sec interval before movement onset. Group means of 8 right-handed adults over averaged consecutive 30 msec intervals, based on the individual BP averages, and on the individual percentage of alpha amplitude averages of the same trial blocks (N between 40 and 55). Arrows indicate mean agonist EMG onsets of the rapid and slow movements. (Modified from Grünewald et al. (17), and completed by unpublished results.)

the case of the period of the late CNV, there are no significant alpha amplitude changes during the BP period. In both paradigms, however,during and shortly after the actual RM the alpha amplitudes show a reduction in all derivations, except of the O_z. These AARs are significant only in the central leads (see fig.6), where they probably reflect the responses of mu activity.

DISCUSSION

To summarize we can say that both AAR and CNV show some similar functional reactivity during the ISI of our warned reaction experiments, but with different spatio-temporal characteristics.

In the group means, statistically significant AARs were found mainly in the first half of the ISI, reaching their maximum within 1 sec after S1 and thereafter declining, sometimes followed by an alpha augmentation prior to S2. These findings correspond to the results of Zappoli et al.[16], who demonstrated that in click-tone-respond experiments an initial alpha decrement in the first post-S1 epoch (400-900 msec) was the most constant event, followed in the remaining part of the ISI by a gross and incoherent intra- and intersubject variability of the alpha amplitude changes. The AARs parallel the 'early CNV'[14,17,27,28], which has a peak latency of about 500-700 msec. Both AAR and early CNV seem to be time-locked to S1; they are essentially identical in proportion and scalp topography to the arousal reaction following non-paired tone stimuli[7,8], and to the 'negative afterwave'[29,30], respectively. The AAR, however, shows a considerably widespread scalp distribution, whereas the early CNV has a frontal maximum. The AAR differentiates between the experimental conditions, while the early CNV does not. On the other hand, if the S1 has no warning function (no reaction task), both phenomena are comparatively less pronounced. The dissociations between the two S1-related EEG changes (often somewhat speculatively regarded as components of an orienting response[7,14,28]) suggest that both phenomena represent different neurophysiological mechanisms.

In the second half of the ISI, a significant AAR, distinct from the post-S1 response and clearly related to S2, could be observed in our group means only in the visual discriminative reaction task, in Ss with good abilities to concentrate. The response was restricted to the occipital region. Reports given by Ss support our interpretation[14] that this pre-S2 AAR is an indication of preparatory visual attention and ocular fixation on the area in which the stimulus will appear. Many studies[31,32,33,34,35,36] have shown that visual attention together with oculomotor functions have a considerable influence on the occipital alpha rhythm. In the simple reaction task, on the other hand, Ss can perform efficiently with relaxed accommodation, since the mere appearance of the light is sufficient to trigger the reaction. In contrast with the results of our experiment II, Pfurtscheller and Aranibar[37] found significant occipital (O_2-A_2) pre-S2 AARs in 6 out of 9 Ss in a click-tone-respond (simple reaction) experiment. One critical point may be that in their study Ss performed with eyes closed, while in our study eyes were open and fixating. The consequences are very different levels of background alpha activity. The details of any occipital alpha activity response, however, are obviously related to some extent to the character of ongoing electrical rhythms in the prestimulus interval and at the instant of stimulus application[38,39,40,41]. So it is necessary to investigate subgroups with different alpha base levels.

Unlike the alpha activity prior to S2, the 'late CNV' is very responsive to the experimental conditions. It shows a functional reactivity similar to the post-S1 AAR[a], but the late CNV has a different scalp topography, being maximal over central regions. These features clearly distinguish late CNV from early CNV and also from post-S1 AAR. Many studies recently pointed to a functional affinity between late CNV in a simple reaction task and BP before self-paced movements[14,17,27,28,42,43,44]. Both slow potential changes probably reflect the same process of response intention. In our experiment II, late CNV and BP showed an equal differentiation between the rapid and slow movement conditions, centrally and parietally (with a remarkable interindividual correspondence[17]). Moreover, in analogy to the late CNV phase no significant AARs were found during the BP. On the other hand, centrally significant AARs could be observed (in both paradigms) with the actual movement (see fig. 6), where they were probably induced by somatic afferent input[45]. The negative finding that no statistically significant AARs occurred during the phase of movement intention conflicts with the positive results of Pfurtscheller and Aranibar[46], who demonstrated significant AARs in most of their 10 Ss not only during uncued short movements, but also during the BP, especially immediately before movement onset. Therefore we controlled our results for both paradigms, using the same methods, in another independent sample with 7 Ss. Here also we found significant AARs in the first half of the ISI, but not prior to S2, and not in the preparatory interval before self-paced movements. Moreover, because Pfurtscheller and Aranibar used bipolar derivations over central regions, we also checked for influences of our common reference electrode by computing

---

[a] The pronounced pre-S2 AAR in the visual discriminative task as compared with the simple reaction task in children with good ability to concentrate has no parallel in the late CNV. On the contrary, these children exhibit centrally a significant larger pre-S2 rise in negativity in the simple reaction task. This may be an indication of a stronger motor set in a condition where Ss in the ISI were certain of the reaction to be made (14,27).

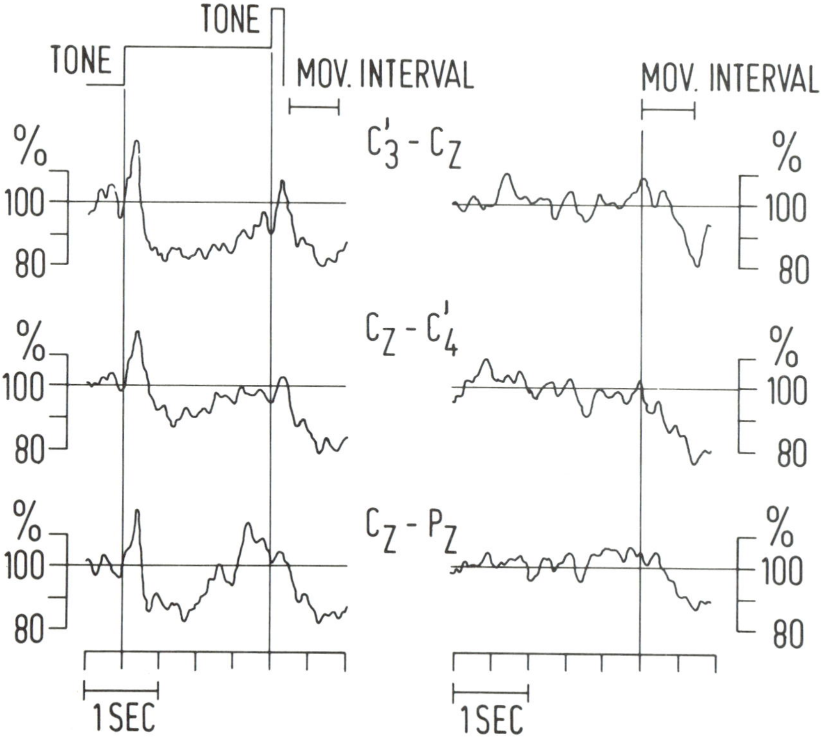

Fig.6. Changes in EEG alpha (8-13 Hz) amplitude in bipolar recordings during the trial epoch of a tone-tone-respond paradigm (left column) and in the period before self-paced movements (right column). Alpha amplitudes expressed as percentages of the pre-S1 amplitude level (left column) and of the amplitude level averaged in the 2.5-2.0 sec interval before movement onset (right column). Group means of 8 right-handed Ss over averaged consecutive 30 msec intervals, based on the individual percentage of alpha amplitude averages (N of trial blocks: 36-56). Horizontal bars indicate the mean movement intervals, including a rapid right-handed horizontal displacement of a rod to a target plate and the return to the starting position. (Data from Grünewald et al. (17).)

the differences between two monopolar derivations, respectively, prior to EEG analysis. The C_3'-C_z, C_4'-C_z and P_z-C_z differences showed essentially the same results as the monopolar recordings (see fig.6). The possible influences of other methodological dissimilarities (especially eyes closed without control for eye movements[46] vs. eyes open, fixating and elimination of all trials with eye movements[17]; button pressing with the thumb[46] vs. flexion-extension at the wrist[17]) remain to be clarified.

ACKNOWLEDGMENTS

Supported by Landesamt für Forschung, Nordrhein-Westfalen.

REFERENCES

1. Walter,W.G. (1961) Electroenceph. Clin. Neurophysiol., 13, 447-448.
2. Morrell,F. (1961) Physiol. Rev., 41, 443-494.
3. Grossman, S.P. (1967) A textbook of Physiological Psychology, John Wiley, New York, pp. 1-932.
4. Nebylitsyn, V.D. (1972) Fundamental Properties of the Human Nervous System, Plenum Press, New York, pp. 1-333.
5. Torres,A.A. (1968) Electroenceph. Clin. Neurophysiol., 24, 297-306.
6. Putney,R.T. (1973) Electroenceph. Clin. Neurophysiol., 34, 485-493.
7. Grünewald,G., Grünewald-Zuberbier,E. and Rasche,A. (1973) Arch. Psychiat. Nervenkr., 217, 315-328.
8. Grünewald-Zuberbier,E., Grünewald,G. and Rasche,A. (1975) Electroenceph. Clin. Neurophysiol., 38, 149-159.
9. Pfurtscheller,G. (1975) Z. EEG-EMG, 6, 194-199.
10. Pfurtscheller,G., Waibel,R. and Schuy,S. (1976) Med. Biol. Eng., 14, 199-204.
11. Pfurtscheller,G. and Aranibar,A. (1977) Electroenceph. Clin. Neurophysiol. 42, 817-826.
12. Grünewald-Zuberbier,E., Grünewald,G., Netz,J. and Klenkler-Wehler,A. (1977) Electroenceph. Clin. Neurophysiol., 43, 517.
13. Zappoli,R., Denoth,F., Navona,C., Rossi,L., Amantini,A. and Ragazzoni,A. (1977) Electroenceph. Clin. Neurophysiol., 43, 494.
14. Grünewald-Zuberbier,E., Grünewald,G., Rasche,A. and Netz,J. (1978) Electroenceph. Clin. Neurophysiol., 44, 37-47.
15. Grünewald,G., Grünewald-Zuberbier,E. and Netz,J. (1978) Electroenceph. Clin. Neurophysiol., 44, 617-625.
16. Zappoli,R., Denoth,F. and Navona,C. (1979) In: H.H.Kornhuber and L.Deecke (Eds.) 5th Int. Symposium on Electrical Potentials Related to Motivation, Motor and Sensory Processes of the Brain. In press.
17. Grünewald,G., Grünewald-Zuberbier,E., Netz,J., Hömberg,V. and Sander,G. (1979) Electroenceph. Clin. Neurophysiol., 46, 538-545.
18. Milstein,V., Stevens,J. and Sachdev,K. (1969) Electroenceph. Clin. Neurophysiol., 26, 12-18.
19. Mulholland,T.B. and Gascon,G. (1972) Electroenceph. Clin. Neurophysiol., 33, 295-301.
20. Elazar,Z. and Adey,W.R. (1967) Electroenceph. Clin. Neurophysiol., 23, 225.
21. Grünewald-Zuberbier,E., Grünewald,G. and Rasche,A. (1973) Electroenceph. Clin. Neurophysiol., 34, 724.
22. Pfurtscheller,G. (1977) Electroenceph. Clin. Neurophysiol., 43, 757-760.
23. Arnolds,D.E.A.T., Lopes da Silva,F.H., Aitink,J.W. and Kamp,A. (1979) Electroenceph. Clin. Neurophysiol., 46, 552-570.

24. Burch,N.R. (1959) Electroenceph. Clin. Neurophysiol., 11, 827-834.

25. Legewie,H. and Probst,W. (1969) Electroenceph. Clin. Neurophysiol., 27, 533-535.

26. Lienert,G.A. (1973) Verteilungsfreie Methoden in der Biostatistik, Bd.1, A.Hain, Meisenheim, pp. 1-736.

27. Gaillard,A. (1978) Slow Brain Potentials preceding Task Performance, Acad. Pers.B.V., Amsterdam, pp. 1-162.

28. Loveless,N. (1979) Event-related slow potentials of the brain as expressions of orienting function, In press.

29. Loveless,N. (1976) In: W.C.McCallum and J.R.Knott (Eds.) The responsive brain, Wright, Bristol, 26-29.

30. Rohrbaugh,J.W., Syndulko,K. and Lindsley,D.B. (1978) Electroenceph. Clin. Neurophysiol., 45, 551-567.

31. Mulholland,T.B. and Evans,C.R. (1966) Nature, 211, 1278-1279.

32. Mulholland,T.B. (1969) In: C.R.Evans and T.B.Mulholland (Eds.) Attention in Neurophysiology, Butterworths, London, 100-127.

33. Mulholland,T.B. and Peper,E. (1972) Psychophysiology, 8, 556-575.

34. Lehtonen,J.B. and Lehtinen,I. (1972) Electroenceph. Clin. Neurophysiol., 32, 139-147.

35. Pollen,D.A. and Trachtenberg,M.C. (1972) Brain Res., 41, 303-314.

36. Grünewald,G., Wita,C., Grünewald-Zuberbier,E. and Kapp,H. (1972) Arch. Psychiat. Nervenkr., 217, 315-328.

37. Pfurtscheller,G. and Aranibar,A. (1978) Electroenceph. Clin. Neurophysiol., 45, 226-235.

38. Morrell,F. (1967) In: G.C.Quarton, T.Melnechuk, F.O.Schmitt (Eds.) The Neurosciences. A Study Program, The Rockefeller University Press, New York, 452-469.

39. Morrell,L.K. (1966) Electroenceph. Clin. Neurophysiol., 21, 552.

40. Grünewald,G., Simonova,O. and Creutzfeldt,O. (1968) Arch. Psychiat. Nervenkr., 212, 46-69.

41. Legewie,H., Simonova,O. and Creutzfeldt,O. (1969) Electroenceph. Clin. Neurophysiol., 27, 470-479.

42. Rohrbaugh,J.W., Syndulko,K. and Lindsley,D.B. (1976) Science, 191, 1055-1057.

43. Loveless,N. and Sanford,A. (1974) Biol. Psychol., 1, 303-314.

44. Syndulko,K. and Lindsley,D.B. (1977) In: J.E.Desmedt (Ed.) Attention, Voluntary Contraction and Event-related Cerebral Potentials, Karger, Basel, 97-131.

45. Kuhlman,W.N. (1978) Electroenceph. Clin. Neurophysiol., 44, 83-93.

46. Pfurtscheller,G. and Aranibar,A. (1979) Electroenceph. Clin. Neurophysiol., 46, 138-146.

© 1980 Elsevier/North-Holland Biomedical Press
Rhythmic EEG Activities and Cortical Functioning
G. Pfurtscheller et al., editors

CHANGES IN ALPHA BAND ACTIVITIES DURING ALPHA BLOCKING INDUCED BY FLASH STIMULATION

JIRO TATSUNO, JUNKO MORI, HIROSHI ASHIDA and EIICHI MARU
Department of Physiology, National Defense Medical College, Tokorozawa, Saitama (Japan)

INTRODUCTION

In response to a sensory stimulus such as a flash the EEG changes in several respects (frequency content, amplitude) as has been noted since Berger's first description[1]. These changes involve various EEG frequency bands but are most conspicuous in the alpha band; therefore one speaks of "alpha blocking" or "alpha attenuation" induced by sensory stimuli. In this report we use the term *alpha band activities* to indicate EEG activities in the band 8 to 13 Hz. Studies of alpha blocking have been carried out using different methodologies; several types of parameters of this phenomenon have been measured: blocking time[2], duration[3], frequency change[4] among others. Measurement of the envelope of alpha activity during alpha blocking, however, began only recently[5,6,7]. In the present investigation we have studied this phenomenon in particular by looking at changes in the envelope of the alpha activity. We have considered here mainly the question of the augmentation of alpha activity which occurs transiently at the end of alpha blocking. This aspect of the phenomenon is called *transient alpha augmentation* or TAA. It is important to settle the question of whether the TAA is a part of the evoked response, a stimulus-triggered change of background activity of part of a recovery process. We discuss here also some data concerning behavioural correlates of this phenomenon since this involves psychophysiological aspects[8,9]. The general problem of explaining the changes in the envelope of alpha activity and the TAA involves processes more complex than those discussed by Goldstein[10].

In this article we have studied only healthy subjects; we have not entered into the clinical uses of the envelope study. Some applications have revealed interesting findings concerning drug effects and mental retardation[11].

METHOD

EEG recording

Healthy medical students participated voluntarily as the subjects of this study. They were male, between 18 and 20 years of age and right-handed. Unless otherwise indicated, five students participated in each experiment. The subjects were investigated while sitting on a chair in a dimly-lit, sound-protected and electrically-shielded room. They were instructed to keep their eyes closed and to stay relaxed but to pay attention to the flash stimulation. The flash light was at 30 cm distance from their eyes; it was a Xenon lamp stroboscope of 0.64 Joule emitting white light of short duration. The noise of the lamp was controlled by the application of white noise (60-80 dB). The flash stimuli were applied in the following fashion: one block of eight stimuli at 2 sec intervals followed by a longer interval of random length (on average ten seconds). The blocks were repeated until the total number of stimuli exceeded 50 (Fig. 1).

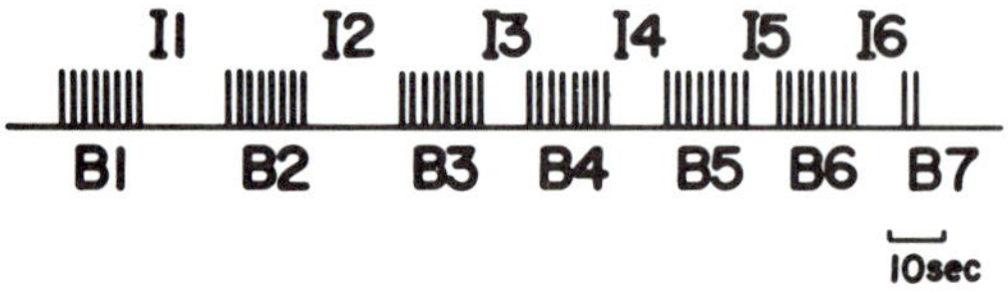

Fig. 1. Trigger pulses for stimulation. B1, B2,...B7: Blocks of 8 stimuli at 2 sec intervals, I1, I2,...I6: Intervals of random duration (on average ten seconds).

EEGs were recorded by non-polarizable silver-silver chloride disc electrodes of 5 mm diameter placed on the scalp according to the international ten-twenty method, at L, R-FP (F_{p1}-F_{p2}); L, R-F (F_3-F_4); L, R-C (C_3-C_4); L, R-P (P_3-P_4); L, R-O (O_1-O_2); L, R-T (T_3-T_4). Linked ear lobe electrodes were used as reference. EEGs which were contaminated by artifacts or tended to show drowsy patterns were not analysed. The EEGs were stored on a FM magnetic tape recorder, whose frequency characteristic was flat from DC to 5 kHz. The EEG signals were played back and filtered in the band 8 to 13 Hz, using an active band-pass filter. The filter attenuated 30 dB at one octave.

Computer processing

Filtered EEG activities were digitalized using an 8 channel (12 bit) A/D converter. The sampling time interval was usually 4 msec. Data processing was carried out by a minicomputer (Hitac 20) or larger computer (Tosbac or Facom system).

1) Square summation and interpolation methods. The former method (square summation) was used to evaluate the *averaged* envelope of alpha activity. This method consisted in calculating the standard deviation at each sample point of

a number of EEG epochs starting from the flash stimulus; usually about 50 trials were measured. In Fig. 2 the absolute value of the standard deviation of 50 trials of alpha activity is shown (SQ). The ripples on the standard deviation curve were eliminated by calculating the moving average on 15 successive points as also shown in Fig. 2 (RS). The latter method (interpolation) was used to study *individual* alpha activity curves; usually three peaks and troughs of alpha waves were employed for the interpolation applying Lagrange polynomials.

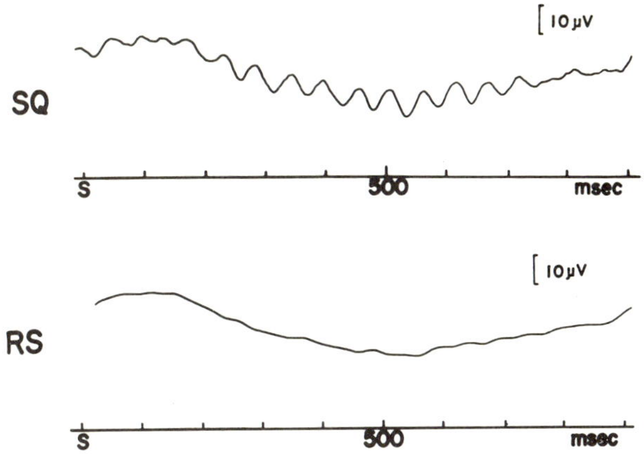

Fig. 2. Example of an alpha blocking curve. Upper curve is drawn by calculating the time course of standard deviations of 50 trials at each sampling point. Lower curve is the result of smoothing of upper curve.

2) Study of "momentaneous frequency" using period analysis. In order to follow rapid frequency changes during alpha blocking, period analysis was used. The EEG was rectified as shown in Fig. 3. The intervals between the peaks of the rectified signals were measured (Li). The time Ti was assigned to the middle of Li. A momentaneous frequency measure was defined as F at time T; as follows: F = 1/2 Li. The values of F(Ti) were classified into 17 classes of time intervals starting from 50 msec after the stimulation. In each of these time intervals, the values of F(Ti) were averaged over 50 trials. This resulted in a series of values of F which gave the time course of the frequency measures.

3) Study of coincidences. In order to answer the question of whether the occurrence of a certain phenomenon, e.g. TAA, coincided in any two derivations, P and Q, we resorted to the computation of the following statistic:

$$\phi = \frac{AD - BC}{\sqrt{(A+B)(C+D)(A+C)(B+D)}}$$

where A was the number of trials in which the phenomenon appeared at both P and Q; B was the number of trials where it was found only at P and C only at Q; D was the number of trials where it did not appear either at P or at Q. If $\phi = 1.0$ there is complete coincidence, if $\phi = -1.0$ there is complete opposition between the occurrences at P and Q; if $\phi = 0$ the phenomenon occurs at random at the two derivations.

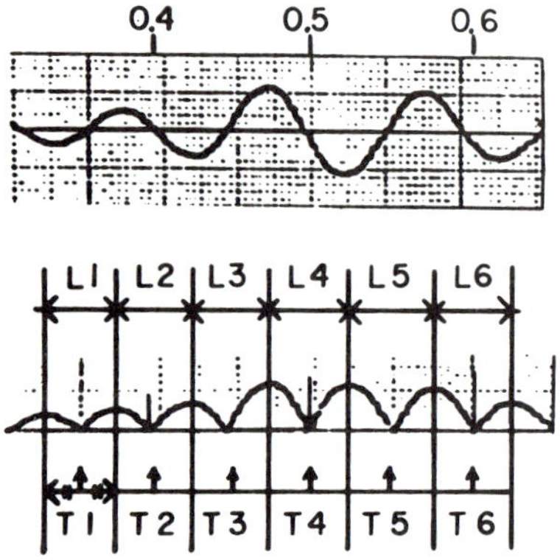

Fig. 3. Peak to peak measurement of frequency. Upper curve: original EEG. Lower curve: rectified EEG. Li: length between peaks. Ti: time assigned.

Definitions

In another publication[12] we determined the envelope of alpha blocking (so called "alpha blocking curve") by the square summation method. We also defined "latency of alpha blocking" meaning the period during which the amplitude of the envelope stayed at a level not significantly different from pre-stimulus level (this "latency" was thus different from the "latent time of blocking" of Bakes[13] and the "blocking time" of Jasper and Cruikshank[2]). In order to measure the amplitude of the envelope we had to define two elements: the zero level and the pre-stimulus amplitude. The former was calculated by averaging 250 potentials (sampled at 4 msec intervals) for one second after the stimulation. The latter was calculated by averaging alpha peak amplitudes for 0.5 sec before the stimuli in 50 trials (Fig. 4).

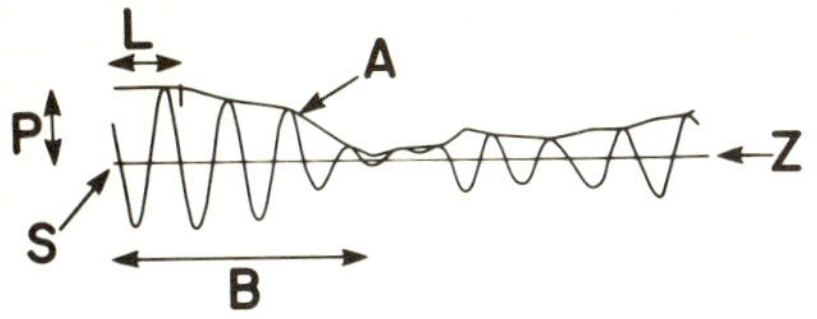

Fig. 4. Explanation of terms
A Alpha envelope or alpha blocking curve.
B Blocking time.
L Latency of alpha blocking.
P Pre-stimulus amplitude.
S Stimulation.
Z Zero level.

Here we define 2 types of trials: "blocking positive"(BL(+)) and "blocking negative"(BL(-)). The former, BL(+), were those trials where the envelope amplitude in the period from 300 to 600 msec after the stimulus decreased by more than 20 % of the pre-stimulus amplitude. The latter, BL(-), were those trials where this was not achieved. This could occur in those trials where the alpha amplitude decrease was lower than 20 % of pre-stimulus amplitude (LV) and those where there was a slow decline of the envelope (more than 500 msec to reach a minimum or zero reaction (SN).

We further considered the recovery phase. We defined as recovery positive [RC(+)] those trials where the envelope amplitude, in the period from 600 to 1000 msec after the stimulus, exceeded pre-stimulus level.

Therefore, we considered the following phases in a trial: DEC - declining stage, BLK - blocking stage and RCV - recovery stage. DEC was the period from the stimulus to 300 msec, BLK from 300 to 600 msec and RCV from 600 to 1000 msec.

The phenomenon of main interest in the present investigation, however, was what we called *transient alpha augmentation* or TAA. This may occur in a period from 500 to 600 msec after the stimulus. According to our definition of TAA(+) the following criteria should be fulfilled:

(i) - the largest peak of the envelope of the alpha activity (TAA max) should lie between 500-600 msec after the stimulus;

(ii) - the maximal amplitude of TAA (TAA max) should exceed 60 % of pre-stimulus amplitude;

(iii) - the amplitude of TAA after attaining TAAmax should decrease (TAAmin 2) to a level below 20 % of TAAmax;

(iv) - the decline (TAAmin 2) should occur not later than 750 msec after the stimulus.

The form of a TAA is shown in Fig. 5 with an indication of a few characteristic parameters. If, in a trial, the criteria for TAA(+) were not fulfilled we called it a TAA(-) trial.

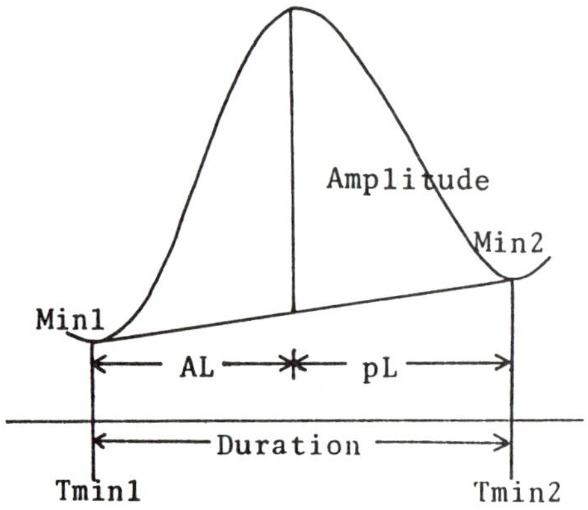

Fig. 5. Description of TAA envelope. AL: anterior limb. PL: posterior limb. Tmin1: time of min 1. Tmin2: time of min 2.

RESULTS

General characteristics of TAA

We noticed that TAA frequently appeared after the stimulus, in the way shown in Fig.6. A TAA consisted of a combination of frequency components, whose phases were not related in any consistent way, despite the fairly stable shape of the TAA envelope. The values of latency to the TAAmax (peak latency), maximal amplitude, duration of anterior limb and time of occurrence of Tmin1 (Tmin1, see also Fig. 5 for definitions) are shown for five subjects in table 1 for different deviations.

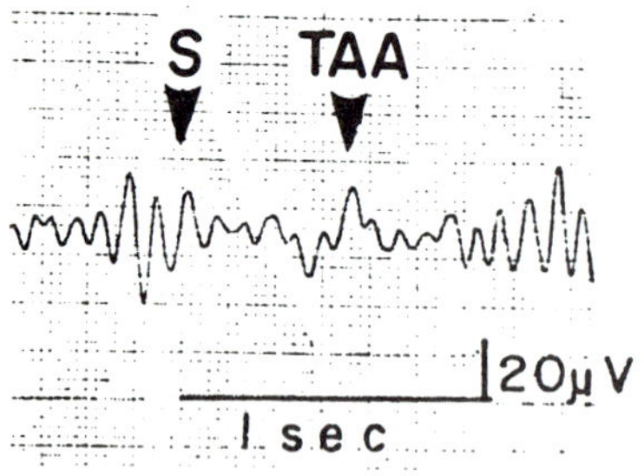

Fig. 6. Example of TAA in a single EEG-trial. S: stimulation. TAA appeared about 0.5 sec after the stimulation.

Topographical aspects of TAA appearance in relation to alpha blocking

These aspects were studied in ten subjects at twelve derivations. Table 2 shows the percentages of BL(+), LV and SN (see Definitions) encountered. Also shown are the percentages of those trials where TAA occurred (TAA(+)) and in which BL(+) TAA(+) was observed; we also show similar percentages for those cases where TAA(+) RC(+) was observed in relation to all trials with TAA(+).

It should be noted (from Table 2) that there were more trials where alpha blocking was negative (BL(-): LV plus SN trials) than positive. A positive alpha blocking (BL(+)) appeared in about 1/3 of the cases, regardless of derivation. Also in about 1/3 of the cases, a transient alpha augmentation (TAA) was encountered. The sequence BL(+) TAA(+) was more often found than the sequence TAA(+) RC(+).

TABLE 1
PEAK LATENCY, AMPLITUDE, DURATION, ANTERIOR LIMB AND TMIN 1 OF TAA

	Peak latency	Amplitude	Duration	Anterior limb	Tmin 1
L FP	546 msec	30.7 μV	282 msec	167 msec	383 msec
L F	549	46.9	319	189	358
L C	555	36.3	298	119	380
L P	550	41.2	301	120	376
L O	548	32.6	303	183	365
L T	550	33.8	289	167	384
R FP	550	37.1	290	172	382
R F	553	53.3	297	185	376
R C	545	43.3	306	180	373
R P	551	41.4	293	164	389
R O	552	35.0	297	167	386
R T	546	37.8	302	190	369
mean	549.58	39.12	298.08	166.92	376.75
±S.D.	3.00	6.47	9.45	23.95	9.20

As regards blocking and TAA, coincidence studies were made in a subgroup who showed a high rate of blocking in occipital (5 subjects) and in a subgroup of low rate (5 subjects). The results are shown in Table 3. It was clear that high ϕ values of blocking and TAA were distributed in approximately corresponding positions in the table. In the high BL(+) rate group the coincidence occurred in restricted areas, namely occipital parietal and frontopolar. In the low BL(+) rate group the areas over which coincidence was found were more wide spread. This finding implied that blocking and TAA became topographically more restricted when alpha blocking occurred frequently, while, when the blocking occurred infrequently, blocking of TAA was found over wide areas.

TABLE 2

PERCENTAGES OF BL(+), LV, SN AND TAA APPEARANCE IN 12 LOCATIONS OF 10 SUBJECTS, AND RATES OF BL(+)TAA(+)/TAA(+) AND TAA(+)RC(+)/TAA(+).

	BL(+)	LV	SN	TAA	BL(+)TAA(+) /TAA(+)	TAA(+)RC(+) /TAA(+)
L FP	37.2%	33.4%	29.4%	39.2%	53.0%	39.5%
L F	39.4	35.0	25.6	30.8	42.9	31.0
L C	38.6	29.6	31.8	38.8	60.5	38.3
L P	34.2	31.6	34.2	34.0	43.2	33.8
L O	35.0	31.0	34.0	32.8	49.1	33.4
L T	37.4	30.4	32.2	38.0	58.9	41.2
R FP	33.4	33.6	33.0	36.4	54.6	34.1
R F	33.2	39.2	27.6	31.2	35.4	31.7
R C	39.2	30.6	30.2	40.0	54.1	39.8
R P	34.4	31.4	34.2	38.8	50.6	39.8
R O	36.2	30.8	33.0	30.0	46.5	31.1
R T	38.2	34.4	27.4	33.2	49.6	36.8
mean	36.4	32.6	31.1	35.3	49.9	35.9
±S.D.	2.3	2.7	3.0	3.7	7.1	3.8

TABLE 3

COINCIDENCE STUDIES IN BLOCKING AND TAA

ϕ values of blocking (left) and TAA (right) in high rate BL(+) (upper side) and low rate BL(+) group (lower side). Data from six locations on left side. Confidence level of 10 % marked by 1) and of 5 % marked by 2).

(BLOCKING)

high BL(+) rate group

	F	C	P	O	T
FP	.188$^{1)}$	.290$^{2)}$	.331$^{2)}$	0.096	.171$^{1)}$
F		.206$^{2)}$	.194$^{1)}$	.153	.088
C			.130	-.024	.152
P				.309$^{2)}$	.094
O					.059

(TAA)

high BL(+) rate group

	F	C	P	O	T
FP	.111	.312$^{2)}$	.262$^{2)}$	.182$^{1)}$	.190$^{1)}$
F		.136	.156	.202$^{1)}$	.084
C			.172$^{1)}$	.026	.167
P				.309$^{2)}$	.148$^{1)}$
O					.120

(BLOCKING)

low BL(+) rate group

	F	C	P	O	T
FP	.367$^{2)}$	.351$^{2)}$	.319$^{2)}$	.178$^{1)}$	.362$^{2)}$
F		.360$^{2)}$	.429$^{2)}$	.168$^{1)}$	.272$^{2)}$
C			.256$^{2)}$	.061	.375$^{2)}$
P				.183$^{1)}$	.268$^{2)}$
O					.165$^{1)}$

(TAA)

low BL(+) rate group

	F	C	P	O	T
FP	.358$^{2)}$	.370$^{2)}$	.297$^{2)}$	.222$^{2)}$	.383$^{2)}$
F		.334$^{2)}$	.407$^{2)}$	.203$^{2)}$	.295$^{2)}$
C			.277$^{2)}$	.040	.340$^{2)}$
P				.182$^{1)}$	.253$^{2)}$
O					.199$^{2)}$

Changes in alpha period durations in relation to different types of trials

We investigated whether there were differences in the duration of alpha wave periods (also called frequency F = 1/2 Li) depending on whether a trial was characterized as TAA(+) or TAA(-); we determined also whether trials which were characterized by simple TAA(+) differed as regards the alpha period or frequency (F = 1/2 Li) from other types where more complex patterns were encountered: BL(+)TAA(+); LV TAA(+); SN TAA(+); BL(+)TAA(-); TAA(+)RC(-). These results are summarized in Table 4. It can be seen that the most consistent difference in alpha period was between TAA(+) and TAA(-), particularly in the first 400 msec after the stimulus. In this case the frequency of TAA(+) remained larger than 11 Hz while that of TAA(-) was smaller.

TABLE 4

SIGNIFICANCE OF DIFFERENCE IN TIME COURSES OF FREQUENCY CHANGES BY T-TESTS.

	TAA(+) VS TAA(-)	TAA(+) VS BL(+)TAA(+)	TAA(+) VS LV-TAA(+)	TAA(+) VS SN-TAA(+)	TAA(+) VS BL(+)TAA(-)	TAA(+)RC(+) VS TAA(+)RC(-)
period (msec)	T	T	T	T	T	T
50-100	2.65 **	0.58	1.19	0.33	3.90 **	0.05
100-150	2.13 **	0.71	0.31	0.81	2.71 **	2.91 **
150-200	2.74 **	0.20	0.31	0.03	1.20	2.82 **
200-250	6.46 **	0.71	0.93	2.31 *	2.21 *	0.75
250-300	7.93 **	1.40	0.82	3.63 **	2.14 *	0.23
300-350	6.25 **	1.74	0.34	3.17 **	0.68	1.32
350-400	3.32 **	0.09	0.50	0.40	1.38	0.44
400-450	0.45	0.96	1.89	0.54	1.08	0.30
450-500	3.25 **	0.24	0.17	0.20	2.70 **	0.37
500-550	1.25	0.42	1.64	1.26	0.18	0.36
550-600	1.14	o.49	0.64	0.08	0.51	1.19
600-650	0.69	0.55	0.23	0.62	0.57	0.49
650-700	3.50 **	0.06	0.93	0.93	2.70 **	0.89
700-750	1.95	0.57	0.90	1.98 *	1.70	0.12
750-800	3.11 **	0.76	1.46	0.29	2.68 *	1.13
800-850	0.50	2.45 *	3.00 **	0.64	0.48	1.99 *
850-900	0.66	0.18	0.90	0.68	0.77	0.78

* significant at 5% level
** significant at 1% level

T t-value

TAA and VEP by averaging discount method.

In order to discount the possibility of TAA being late components of visually evoked potential, summations were made to get averaged responses from data of TAA(+) and TAA(-). By careful examination of 12 locations in five subjects, it was assured that in the evoked potentials there were no peaks corresponding to TAA envelope. TAA was probably cancelled by the variable phase of the EEG activity. One example is shown in Fig. 7. It is easily conceivable that TAA appearance might have affected the evoked response itself, but it will be left to detailed investigations in the future to determine this.

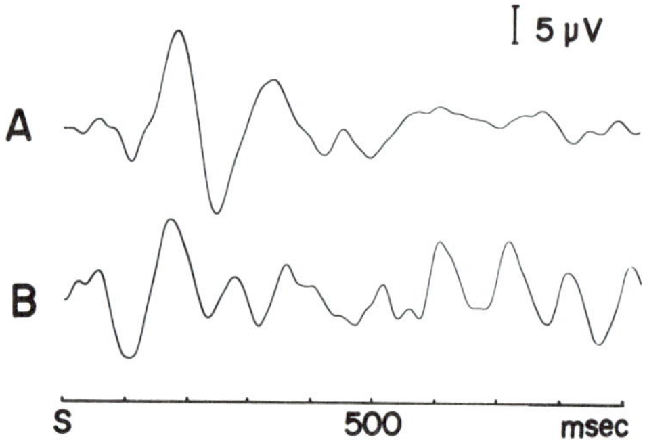

Fig. 7. Averaged visually evoked potentials. A: obtained by summation of TAA(+) data. B: obtained by summation of TAA(-) data. Upward deflection indicates negativity.

Appearance rates of blocking and TAA in relation to stimulation procedure.

The relation between BL(+) and TAA and the stimulation procedure should be mentioned briefly. Figure 8A shows appearance rates of alpha blockings in response to eight successive stimuli. For the first stimulus of each block, the blockings were elicited most frequently, but for the second the frequency of occurrence was the lowest. Afterwards, the appearance rates rose and fell repeatedly, though the range of these oscillations decreased. The course of the decrease in the rates to stimulus 1, 3, 5 probably reflected habituation to the stimulation. Rebound of the rate to stimulus 7 might be due to anticipation of the end of stimulation.

The changes in TAA appearance in the eight successive stimuli are shown in Figure 8B. The rates increase gradually from the first to the sixth and then decrease towards the eigthth. This largest rate in the sixth might imply

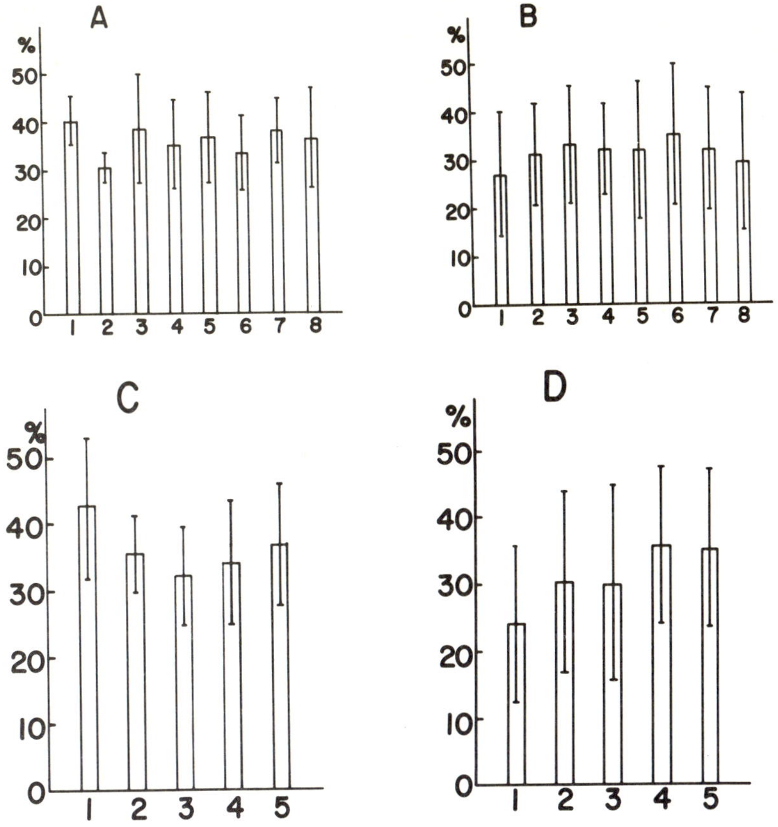

Fig. 8. Appearance rates of blocking and TAA to eight succeeding stimuli and every ten stimuli in temporal order. Bars represent standard deviations. A: BL(+) rates to eight succeeding stimuli, B: TAA(+) rates to the same. C: BL(+) rates to every ten stimuli, D: TAA(+) rates to the same. Significant differences A: 1 and 2 at level of 1 % and 2 and 7 at level of 5 %, B: 1 and 6 at level of 10 %, C: 1 and 3 at level of 10 %, D: 1 and 5 at level of 5 % and 1 and 2 at level of 10 %.

that a relaxed state favours a high TAA rate. It is probable that the expectation of the end of the block lowered the TAA rate.

Figure 8C shows the rates of BL(+) in response to each ten stimuli of a total of 50 stimulations. An exceedingly high rate was observed with the first ten stimuli and a rapid decrease in rate then occurred, reaching a minimum within the third ten stimuli. Afterwards the rates increased and the rate in response to the fifth ten stimuli returned to the level of that to the second ten.

These changes were also presumably due to habituation and expectation of the end of the experiment. Figure 8D shows TAA appearance rates to each ten stimuli of total 50 stimulations. The rates increased clearly in the later part of the experiment. The rise of the rate suggests that the rate increased in more relaxed conditions.

DISCUSSION

In the envelope study two methods developed by us were employed, one of which was square summation. This consisted of summation of squared potentials at each sampling point. The squaring prevented the on-going background activities from being cancelled. In other words, the method was nothing but the calculation of standard deviations. By definition, alpha envelope is the curve connecting peaks of alpha activities. Since peaks of alpha activities should be expressed by negative and positive extremities of alpha amplitude distribution, the envelope must be in relation to the *range*, that is, the distance between negative and positive extremities. The linearity between the range and the standard deviation was already confirmed[14,15]. Therefore the envelope should be obtained by calculating the time course of standard deviations of alpha amplitude distributions at each sampling point.

The other method used was an application of Lagrange interpolation. This was used for obtaining the alpha envelope in individual trials. Compatibility of the two methods was confirmed[11]. Pasquali[16] detected alpha envelope in waxing and waning EEG. Lansing and Barlow[6] pursued alpha envelope study on alpha blocking; especially in the recovery state. We began our alpha envelope studies from the initial stage of alpha blocking.

Envelope studies revealed various facts about alpha blocking. Firstly, the latency of alpha blocking was clarified and measured[17]. Topographical differences in the latencies for the various locations studied were found. An increase in latency along the occipito-frontal direction was observed. Secondly, the fitting of the alpha envelope by way of Gompertz model was found to be quite satisfactory. This led to the proposal of a model of alpha blocking (details will be reported in a separate article).

From the envelope study, transient alpha augmentation (TAA) was noted[18]. TAA appeared frequently and widely. TAA envelope could not be observed in average evoked potentials by the ordinary summation method, probably due to the instability of the phases of the activities that caused TAA cancellation in the summation. Though TAA might be regarded as a step in the recovery after alpha blocking, functional connection to the recovery process was rather weak, as revealed by studying the ratio of TAA(+)RC(+)/TAA(+). TAA was transient or phasic, and thus could be distinguished from the subsequent genuine recovery. By the timing of its occurrence, TAA could be differentiated from Morell's stimulus-provoked alpha activity[19]. The intervention of TAA must be taken into consideration in phase studies such as Goldstein's[10].

From the ratio of TAA appearance in relation to blocking,that is,BL(+)TAA(+)/TAA(+), it was impossible to conclude functional connection of TAA to preceeding blocking.

By complex demodulation technique, Childers[20] and Nogawa et al.[7] mentioned that the envelope increased around 0.5 sec after the stimulation. Childers showed that this increase had no corresponding peak in VEP. Nogawa et al. concluded that blocking and augmentation of alpha activities could appear successively in the same response to flash stimulation. Our data agree with their conclusions.

One of the interesting findings about TAA was its relation to the time course of frequency change in alpha blocking. The frequency changes in alpha blocking were reported by Kawabata[4] and by Pfurtscheller and Aranibar[21]. But their measurement was every half-second. In order to identify rapid changes in frequency, peak-to-peak measurements was used in this study. It revealed the difference of the time courses between TAA(+) and TAA(-). In TAA(+) there was a characteristic frequency decrease around the time when TAA began to rise. TAA envelope and frequency changes seemed to run parallel. Therefore it could be concluded that TAA was a phenomenon which brought about changes in both amplitude and frequency. It must be noted that BL(+)TAA and LV-TAA(+) showed more or less the same time course, whereas SN-TAA(+) had a different time course, which was similar to TAA(-). The underlying mechanism of the frequency fall around 300 msec in TAA(-) was as yet unknown.

A study of coincidences of the occurrence of TAA over different areas revealed approximately the same topographical distributions in blocking and TAA, which suggested the intimate relation between the two. It is also worth noting that coincidence occurred in very restricted areas in a high rate group, while it occurred in wider areas in a low rate group. It was observed that TAA in occipital areas occurred in rather isolated fashion, while TAA in the central areas was more widespread. This fact suggested that TAA in these two locations had different characteristics.

Appearance rates of blocking and TAA in relation to stimulation procedure indicated that the blocking and TAA did not always appear in parallel. TAA seemed to appear more readily in a relaxed state.

SUMMARY

Changes in alpha band activities during alpha blocking induced by flash stimulation were investigated by way of the envelope obtained by square summation or interpolation method. Subjects were healthy medical college students. EEG was recorded monopolarly from twelve locations with reference electrode on earlobe. Transient alpha augmentation (TAA) was mainly investigated in this report.

1. TAA appeared in about 35 % of the trials, widespread in 12 recording locations.
2. TAA had peak latency of 550 msec, duration of 300 msec and peak amplitude of about 40 μV.
3. During TAA there were not only changes in amplitude but also in alpha frequency.
4. TAA bore closer relation to preceeding blocking rather than to subsequent recovery.
5. Wherever TAA occurred in central areas it was found to coincide with TAA occurring over a large area of the scalp. TAA in occipital area was topographically more restricted.
6. There was no peak in the averaged VEP which corresponds to the peak of TAA envelope.
7. TAA seemed to be related to a relaxed state.

ACKNOWLEDGEMENTS

The authors wish to thank Mrs. Junko Umehara and Miss Kazuko Onuki for their expert technical assistance.

REFERENCES

1. Berger, H. (1930) J. f. Psychol. u. Neurol., 40, 160-179.
2. Jasper, H.H. and Cruikshank, R.M. (1939) Psychol.Bull.,24,406-418.
3. Loomis, A.L., Harvery, E.N. and Hobart, G. (1936) J. exp. Psychol., 19, 249-279.
4. Kawabata, K. (1973) IEEE Trans.Biomed.Eng., 20, 444-452.
5. Tatsuno, J. (1972) Clin. Electroenceph., 14, 525-531.
6. Lansing, R.W. and Barlow, J.S. (1972) Electroenceph.clin.Neurophysiol., 32, 149-160.
7. Nogawa, T., Katayama, K., Tabata, Y., Ohshio, T. and Kawahara, T. (1976) Electroenceph.clin.Neurophysiol., 40, 78-83.
8. Ulett, G.A., Gleser, G., Winokur, G. and Lawler, A. (1953) Electroenceph. clin. Neurophysiol., 5, 23-32.

9. Fenton, G.W. and Scotton, L. (1969) Brit. J. Psychiat., 113, 1283-1289.
10. Goldstein, S. (1970) Electroenceph.clin.Neurophysiol., 29, 127-136.
11. Tatsuno, J. (1976) Clin. Electroenceph., 18, 743-750.
12. Tatsuno, J. (1973) Electroenceph.clin.Neurophysiol., 34,726-727.
13. Bakes, F.P. (1939) J. exp. Psychol., 24, 406-418.
14. Tatsuno, J. (1977) Electroenceph.clin.Neurophysiol., 43, 526.
15. Tatsuno, J. and Karasawa, R. (1976) J. Natl. Def.Med.Coll., 1, 115-121.
16. Pasquali, E. (1969) Electroenceph.clin.Neurophysiol.,26, 106-109.
17. Tatsuno, J. (1973) Electroenceph. Electromyograph., 2, 113-115.
18. Tatsuno, J. (1978) Eighth Japan EEG and EMG Meeting Abstract., 154.
19. Morrell, L.K.(1966) Electroenceph. clin. Neurophysiol.,21,552-561.
20. Childers, D.G. (1973) Electroenceph. clin. Neurophysiol.,34,446-447.
21. Pfurtscheller, G. and Aranibar, A. (1977) Electroenceph. clin. Neurophysiol., 42, 817-826.

© 1980 Elsevier/North-Holland Biomedical Press
Rhythmic EEG Activities and Cortical Functioning
G. Pfurtscheller et al., editors

THE INFLUENCE OF FREQUENCY-MODULATED TONE AND TONE BURST ON THE ON-GOING EEG

J.L. BLOM
Netherlands Institute for Preventive Health Care - TNO, Wassenaarseweg 56, P.O. Box 124, Leiden (The Netherlands)

SUMMARY

A study was performed to investigate the influence of frequency modulated and tone burst stimulation on the ongoing EEG over different scalp areas.

The stimulus for the FM-AER was a 1000 Hz tone, modulated downwards, of 55 dB HL. The stimulus for the tone-AER was a 1000 Hz 10 msec 90 dBA tone. Both stimuli were, with a random time interval between 2 and 3 seconds, delivered to both ears. Spectral analysis of the period immediately before and after the stimulus showed no significant influence on the power in the α and θ band with the FM-stimulus neither with a long nor a short duration of the stimulus. The tone influenced the relative α power over the occipital and central areas which diminished more than 3 percent. The relative θ power was increased over the central area more than 4 percent. It is discussed that this change in power should partly be caused by general alerting mechanisms and not by processing of the signal itself.

INTRODUCTION

It has been long known that stimuli may change background activity in the EEG resulting in desynchronisation, caused by reticulo-cortical activation as first described by Moruzzi and Magoun in 1949[1].

In evoked response studies the influence of the stimulus on the background EEG is often neglected. Most studies are centered in the elimination of the background activity, regarding it as an unwanted noise component and virtually unrelated to the stimulus event. Recently some studies have been done on background activity[2,3] using CNV paradigms or click and/or flashes.

In the auditory system nearly all studies on evoked or event related potentials are done using click stimulus or tone bursts. It is known from animal studies that sound stimuli give rise to arousal with blocking of the α rhythm, and when the stimulus is novel, i.e., the first from a series, a complex behavior known as the orienting reflex develops. It is therefore very likely that these changes in background activity in the EEG will also occur in ER research using this kind of sound stimuli.

Another method of stimulating the auditory system is the use of a frequency modulated tone of constant intensity. This type of stimulus, first described by Spoor et al.[4] gives a clear response over the primary auditory cortex as well as over the central area, just like the amplitude modulated stimuli but of a different form and with different amplitude and latency components. Both types of stimuli differ appreciably in the way they excite the auditory system. In a study to evaluate some properties of the frequency modulated auditory evoked response (FM-AER) we became interested in the changes of the background activity of the EEG.

MATERIAL AND METHODS

Earphones (B & O type N70) delivered to both ears a continuous tone with a frequency of 1000 Hz from an instrument made in our laboratory. The individual hearing level (HL) was determined using a procedure developed in our laboratory[5]. The 1000 Hz tone was modulated downwards 120 Hz to 880 Hz with a fixed down and return slope duration of 10 msec. The plateau duration varied in each session (see fig.2). The intensity of the tone remained constant at 55 dB HL (fig.1) (approx. 75 dBA).

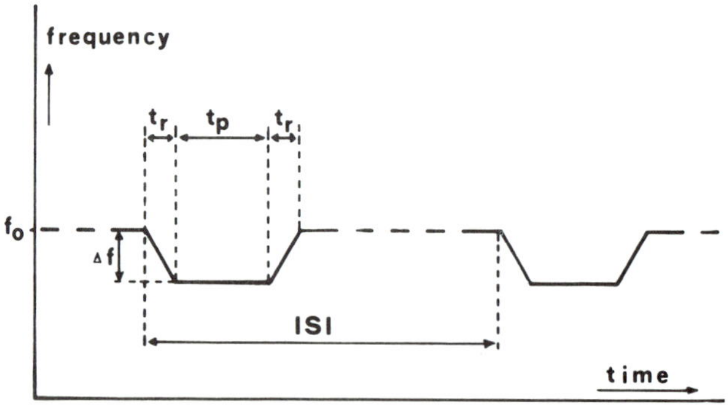

Fig. 1. Identification of the different stimulus parameters for frequency modulation. f_o = main frequency 1000 Hz in this study. Δf = modulation frequency. t_r = ramp time: in this study constantly kept at 10 msec. t_p = plateau time. ISI = Inter Stimulus Interval; this varied in this study between 2 and 3 seconds.

As for amplitude modulation a tone was delivered via a loudspeaker located at 25 cm in front of the subject with an intensity of 90 dBA at the level of the ears. Its frequency was 1000 Hz with an up and down slope duration of 2 msec and a duration of 10 msec (Ahrend v. Gogh type 516503). The 10 following montages were used according to the international 10-20 system:

1. C3-Fz	2. C4-Fz	3. T3-Fz	4. T4-Fz	5. Cz-A1
2. Cz-Fz	7. O1-A1	8. O2-A2	9. T3-A1	10. T4-A2

Standard Ag/AgCl electrodes were used and recordings were made with a conventional electroencephalograph (Ahrend v. Gogh ET16) with time constant 0.6 sec and lowpass filter 3 dB cut-off point at 75 Hz. All signals were recorded on analog tape (Ampex PR 2200) and processed on-line on a PDP 11/34 for evoked response analysis. ER were stored for later statistical analysis.

The stimuli were randomly delivered with a minimal stimulus interval of 2 seconds and a maximal interval of 3 seconds. One hundred sweeps were used to construct the AER.

From the signals recorded on analog tape spectra were calculated for 50 stimuli. Each two-second period was split into a one-second period immediately preceding the stimulus (period 1) and a one-second period following it (period 2). From these 100 one-second periods spectral analysis was done using a FFT, smoothed with an elliptical 3-point filter, and the average power in the δ band (1-3 Hz), θ band (4-7 Hz), α band (8-12 Hz) and β band (13-25 Hz) was calculated. Thereafter the powers in the different bands were averaged over the 50 events giving the averaged power in the δ, α, θ and β bands before, and after a stimulus.

In a first approach to analyse the spectral data statistically a t-test was applied on the differences in mean spectral values before and after the stimulus.

SUBJECTS

Twelve subjects, 4 men and 8 women, in age varying between 19 and 38 years participated in this study. They were recorded in a semi sound-proof electrically shielded room. They were comfortably seated in a modified dentist chair with the eyes closed. All had normal hearing capabilities and a normal medical record.

RESULTS

As this study was started to analyse the properties of the FM-AERS, some of the main properties of these will be described after which the changes in spectral parameters will be shown. Also some of the differences between the FM-AER and the tone AER will be given.

Auditory Evoked Responses

In fig.2 the form of the different AERS from one subject are shown. For the study of the characteristics of the FM-AER plateau durations of 10, 20, 50, 100 and 500 msec were used.

An EEG recording in the presence of a continuous 1000 Hz tone of 55 dB HL was used to assess the background activity. During this recording pulses were delivered at random intervals which were used as triggers for the averaging program. One recording with the tone AER was made. Analysing the ongoing EEG afterwards revealed that in some subjects the attention during one recording changed slightly manifesting itself in a reduction of the α rhythm. This instability in attentive state may have some influence in the evoked responses and the spectral components. During the recording therefore various methods were tried to keep the subject in the same attentive state. This was mostly done by speaking to the subject.

Clear FM-AERS and tone-AERS were seen (fig.2) in the following derivations: Cz-A1, T3-Fz and T4-Fz. In the other derivations the amplitude was smaller and the variability larger. The form of the FM-AER is fairly simple compared to the tone-AER recorded from the same derivations. Two types of the former were encountered: one at the temporal derivations and another at Cz-A1. The FM-AER recorded at the temporal derivation was characterized by an initial positive deflection with a peak-latency between 100-120 msec followed by a negative trough which reached its lowest value at about 160-200 msec. In some cases it was preceded by a small negative wave at about 60 msec and followed by a positive inflection of variable latency. The FM-AER recorded at Cz-A1 initiated in most cases with a small positive wave at 60 msec followed by a large negativity reaching its maximum between 90 and 120 msec and followed by a large positive wave between 180-220 msec. Thereafter a negativity of variable latency and amplitude developed.

In some cases, the detection of the AER was hampered by large muscle artefacts which were picked up especially by the temporal electrodes.

The tone-AER was very clear over all derivations - except the occipital -. The largest amplitudes were recorded, as has been mentioned before, over the vertex, while the difference in amplitude between the central and the temporal regions was not always the same. In some subjects the largest ER were obtained from temporal, in others, from central derivations. No clear hemisphere asymmetry was encountered.

As this paper deals only with the influence on the background activity of these evoked responses no further results will be presented. They will be published elsewhere.

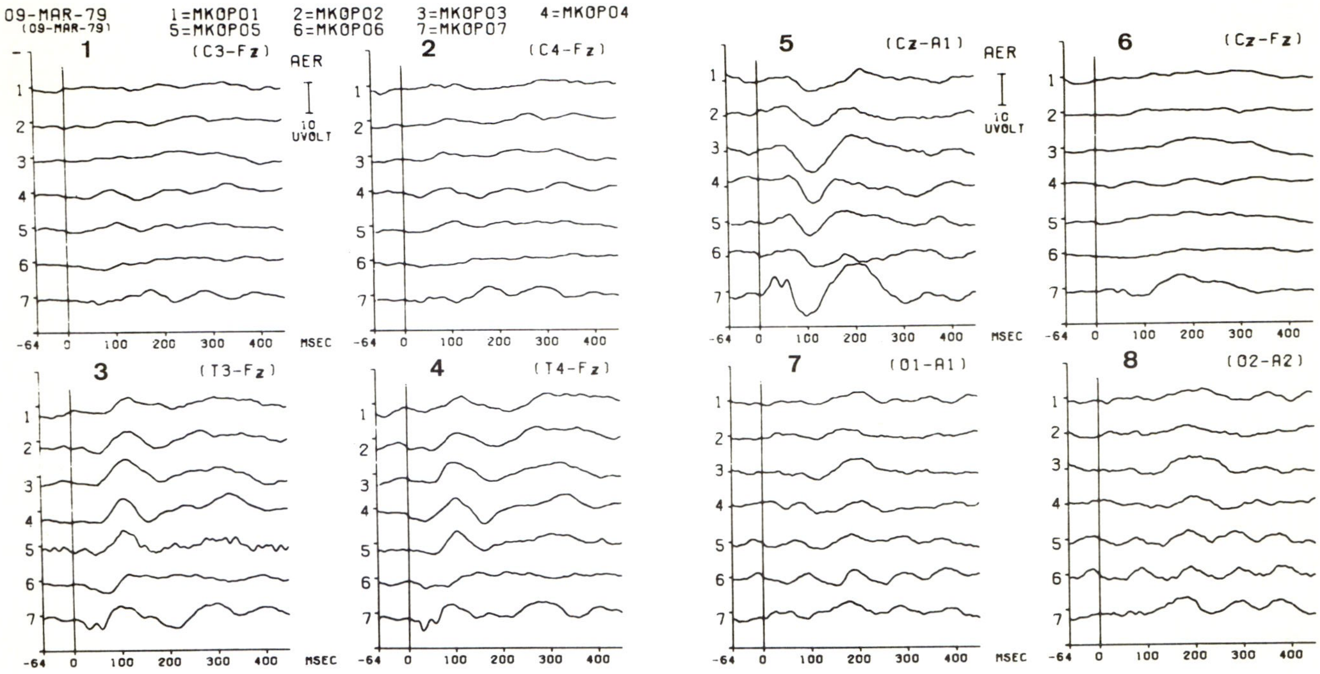

Fig. 2. Example of the auditory evoked responses of one subject (AERS). Seven different AERS are displayed for 8 channels. The first 4 sets show the central (C3 and C4) and temporal (T3 and T4) electrodes, referred to the frontocentral electrode (Fz). Sets 5 and 6 show the vertex electrode (Cz) referred resp. to the left earlobe and the frontocentral electrode and sets 7 and 8 the occipital electrodes referred to the homolateral earlobes. The 7 AERS per set are recorded with different plateau length. Record 1 has a 10 msec plateau duration, record 2: 20 msec, record 3: 50 msec, record 4: 100 msec and record 5: 500 msec plateau duration. Record 6 shows the averaged EEG using random triggers when a continuous tone is presented to both ears and record 7 gives the responses for tone-AER 90 dBA. Sampling time is 1 msec, total nr. of samples is 512. The line indicates the start of the stimulus (except for record 6 where the stimulus was continuously presented). Negativity is downwards.

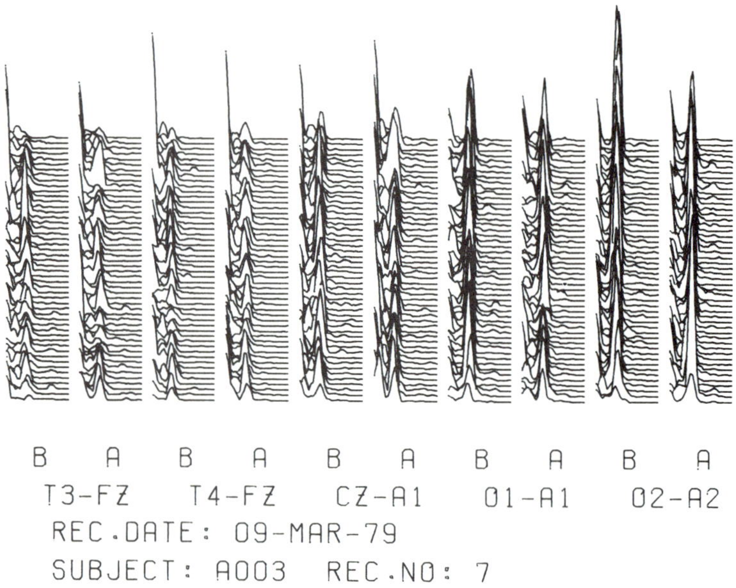

Fig.3: Example of the compressed spectral arrays of 50 sweeps for 5 channels recorded for the tone-AER. B = before the stimulus, A = after the start of the stimulus. Each spectrum is calculated over a one second period. The amplitude is taken in arbitrary units. This is an example showing the large variability between successive sweeps in these leads. The effect of the stimulus can be observed especially in the right occipital lead.

Spectral analysis

These analyses were done for all 10 montages as given in the section on methods. Fig.3 shows an example of the spectra calculated over the periods before and after a stimulus on a linear scale and expressed in μV^2. The decrease in α power with the tone-AER in the occipital areas can be seen. All our subjects showed a maximal α power in the occipital derivations as may be expected. However, as can be seen in fig.3, α power is clearly visible in the vertex and in the temporal derivations.

For the spectral analysis three different sessions were used to study the effects of the stimuli upon background EEG. They were the sessions with a plateau duration of 50 and 500 msec of the FM-AER and the tone AER.

Fig.4 is an example of the averaged power for the three sessions before (left) and after (right) the stimulus.

As can be seen the peak α power is lower for the 500 msec plateau while the 50 msec plateau and the tone-AER display about the same height in amplitude.

This lower amplitude is seen in 5 of the 12 subjects before as well as especially after the stimulus. This effect has not been tested yet for its significance as it may well reflect the difference in overall vigilance in these subjects, which will be discussed later.

The differences in absolute and relative power between period one (before) and period two (after) for the three different sessions in the α band were calculated from these spectra.

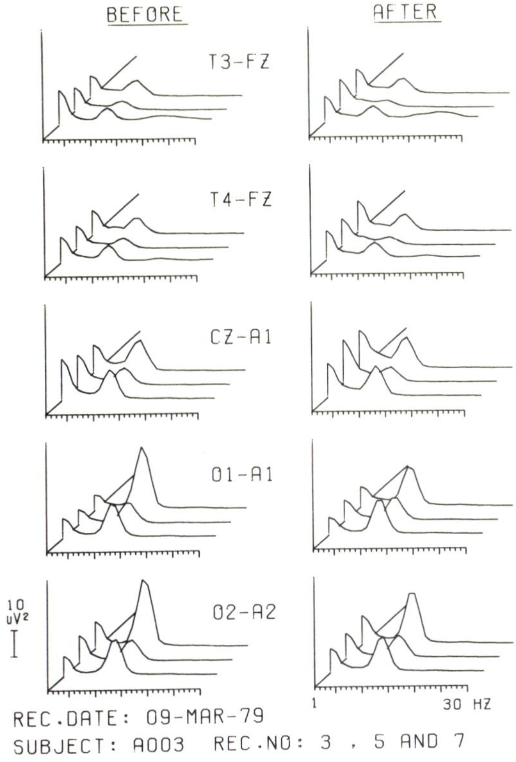

Fig.4: Mean spectra averaged over 50 sweeps for 5 channels for subject A003. BEFORE: main spectra calculated over the one-second period before the stimulus, AFTER: main spectra calculated over the one-second period after the stimulus. The foremost spectrum is calculated for the FM-AER with a 50 msec plateau, the middle is calculated for the 500 msec plateau and the third is calculated for the tone-AER. The activity in the β band in the temporal channels is caused by muscle artifacts. For further discussion see text. Calibration is 10 $\mu V^2/Hz$.

Table 1 shows the differences in relative power in the α band and table 2 these differences in the θ band. A minus sign means that the power in period 2 was lower than in period 1, a plus sign that is was higher. A two-sided t-test on differences was used to test the significance of the difference.

The FM-AER

As can be seen in column one in table 1 a decrease in relative α power of about 2% of fairly low significance for the 50 msec plateau duration can be found only in the midline montage, while the 500 msec plateau duration does not have any influence on the background EEG, although (see fig.2) a distinct AER can be recorded which does not differ much from the 50 msec AER. It is notable however that in the case of the 50 msec plateau duration a decrease in α power, however small and not attaining significance, is noted in all montages which is not the case for the 500 msec plateau.

In the θ power (table 2) the same can be said but for the opposite direction. A small significant increase in θ power may be seen over the left temporal region, but not elsewhere, for the 50 msec plateau, while the other channels also show a tendency to a small increase in θ power. In the 500 msec session no significant increase or decrease of the θ power can be seen and no montage shows any change in this spectral component caused by the stimulus.

The tone AER

A much different picture exists for the tone-AER. Here in practically all derivations a significant decrease in relative α power can be seen which is the largest in the central and occipital channels, being about 5% in the vertex duration, and 4.3 and 3.0 percent resp. for the left and the right occipital derivation. But temporal and central regions with about 3% are also very noticeable. Most changes have a significance of less than 1 percent when the double-sided t-test is used. It must be noted that in the data a small difference between left and right derivations can be seen. However this difference is statistically not significant.

As regards the relative θ power again the opposite is visible. Here, however, the largest increase is in the vertex derivations and over the temporal region. This change in θ power over the vertex is probably caused by the fact that the evoked response itself has frequency components in this range which will be discussed later. A small increase is visible over the occipital region which, however, is only significant at a 5% level over the right hemisphere. Interestingly, over the central areas (C3 and C4) the θ power is not increased although a very distinct tone-AER can be recorded there. Also it must be noted that the increase is less over all derivations than the decrease in α power.

DISCUSSION

This study is part of a larger one dealing with the characteristics of the FM-AER. The main difference between the FM-AER and the more familiar tone-AER or click AER lies in the way the peripheral receptor is stimulated. Much work has been done on the influence of ramp duration of the stimulus on the latency

TABLE 1

DIFFERENCES IN RELATIVE α-POWER BEFORE ($P\alpha^1$) AND AFTER ($P\alpha^2$) THE STIMULUS EXPRESSED IN PERCENTAGES OF THE TOTAL POWER, AVERAGED OVER ALL 12 SUBJECTS FOR THE THREE DIFFERENT STIMULUS SITUATIONS

derivations	FM-tone duration 50 msec $P\alpha^1$-$P\alpha^2$	signif. level	FM-tone duration 500 msec $P\alpha^1$-$P\alpha^2$	signif. level	tone-burst duration tone 10 msec $P\alpha^1$-$P\alpha^2$	signif. level
C3 - Fz	- .6	n.s.	+1	n.s.	-3	<0.001
C4 - Fz	- .3	n.s.	- .2	n.s.	-2	< .01[a]
T3 - Fz	- .7	n.s.	+ .2	n.s.	-3	< .01
T4 - Fz	-1.7	n.s.	-1.2	n.s.	-2.3	<0.001
Cz - A1	-2.1	<0.10	+ .6	n.s.	-5.2	<0.01
Cz - Fz	- .7	n.s.	- .6	n.s.	-5.1	<0.001
01 - A1	- .4	n.s.	- .2	n.s.	-4.3	<0.01
02 - A2	-1.2	n.s.	- .3	n.s.	-3	<0.02
T3 - A1	-1.7	n.s.	- .8	n.s.	-2.5	<0.10
T4 - A2	- .6	n.s.	- .6	n.s.	-2.3	n.s.

TABLE 2

DIFFERENCES IN RELATIVE θ-POWER BEFORE AND AFTER THE STIMULUS EXPRESSED IN PERCENTAGES OF THE TOTAL POWER, AVERAGED OVER ALL 12 SUBJECTS FOR THE THREE DIFFERENT STIMULUS SITUATIONS

derivations	FM-tone duration 50 msec $P\theta^1$-$P\theta^2$	signif. level	FM-tone duration 500 msec $P\theta^1$-$P\theta^2$	signif. level	tone-burst duration tone 10 msec $P\theta^1$-$P\theta^2$	signif. level
C3 - Fz	- .8	n.s.	+ .2	n.s.	+ .4	n.s.
C4 - Fz	+ .2	n.s.	+ .6	n.s.	+ .4	n.s.
T3 - Fz	+1.4	< .10	+0.3	n.s.	+1.6	<0.01
T4 - Fz	+ .8	n.s.	+1.1	n.s.	+ .8	n.s.
Cz - A1	+ .9	n.s.	-0.5	n.s.	+4.3	<0.002
Cz - Fz	+ .1	n.s.	+1.5	n.s.	+3.2	<0.01
01 - A1	- .1	n.s.[a]	- .5	n.s.	+ .5	n.s.
02 - A2	+ .3	n.s.	-0	n.s.	+1.2	<0.05
T3 - A1	+1	n.s.	- .5	n.s.	+2	<0.02
T4 - A2	+ .7	n.s.	+ .5	n.s.	+2.2	<0.02

[a] Calculated over 11 subjects in that derivation. One value is indicated as an outlier using the outlier test of DIXON[6]

and amplitude of the response by Arlinger[7]. With the tone-AER a specific small part of the cochlea is stimulated with the sudden occurrence of a stimulus with a certain intensity and duration, while with a click a large number of elements especially in its lower part are activated. With FM-AER there is only a shift in activation from one small part which is practically continuously stimulated to another which is only for a very short period activated. Although the difference, as regards processing in the periphery, seems to be not very impressive, apparently the influence on the central parts is of a different nature. Also the work by Arlinger[7], which shows that disturbances in the central processing systems affect the FM-AER more generally than cochlear lesions, supports this notion. As can been seen from figure 2, the differences in the waveform of the evoked responses are very clear, although differences in intensity of the stimuli may play some part. A more elaborate analysis of the FM-AER will be published elsewhere.

The changes caused in the background activity show that profound differences exist between the tone-burst and the FM tone-stimulation.

It was shown[8] that a relation existed between the α rhythms and components in the secondary part of the tone-AER when the vigilance level changed. As we compared the changes in spectral components immediately before and after a stimulus, it is highly unlikely that the differences could be caused by vigilance changes in our measurements.

An important difference between the FM-AER and the tone-AER is the fact that in the former, at cortical level according to the hypothesis of tonotopy, only a shift will occur in the localisation of the response in the cortex while the latter initiates a response every time a stimulus arrives. The hypothesis of tonotopy relies on, e.g. the work of Tunturi[9, 10]. This hypothesis supposes a local projection on the cortex of the neurons responding to certain frequencies in a similar way to the cochlea. Although it is doubtful whether a rigid pattern as described by Tunturi[11] actually exists, there certainly is a similar pattern of frequencies as in the cochlea. It can therefore be concluded that the FM-AER is the result of a process different from the tone-AER.

It may be argued that the tone-AER combines more frequency components and thus affects a larger number of neurons. However, the use of a sloped amplitude envelope minimizes this effect.

The difference in spectral changes between the FM and tone-stimulation may be caused by differences in the two sensory processes or be an expression of differences in attentive value between the two types of stimuli. The intensity difference between the two stimuli seriously hampers the interpretation of these data. It was caused by the fact that these stimuli are used in a different investigation in which other applications determined these intensity parameters.

The larger intensity of the tone could affect the overall vigilance level during the recording session, resulting in a different amount in α activity when compared to vigilancy during the FM-AER. Important to note is the difference in relative α power between the two sessions of the FM-AER with different plateau duration. This difference exists as much before as after the stimulus. It is seen in half of the subjects. An example is given in figure 4. It may well be that in the 500 msec plateau sessions the overall vigilance level of these subjects was lowered. However, this was not visible in the AER, nor in the spontaneous EEG on visual inspection. From animal experiments it is suggested that in cortical processing of auditory signals, frequency separation as well as length of tone appraisal plays an important role[12,13]. These data, however, do not permit us to more than point at the possibility of the length of tone as an explanation for this finding.

The changes in α power for the tone-stimulus can well be explained by changes in cortical excitability caused by the stimulus processing.

It is known from other experiments[14] that information processing changes the excitability in the cortex. It is proposed that this is not only a specific process but also a more general process occurring at subcortical levels, presumably the reticular formation, which has a general effect on the excitability cycle of the cortex. Since Moruzzi and Magoun[1] proved its existence, many authors have pointed to the ascending reticular activating system (ARAS) as the mediator of the more general processes involved in the cortex accompanying sensory activity.

More recent literature[15, 16] indicates that this activating system is more complex than originally stated and consists at least of two different components, one the mesencefalic reticular formation related to the more general arousal process, the other related to different activity connected with cognitive processes and mediated by a medio-thalamic-fronto-cortical system in which the thalamic reticular nuclei play an important role.

This more general process may well be reflected in the change in relative power over all regions including the occipital areas as found by tone-stimulation. It is remarkable that over the latter the reduction is only slightly smaller than over the central area (Cz), while the difference in AER amplitude is much larger. This implies that the changes are not only due to the special effects caused by the processing of the signal itself but are also the expression of a more general effect, as has been explained above.

Apparently this more general process is to a lesser degree involved in the FM- than in the tone-stimulation. More studies are required to elucidate this point.

LITERATURE

1. Moruzzi, G., and H.W. Magoun. Brain Stem reticular formation and activation of the EEG. Electroenceph.clin.Neurophysiol., 1949, 1: 455-73
2. Pfurtscheller, G., and A. Aranibar. Evaluation of event-related desynchronization (ERD) preceding and following voluntary selfpaced movement. Electroenceph.clin.Neurophysiol., 1979, 46: 138-47
3. Pfurtscheller, G., A. Aranibar and H. Maresch. Amplitude of evoked potentials and degree of event-related desynchronization (ERD) during photic stimulation. Electroenceph.clin.Neurophysiol., 1979, 47: 21-30
4. Spoor, A., F. Timmer and D.W. Odenthal. The evoked auditory response to intensity modulated and frequency modulated tones and tone bursts. Int-Audiol., 1969, 8: 410
5. Lindeman, H.E. Personal communication, 1978
6. Jonge, H. de. Inleiding tot de medische statistiek. 1962, Leiden, NIPG
7. Arlinger, S.D. Auditory responses to frequency ramps. 1976, Thesis. Linköping University Medical Dissertation, 40
8. Pfurtscheller, G., and B. Pfurtscheller. Korrelation zwischen α-Rhythmus und gemittelten kortikalen Reizantworten. Z. EEG-EMG, 1970, 1: 197-204
9. Tunturi, A.R. Physiological determination of the arrangement of the afferent connections to the middle ectosylvian auditory area in the dog. Am.J.Physiol. 1950, 160: 395-401
10. Tunturi, A.R. Statistical analysis of afferent potential in the auditory cortex. Am.J.Physiol., 1967, 213: 597-612
11. Tunturi, A.R. Analysis of cortical auditory responses with the probability pulse. Am.J.Physiol., 1955, 181: 630-8
12. Scharloch, D.P., W.D. Neff and N.L. Stroninger. Discrimination of tone duration after bilateral ablation of cortical auditory areas. J.of Neurophysiol., 1965, 28: 673
13. Nieder, P.C., and N.L. Stroninger. Evoked potentials in auditory cortex after bilateral transaction of brachium of the inferior colliculus in the cat. Journ.Neurophysiol., 1965, 28: 1185
14. Blom, J.L. L'influence de la charge mentale sur les potentiels évoqués. Le Travail Human, 1974, 37: 139-214
15. Skinner, J.E., and C.D. Yingling. Regulation of slow potential shifts in nucleus reticularis thalami by the mesencephalic reticular formation and the frontal granular cortex. Electroenceph.clin.Neurophysiol., 1976, 40: 288-96
16. Yingling, C.D., and J.E. Skinner. Regulation of unit activity in nucleus reticularis thalami by the mesencephalic reticular formation and the frontal granular cortex. Electroenceph.clin.Neurophysiol., 1975, 39: 635-42

© 1980 Elsevier/North-Holland Biomedical Press
Rhythmic EEG Activities and Cortical Functioning
G. Pfurtscheller et al., editors

HYPOTHESIS OF BACKGROUND EEG

THOMAS MULHOLLAND and DAVID GOODMAN
Edith Nourse Rogers Memorial
Veterans Administration Hospital, Bedford, MA 01730 (U.S.A.)

The methods for estimating the average evoked potential (AEP) require assumptions about (1) the statistical texture of the EEG, e.g., that the EEG is a continuous random variable which is statistically stationary, and (2) the reliability of the evoked potential, e.g., that it is an invariant function, i.e., it is time-locked to the stimulus with a very high replicative reliability. The interpretation of average evoked potentials (AEP) also requires assumptions about (3) the specificity of the effects of various conditions (independent variables) on the EP. For instance, it is often assumed that the effect of an independent variable does not cause a deviation from stationarity of the EEG.

Stationarity of background EEG. The assumption of stationarity of "background" EEG is fundamental for the optimal application of averaging methods and there is widespread recognition in the EP research literature that EEG is not stationary[1]. The assumption that the EEG is a continuous random variable which is stationary is crucial for replacing ensemble averages by time-averaging[2]. Moreover, a deviation of the EEG from stationarity toward a periodic function with reduced random variance poses problems regarding the sample size required for an acceptable level of confidence in the average EP. For instance Becker (1973)[3] noted that if a stimulus were presented at random times in relation to phase-angle of an "alpha" background of 20 μV, a sample size of about 200 would be necessary to measure a 5 μV non-random potential with a signal/noise ratio of 4.

Deviations from stationarity in the EEG occur often, i.e., an EEG epoch is stationary only for a brief time[4]. Recently, Praetorius et al. (1977)[5] and Michael and Houchin (1979)[6] have provided convincing evidence for a rapid sequence of deviations from stationarity in the EEG.

The method described by Praetorius et al.[5] is summarized as follows: (1) EEG is digitized (100/sec) and passed through a digital band pass filter (0.8 c/sec to 25 c/sec but offs); (2) the first 20 seconds more or less are used to obtain the first 10 values of autocorrelation function which are (3) used to calculate the coefficients of an autoregressive filter; (4) the EEG is then fed through the filter and the output is a prediction error record(s). Large peaks in its prediction error record are eliminated; (5) a "window" of 2 sec length is moved over the prediction error record; (6) from the portion of the prediction error record seen through the "window" a spectral error measure (SEM) as a function of time is calculated; (7) once a pre-set threshold SEM is reached, a pattern boundary is signalled at a time corresponding to the center of the "window"; (8) at this time-point, coefficients of the autoregressive filter, total power of the segment just detected (taken from its autocorrelation function), and its duration are stored; (9) from these data, power spectral estimates of the segments can later be calculated; (10) the procedure is then repeated with the next 20 sec sample.

They found that the number of deviations from stationarity ranges from 4 in 24 sec to 9 in 24 sec or from 10 per min to 23 per min. Thus, the spontaneous EEG is not stationary for longer than 6 sec, even without a disturbing stimulus.

Invariance of the evoked potential. The assumption of strict invariance in the EP research literature is not defensible either[1,7] but that is not our primary concern here.

Specificity of effects on evoked potentials. When sensory stimuli, receptor adjustments, or motor tasks are studied in relation to the EP there are also simultaneous and subsequent effects on the EEG, i.e., those independent variables are associated with non-random effects on the EEG. The effects may be a definite, rapid transition from synchronous activity to a lower-power, mixed-frequency EEG or, a less definite, but non-random change in the power spectrum of the EEG.

The response of posterior rhythms in the alpha band following a visual stimulus as a shift from "alpha" rhythms to a signal composed of mixed frequency and a lower RMS amplitude. This is the classical "alpha blocking" which is part of the history of EEG research. Evidence for non-random effects on the EEG following sensory stimulation, e.g., alpha attenuation, form part of the large body of literature on the changes in alpha rhythms following stimulation (for reviews see[8,9,10]).

More recently, computer-assisted on-line sampling, high-resolution measurement, speedy spectral analysis, and rigorous quantitative definition of EEG responses, have revealed in detail the particular deviation from stationarity of the EEG which is reliably associated with stimulus and task variables[11,12,13,14]. In keeping with improved quantitative methods, a new improved nomenclature has been introduced. Alpha is now "rhythmic activity in the alpha band" (RAAB) and alpha attenuation following a stimulus is "event-related desynchronization" (ERD). These researchers and others (e.g., Kuhlman, 1978[15]) have shown that the EEG effects following stimulation or in association with tasks were modality-specific, EEG-location specific and task specific.

Indirect evidence that there is a reliable shift in the EEG following a visual stimulus is also given by the fact that the variance of duration of alpha and of no-alpha (EEG intervals which do not meet criteria for alpha) is reduced by a feedback from the occurrence of occipital alpha to the occurrence of a visual stimulus[16].

Despite the reliability of occurrence of an alpha-"blocking" response, quantitative aspects of this response, for example, its duration, are extremely variable. This variability decreases the usefulness of alpha-blocking measures as on-line indices of orienting to a visual stimulus. Feedback from the EEG response back to the stimulus was first used to reduce unwanted, unsystematic variation of alpha suppression and to improve the economics of sampling[16]. With feedback, each alpha could be associated with a stimulus, which decreased the time required to obtain an adequate serial sample of responses.

The primary descriptive statistics which are calculated separately for alpha and no-alpha events for each EEG channel for each condition (divided into before and during feedback subsections) are: the mean time duration ($\bar{X}$), the standard deviation of the time durations (S), and the ratio formed by dividing the calculated mean by the corresponding standard deviation. This control ratio ($\bar{X}/S$) is the reciprocal of the coefficient of variation and is a sensitive discriminator of the relative variability of the EEG response. Both alpha control ratio and no-alpha control ratio are increased for (1) reliable feedback compared to (2) less reliable feedback (simultaneous, contralateral EEG not externally connected to the stimulus) compared to (3) zero-reliability, "sham" feedback (random stimulation) and compared to (4) no-stimulation conditions[17,18,19] (Fig.1).

In a more recent experiment[20] lateralized functional specialization of cortical processing of linguistic and spatial information and anatomical lateralization of processing of hemiretinal stimuli were investigated with a randomized factorial design in 14 right-handed male undergraduates using EEG alpha-contingent visual stimulation.

Four monopolar EEG recordings were referenced to linked earlobes with the right mastoid grounded. Electrodes were placed on the occipital-parietal border in each hemisphere (midway between O_1 and P_3 and between O_2 and P_4), over Wernicke's area in the left hemisphere (defined as the center of the triangle

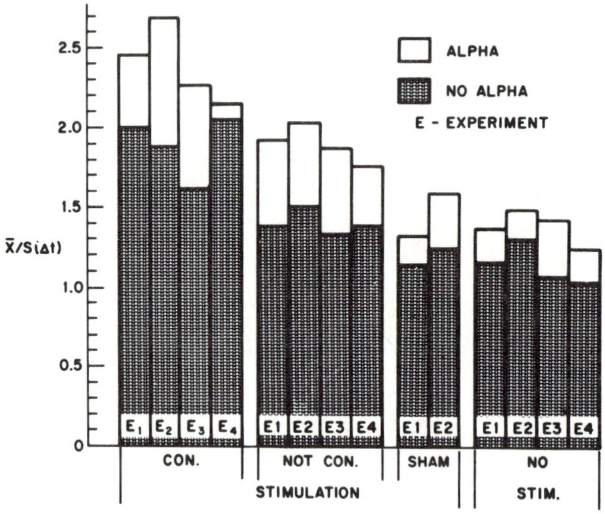

Fig. 1. Control ratios for contingent, connected EEGs; non-connected EEGs; non-contingent random stimulation and baseline, no stimulation conditions. Results from four experiments are shown[17,18,21].

formed by locations P_3, T_3 and T_5) and at the homologous electrode placement over the right hemisphere.

Words or geometric designs of equal luminosity and size were presented by computer to randomly selected visual half-fields contingent on the occurrence of alpha at one of four placements (left and right occipital and left and right temporal).

The replicative reliability of alpha block and, by implication, the control of the EEG by each type of stimulus at different cortical sites, was assessed with the alpha control ratio.

The alpha control ratio was significantly greater for alpha-contingent vs. random stimulation. For alpha-contingent data alone, cerebral lateralization was significantly demonstrated in the two-way interactions of EEG hemisphere x visual field of stimulation, and of EEG hemisphere x type of stimulus (verbal or spatial). The alpha control ratio was higher for contralateral visual half-field stimulation compared to ipsilateral stimulation (Fig. 2). Control ratio was also significantly greater for cognitively appropriate stimulus-hemisphere combinations, i.e.,verbal stimuli and left cerebral hemisphere, or spatial stimuli and right hemisphere vs. inappropriate combinations, i.e., verbal and

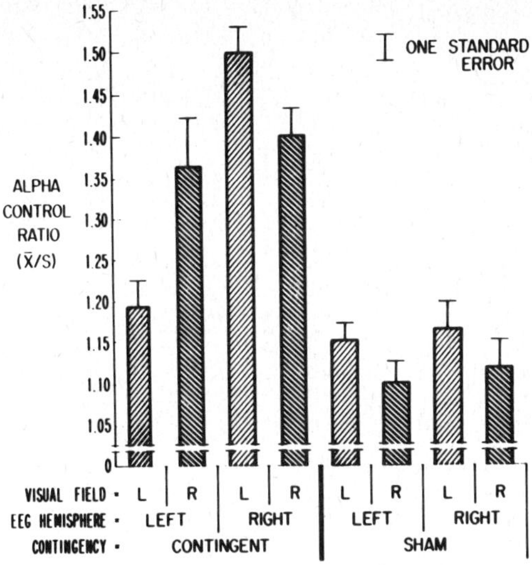

Fig. 2. Hemiretinal lateralization effect for alpha control ratio. Three-way interaction of contingency (alpha contingent stimulation vs sham feedback), visual field, and EEG hemisphere is illustrated[20].

right hemisphere, or spatial and left hemisphere (Fig. 3). No significant effects were obtained for noncontingent stimulation, i.e., "sham feedback".

In another study to identify the EEG location of the most reliable ERD to visual stimulation, Mulholland (1977)[21] showed that the ratio of $\bar{X}$/S of alpha and of no-alpha durations in the EEGs which were connected to the visual stimulus by an external path was greatest for posterior EEGs compared to central-parietal EEGs, i.e., the replicative reliability of the response to visual stimulation was greatest in the posterior EEG. The decrease of the control ratio ($\bar{X}$/S) from posterior to central EEG locations was not found for contralateral EEGs which received the same schedule of stimulation but were not connected to the stimulus by an external path.

The reduction of unwanted variance of alpha and "no-alpha" durations by alpha-contingent stimulation has also proved useful for achieving a standard, reliable quantitative estimate of the EEG component of the so-called human orienting-response[22,23,24,25].

Independence of successive EEG samples. Another set of assumptions concerns the independence of the EEG epochs which "contain" the evoked potential. For instance it is often assumed that successive samples are not correlated, an

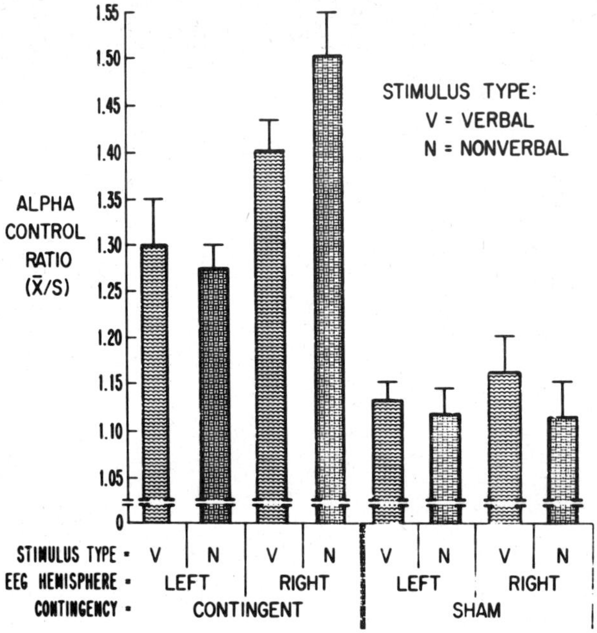

Fig. 3. Linguistic lateralization effect for alpha control ratio. Three-way interaction of contingency (contigent vs sham feedback) verbal vs non verbal stimuli and left vs right-sided EEGs[20].

assumption often criticized[1] or, that the EPs and EEGs do not vary systematically as a function of the prior number of stimulus events.

However, EEG effects produced by stimulus events and by receptor adjustments do change in a systematic way when the stimulus is repeated or continued. An example of such "disturbance and recovery phenomena" is the well-known habituation of alpha suppression following repeated stimulus-events[22,26] and, following a step increase of stimulation. Also, a step-change of stimulus intensity has effects on a subsequent series of alpha and no-alpha events[27,28].

It is well-known that an attenuation of posterior alpha rhythms occurs in association with voluntary ocular fixation and pursuit tracking which are part of the response of "looking-at-a-stimulus"[29,30,31]. Simply opening the eyes in the dark produces a reliable shift of the EEG which persists over a series of alpha and no-alpha events then decreases[12,18,32].

Instructions to the subject such as "pay attention to the 15th stimulus" have a reliable effect on subsequent desynchronization of alpha following the

15th visual stimulus compared to stimuli before and after[33].

Thus, there are desynchronization responses (ERDs) which: (1) are related to a single event or; (2) show trends over a series of stimulus-response-events or; (3) are related to a prior event. All of these event-related effects on the EEG are relevant to the methods and procedures of EP, because the various experimental variables, tasks, and cognitive and perceptual processes which are hypothesized to modify the EP also cause non-random modification of the EEG. Since the valid estimation of an EP depends on the stationarity of the EEG background and since the stationarity of the EEG is lost in association with the conditions and variables under investigation, there is an obvious conflict of methodological requirements for valid estimates of EPs and testing of experimental hypotheses with regard to EPs.

Some methodological solutions. A partial solution to this problem would be to present a stimulus only when the EEG had minimum power and was desynchronized. Modern studies of the visual evoked potential use methods which decrease the power of alpha in the EEG. For example, a subject may be instructed to "be alert" and to look at a checkerboard pattern which will reverse black and white areas. Since it is well-known that occipital alpha is diminished by attentively looking at a patterned visual stimulus[34], this is a reasonable methodological solution.

In a study of EP amplitude and parallel ERD, Pfurtscheller et al.[34] compared EPs obtained under two conditions of visual stimulation: irrelevant flashes and relevant flashes. A relevant flash was preceded by the press of a button; an irrelevant flash was not associated with a motor-attentional response. The purpose of a "relevant flash" was to increase the probability and magnitude of an event-related desynchronization (ERD). They found that the power in the alpha band decreased more for the relevant flashes compared to the irrelevant flashes; EP amplitudes were greater for the relevant flashes compared to irrelevant flashes.

As described previously, the alpha rhythm tends to increasingly recur as the stimulation conditions are repeated or continued (habituation). Thus, pragmatic methodological solutions like those described above which decrease the occurrence of alpha in the EEG may not be effective throughout an experimental trial or session. A further methodological solution would be to control the presentation of a stimulus automatically so that its presentation is contingent upon the occurrence or non-occurrence of alpha in the EEG.

Perhaps the concept of a stationary, quasi-random, "background" EEG is unnecessary. By means of contingent stimulation and sampling operations EPs might be even more useful as indicators of the sequence of functional reactions in the brain, and of their distribution.

ACKNOWLEDGEMENT

Supported by funds from the Veterans Administration Research Program.

REFERENCES

1. Regan, D. (1972) Evoked potentials in Psychology, Sensory Physiology and Clinical Medicine. Wiley, New York, 244-248.
2. Gasser, T. (1977) General characteristics of the EEG as a signal. In: Remond A. (Ed.). EEG Informatics, Elsevier, New York, 37-55.
3. Becker, W. (1973) Discussion. In: McCallum, W.C. and Knott, J.R. (Eds.) Event-related slow potentials of the brain, Electroenceph.clin.Neurophysiol., Suppl. 33, 127-130.
4. Elul, R. (1972) The genesis of the EEG. Internat. Rev. Neurobiol. 1972, 15, 227-272.
5. Praetorius, H.M., Bodenstein, G., and Creutzfeldt, O.D. (1977) Adaptive segmentation of EEG records: a new approach to automatic EEG analysis. Electroenceph.clin.Neurophysiol., 42, 84-94.
6. Michael, D. and Houchin, J. (1979) Automatic EEG analysis: A segmentation procedure based on the autocorrelation function. Electroenceph.clin.Neurophysiol., 46, 232-235.
7. Jarvilehto, T., Hari, R., and Sams, M. (1978) Effect of stimulus repetition on negative sustained potentials elicited by auditory and visual stimuli in the human EEG. Biol. Psychol. 7, 1-12.
8. Hill, D., and Paar, G. (1963) Electroencephalography: A symposium on its various aspects. MacMillan, New York.
9. Evans, C.R., and Mulholland, T.B. (1969) Attention in Neurophysiology, Butterworths, London.
10. Remond, A. (1976) The EEG of the waking adult. Handbook of Electroenceph. and Clinical Neurophysiol., Vol. 6, Part A, Elsevier, Amsterdam.
11. Kawabata, N. (1973) A nonstationary analysis of the electroencephalogram. IEEE Trans. Biomed. Eng. Vol. BME-40, 444-452.
12. Kawabata, N. (1974) Dynamics of the electroencephalogram during performance of a mental task. Kybernetik, 15, 237-242.
13. Pfurtscheller, G., and Aranibar, A. (1977) Event-related cortical desynchronization detected by power measurements of scalp EEG. Electroenceph. clin. Neurophysiol., 42, 817-821.
14. Aranibar, A., and Pfurtscheller, G. (1978) On and off effects in the background EEG activity during one-second photic stimulation. Electroenceph. clin. Neurophysiol., 44, 307-316.
15. Kuhlman, W. (1978) Functional topography of the human mu rhythms. Electroenceph. clin. Neurophysiol., 44, 83-93.

16. Mulholland, T., and Runnals, S. (1962) Evaluation of attentioan and alertness with a stimulus-brain feedback loop. Electroenceph. clin. Neurophysiol., 14, 847-852.

17. Eberlin, P., and Mulholland, T. (1976) Bilateral differences in parietal occipital EEG induced by contingent visual feedback. Psychophysiol. 13, 212-218.

18. Mulholland, T., and Eberlin, P. (1977) Effect of feedback contingencies on the control of occipital alpha. Biofeedback Self-Regulat. 2, 43-57.

19. Goodman, D. (1978) Detection of cerebral lateralization of function using EEG alpha feedback. PhD Dissertation. University of California, Los Angeles.

20. Goodman, D., Beatty, J. and Mulholland, T. (1980) Detection of cerebral lateralization of function using alpha contingent visual stimulation. Electroenceph.clin. Neurophysiol., in press.

21. Mulholland, T. (1977) Biofeedback method for locating the most controlled response to visual stimulation. In Beatty, J. and Legewie, H. (Eds.) Biofeedback and Behavior, Plenum, New York, 95-106.

22. Sokolov, Y. N. (1963) Perception and the Conditioned Reflex. Pergamon, New York. Translated from: Vaspriyatiye Uslovnyl Refleks. Moscow University Press, Moscow, 1958.

23. Mulholland, T., McLaughlin, T., and Benson, F. (1976) Feedback control and quantification of the response of EEG alpha to visual stimulation. Biofeedback Self Regul. 1, 411-421.

24. Mulholland, T., McLaughlin, T., and Benson, F. (1979) Quantification of the human orienting response to another human. J. Nerv. Ment. Dis. 167, 166-170.

25. Boudrot, R., Goodman, D., and Mulholland, T. (1979) An EEG alpha detection, feedback stimulation and data analysis system. Behavior Res. Meth. Instr. 10, 646-651.

26. Mulholland, T. (1973) Objective EEG methods for studying covert shifts of visual attention. In: McGuigan, F. and Schoonover, R. (Eds.) The Psychophysiology of Thinking, New York, Academic Press, 109-151.

27. Mulholland, T., and Runnals, S. (1964) Cortical activation during steady and changing visual stimulation. Electroenceph. clin. Neurophysiol. 17, 371-375.

28. Mulholland, T.,and Gascon, G. (1972) Quantitative index of the orienting response in children. Electroenceph. clin. Neurophysiol.,33, 295-301.

29. Durup, G., and Fessard, A. (1935) L'electroencephalogramme de l'homme. Observations psychophysiologiques relatives a l'action des stimuli visuels et auditifs. L'Annee Psychol. 36, 1-35.

30. Mulholland, T., and Peper, E. (1971) Occipital alpha and accommodative vergence, pursuit tracking and fast eye movements. Psychophysiol. 8, 556-575.

31. Goodman, D. (1976) The effect of oculomotor activity on alpha blocking in the absence of visual stimuli. Psychophysiol. 13, 462-465.

32. Bryan, J., and Luria, Z. (1978)Sex-role learning: a test of the selective attention hypothesis. Child Development,49, 13-23.

33. Mulholland, T., and Runnals, S. (1963) The effect of voluntarily directed attention on successive cortical activation responses. J. Psychol. 55, 427-436.

34. Pfurtscheller, G., Aranibar, A., and Maresch, H. (1979) Amplitude of evoked potentials and degree of event-related desynchronization (ERD) during photic stimulation. Electroenceph.clin.Neurophysiol., 47, 21-30.

EEG RHYTHMS DURING INTERNAL EXPERIENCES

Rhythmic EEG Activities and Cortical Functioning
G. Pfurtscheller et al., editors

CLASSES OF SPONTANEOUS PRIVATE EXPERIENCES, AND ONGOING HUMAN EEG ACTIVITY.

D. LEHMANN and M. KOUKKOU
Dept. of Neurology, University Hospitals, Zurich, and Research Dept., University Hospital of Psychiatry, Zurich (Switzerland)

By introspection we know that frequently, mental activity occurs without externalization of the results via motor acts. Such private mental, perceptual and cognitive activity might belong to many different classes, as e.g. intended, abstract, mathematical operations, intended visual imaging, spontaneous narrative thought chains, spontaneous dreams, etc. Frequently such private mental activity without externalized output occurs spontaneously, without specific outside instructions or pre-set plans. In fact, spontaneous, uncontrolled cognition and perception of externally and internally originated information may be more frequent in everyday life than systematic mental activity which follows instructions, or follows detailed pre-set plans. The characteristics of spontaneously occurring mental activity differ widely, as e.g. between dreams with strong emotions, and numerical or spatial problem solving. These spontaneous, private or internal, cognitive or perceptual events and experiences which are observable by introspection may be investigated in EEG studies, using the techniques of (1) spontaneous, and (2) elicited reports of the subjects' experiences. Classes of spontaneous private events, and classes of ongoing, spontaneous EEG activity, and their mutual correlations may be established. Such correlations may be related to different (spontaneous or induced) functional states of the brain, which are, in turn, suggested to determine the functionally available processing strategies and memory stores of the brain (see Koukkou et al., in press[1]).

The present paper will review briefly some of our studies in which we attempted to relate the EGG data to the subjects' spontaneous and elicited reports of spontaneous, private mental events. The review will demonstrate the different experimental approaches which we used, and the sort of information which one might obtain.

METHODS

Method of spontaneous reports: By pressing or releasing a microswitch, subjects report the spontaneous occurrence of a previously specified type (class) of private events. The EEG epoch preceding (or during) the subject's signal is analysed off-line (classed), typically starting 1 to 5 sec before the report signal, and analysing backwards in time. These EEG epochs immediately before (or during) the specified reports are compared to epochs without reports, or to epochs which are associated with a different type of reports.

Method of elicited reports: The subject is interrogated by the experimentor during the occurrence of one, or several, previously specified types of EEG activity; the subject's reports are classed. Classes of private events are correlated with classes or types of ongoing EEG activity. In another version of this approach, the fate of information which was presented to the subject is related retrospectively to the type of EEG activity which existed immediately after the input of the information (see learning study, below). - The most popular application of the method of elicited reports may be the study of mentation during REM and NREM sleep states (e.g. see Schwartz et al., 1978[2]); in these studies, however, the classification of the EEG typically is based on a distinction between long-lasting states of sleep EEG. Other studies were concerned with mentation in relaxed wakefulness (see Foulkes and Fleisher, 1975[3]).

The experimental methods described above aim at brief psychological and physiological states which cannot be attained at will at a given moment in time; also, these experimental situations avoid simultaneous execution of symbolization/cognition functions and motor command functions. Thus, the outlined approaches for the study of spontaneous, private events in relation to brief EEG activity states might extend the range of types of mentations which have been investigated in experiments on the relationships between the execution of cognitive or cognitive-motor tasks which were performed according to instructions, and characteristics of ongoing EEG activity in the human (e.g. Berkhout et al., 1969[4]; Creutzfeldt et al., 1969[5], and see Rebert and Low, 1978[6]; however see Yingling, this volume[7]). We note that in the cat, EEG states were identified which are associated with specific behaviour (e.g. Rougeul and Buser, 1978[8]; Bouyer et al., this volume[9]).

STUDIES WITH SPONTANEOUS REPORTS

Vigilance (Kuhlo and Lehmann, 1964[10]): During pre-sleep drowsiness, healthy volunteers reported fluctuations of vigilance, i.e. the re-emergence of self-awareness or consciousness after a brief, retrospectively recognized epoch of the feeling of "floating". In visual analysis of the EEG, the subjects' repeated reports correlated well in time with preceding episodes of slowing of EEG alpha waves by 1-2 Hz, or preceding disappearance of alpha waves, having a minimum duration of 1.5 sec. These findings support earlier observations by Davis et al. (1938)[11].

Kamiya (1969)[12] described subjects who operationally learned to report repeatedly the occurrence of an internal state which was associated with ongoing alpha EEG activity (vs non-alpha EEG), but who were unable to give a verbal description of this internal state; this observation indicates that the method of spontaneous reports might be used to define EEG states which are associated with operationally defined subjective moods or feelings.

Perceptual threshold (Lehmann et al., 1965[13]): Stabilized retinal images (contact lens method) of low luminance were observed by the subject. The subject reported by key press or key release the spontaneously alternating epochs of visibility and invisibility (fading) of the stabilized visual target. Visual assessment of the EEG showed that perceptual target disappearance was reported on the average 650 msec after onset of an EEG alpha spindle. The subject's mean reaction time to experimentor-induced switch-off of the target was 607 msec, and the mean alpha-onset latency after target-off was 579 msec. Thus, the private, perceptual experience of target fading could not have been caused by a retinal event, since in that hypothetical case, the difference between reaction time and alpha latency should have been 28 msec (the obtained difference of 650 msec disagread with the expected difference at $p > 99.9$ %). Similar results were found in three subjects; and related observations were made concerning the end of alpha spindles, and the corresponding reports of subjective reappearance of the target. These results were confirmed and extended in studies by Keesey and Nichols (1967[14], 1969[15]).

Our observations are indicative of central control of fluctuations of perceptual threshold during wakefulness ("fluctuations of attention", see Woodworth and Schlosberg, 1954[16]), which are reflected by different classes of the EEG activity, where epochs of rhythmic alpha correspond to a brain functional state with increased perceptual threshold, and absence of alpha corresponds to a state with lowered threshold. This result is in good agreement with observations of changes of alpha amplitude and motor reaction time as a function of the respiratory cycle, the inspiratory phase corresponding to lower alpha amplitude and faster reaction time than the expiratory phase (see Lehmann, 1977[17]).

Cannabis-induced visual hallucinations (Koukkou and Lehmann, 1976[18]): Volunteers took low doses of Delta-9-tetra-hydro-Cannabinol. The subjects (n=12) lay down with closed eyes on a comfortable couch, and were instructed to observe and report the occurrence of hallucinatory experiences. The reports were classed by independent raters. The parieto-occipital EEG during the suspected epochs of hallucinations, i.e. during the 5 - 80 sec prior to the subjects' reports, was analysed (base line crossing analysis), and the resulting wave spectra were compared with those of resting EEG epochs without intervening reports of hallucinatory experiences. The wave spectra during visual hallucinations showed significantly more theta activity, and a significantly lowered mean frequency of the theta-alpha band (lowered mean values of the frequency distribution between 2 and 13 Hz) in comparison with resting epochs without hallucinations. Thus, different strategies of processing and evaluation of internal information are used by the subjects during brain states which differ in terms of EEG frequency distributions.

STUDIES WITH ELICITED REPORTS

Daydreaming mentation (Lehmann et al., 1979 in press[19]): The parieto-occipital EEG was recorded from 26 healthy volunteers during about one hour of "daydreaming" (mind wandering), while they lay with closed eyes on a comfortable couch during the late afternoon. They were told that from time to time reports of their mentation will be elicited. Fifteen reports were collected from each subject, during two classes of spontaneous EEC activity: (1) after more than 20 sec of alpha EEG, or (2) after more than 4 sec of EEG flattening without alpha (the mean duration was 9 sec), diagnosed by visual inspection of the paper record. The subjects' reports of their mentation were classed blindly by two raters (who agreed in 90 %

of the cases) into visual imagery, non-visual thinking, and other experiences. The subjects did not show sleep EEG activity. The data of the first and last two reports were analysed; the relationship between classes of experiences and EEG patterns were very similar for the first and last two reports. During EEG flattening, significantly more of the 52 reports were visual experiences than during alpha (39 % vs 12 %, chi square $p=0.002$), whereas during alpha, significantly more of the 52 reports were classed as thinking than during flattening (49 % vs 29 %, chi square $p=0.05$).

The results show that the two different functional brain states which are distinguished by presence or absence of alpha EEG patterns, and which occur spontaneously in fairly short intervals during daydreaming, are associated with definite predominances of clearly different types of mentation strategies. In the present case, one may argue that either, spontaneous visual imagery blocks EEG alpha, similar to the alpha-blocking effect of visual stimuli; or else, that the spontaneously occurring blocking of alpha EEG facilitates visual imagery. The results of the study described below will show that the first hypothesis does at least not apply to all brain states (visual imagery was associated with alpha EEG at the onset of drowsiness in the study described below).

Hypnagogic visual experiences (Koukkou et al., 1975[20]): Mentation reports during sleep onset ("hypnagogic hallucinations") were collected with an experimental design similar to that used in the daydreaming study. However, each of the 36 subjects was alerted and interrogated only three times: (a) after more than 4 sec of flattening (low aplitude theta), (b) during alpha immediately following more than 4 sec of flattening, and (c) after 20 sec of alpha following more than 4 sec of flattening. The sequence of the 3 conditions was randomized over subjects. The obtained reports were classed into visual and other hypnagogic experiences. The frequency of reports of visual experiences turned out to be highest in condition (c), during alpha, for the first awakenings ($p < 0.07$), and highest in condition (a), low amplitude theta, for last (third) awakenings ($p < 0.05$). The latter result confirmed the qualitative report of an earlier study (Kuhlo and Lehmann, 1964[10]).

Obviously, there are ruleful relationships between frequency of occurrence of a given type of mentation (visual imagery, i.e. so-called hypnagogic hallucinations) during sleep onset, and the ongoing type of EEG activity as assessed with visual analysis. The sleep onset data show more complexity than the daydreaming data, since the relationship appears to differ as a function of time after the onset of drowsiness. This may be accounted for by state-dependent recall: The functional state after the third awakening may differ from that after the first awakening, favouring recall after the third awakening of experiences which were stored in a less alert state (state dependent recall, see Overton, 1979[21]). On the other hand, it remains to be seen whether the EEG patterns which were diagnosed visually as e.g. low amplitude theta, are identical for the first and third awakenings in more sophisticated computer EEG analysis. At any rate, the predominance of visual imagery in first awakenings during alpha epochs suggests that at least during drowsiness, visual imagery does not block alpha. This is of interest for the discussion of the daydreaming data reported above.

Quality of recall of learned material (Lehmann and Koukkou, 1974[22]): Learning designs are particularly suited to test the hypothesis that the functional brain state as expressed by the EEG determines the fate of information in the brain. Emotionally neutral, short sentences were presented acustically to subjects during sleep with EEG slow waves. Later, after final awakening, the quality of recall of the sentences was tested, and three classes of quality of recall were established: (1) spontaneous, correct recall; (2) recognition; and (3) no recall, no recognition. The EEG epochs which immediately had followed the presentation of the sentences were computer-analyzed off-line into wave spectra; it was found that quality of later recall of the sentences was systematically better if the wave spectra contained more alpha and less theta during the 30 sec after the presentation of the sentences; this was true for novel and for familiar sentences. In addition, equal quality of recall was associated with systematically more alpha and less theta for novel than for familiar sentences after the presentation. These results confirmed and extended those of a similar, earlier experiment (Koukkou and Lehmann, 1968[23]), where visual assessment of the post-presentation EEG had shown that the quality of subsequent recall increased systematically with increased duration of a presentation-induced post-presentation alpha EEG pattern. Thus, the elicited reports of the recall quality of learned material in comparison with the characteristics of the ongoing EEG during the time when the material is being stored support the hypothesis that brain functional states which are manifested in the scalp EEG determine the fate of information in the organism.

CONCLUSION

In summary, the examples reviewed above indicate that the experimental strategies of (1) spontaneous reporting of pre-specified classes of private, cognitive or sensory, internal experiences during EEG recordings, or of (2) eliciting reports of private, introspectively recognized events during pre-specified types of ongoing EEG activity yield data which permit us to describe, in EEG terms, different short-lasting brain functional states which correlate with general types or strategies of spontaneous human mentation. Some of the experimental designs (e.g. the study of perception of stabilized retinal images, and the study of quality of learning) lead to the conclusion that in fact, the different brain states as manifested in the ongoing EEG are not only the possible consequences, or concomitants of the brain's data processing. Rather, the momentary functional brain state which is observable in the ongoing EEG - after having been readjusted in response to the input information, i.e. in response to arousal stimuli or habituated stimuli - might be concluded to determine the processing strategies and storage spaces which are functionally available to the brain at the moment. This would mean that the momentary class of ongoing EEG activity which reflects the brain functional state is an indicator of the fate of the information in the brain.

The study of spontaneous, private (introspectively recognized) events with EEG techniques appears to be a fascinating field where physiological measures might help to sort psychological phenomena of a very wide range.

ACKNOWLEDGEMENT

Part of this work was supported by the Swiss National Science Foundation, and by the Smith-Kettlewell Institute of Visual Sciences, San Francisco, California.

REFERENCES

1. Koukkou, M., Lehmann, D. and Angst, J., eds., (in press) Functional States of the Brain: Their Determinants. Elsevier, Amsterdam.

2. Schwartz, D.G., Weinstein, L.N. and Arkin, A.M. (1978) Qualitative aspects of sleep mentation. In: A.M. Arkin, J.S. Antrobus and S. J. Ellman (eds.), The Mind in Sleep. Erlbaum, Hillsdale N.J., pp. 143-241.

3. Foulkes, D. and Fleisher, S. (1975) Mental activity in relaxed wakefulness. J. abnorm. Psychol., 84, 66-75.

4. Berkhout, J., Walter, D.O. and Adey, W.R. (1969) Alterations of the human electroencephalogram induced by stressful verbal activity. Electroenceph. clin. Neurophysiol., 27, 457-469.

5. Creutzfeldt, O.D., Grünewald, G., Simonova, O. and Schmitz, H. (1969) Changes of the basic rhythms of the EEG during performance of mental and visuomotor tasks. In: C.R. Evans and T.B. Mulholland (eds.), The Concept of Attention in Neurophysiology. Butterworth, Oxford, pp. 148-168.

6. Rebert, C.S. and Low, D.W. (1978) Differential hemispheric activation during complex visuomotor performance. Electroenceph. clin.Neurophysiol., 44, 724-734.

7. Yingling, C.D. Cognition, action, and mechanisms of EEG asymmetry. This volume.

8. Rougeul-Buser, A., Boyer, J.J. and Buser, P. (1978) Transitional states of awareness and short term fluctuations of selective attention: neurophysiological correlates and hypotheses. In: P. Buser and A. Rougeul-Buser (eds.) Cerebral Correlates of Conscious Experience, Elsevier, Amsterdam, pp.215-232.

9. Boyer, J.J., Rougeul-Buser, A., Montaron, M.F. and Buser, P. Fast cortical mu rhythms and directed attention in the awake cat. This volume.

10. Kuhlo, W. and Lehmann, D. (1964) Das Einschlaferleben und seine neurophysiologischen Korrelate. Arch.Psychiat.Nervenkr., 205, 687-716.

11. Davis, H., Davis, P.A., Loomis, A.L., Harvey,E.N. and Hobart, G. (1938) Human brain potentials during the onset of sleep. J. Neurophysiol., 1, 24-38.

12. Kamiya, J. (1969) Operant control of the EEG alpha rhythm and some of its reported effects on consciousness. In: C.T. Tart (ed.), Altered States of Consciousness. Wiley, New York, pp. 507-517.

13. Lehmann, D., Beeler, G.W. and Fender, D.H. (1965) Changes in pattern of the human electroencephalogram during fluctuations of perception of stabilized retinal images. Electroenceph. clin. Neurophysiol., 19, 336-343.

14. Keesey, U.T. and Nichols, D.J. (1967) Fluctuations in target visibility as related to the occurrence of the alpha component of the encephalogram. Vision Res., 7, 859-879.

15. Keesey, U.T. and Nichols, D.J. (1969) Changes induced in stabilized image visibility by experimental alteration of the ongoing EEG. Electroenceph. clin. Neurophysiol., 27, 248-257.

16. Woodworth, R.S. and Schlosberg, H. (1954) Experimental Psychology. Holt, New York.

17. Lehmann, D. (1977) Cortical activity and phases of the respiratory cycle. Proc. 18th Int. Congr. Neuroveg. Res., Tokyo, pp. 87-89.

18. Koukkou, M. and Lehmann, D. (1976) Human EEG spectra before and during Cannabis hallucinations. Biol. Psychiat., 11, 663-677.

19. Lehmann, D, Koukkou, M. and Andreae, A. (1979) Daydreaming mentation and EEG patterns. Sleep Res., 8, (in press).

20. Koukkou, M., Dittrich, A. and Lehmann, D. (1975) Hypnagogic experiences and EEG. Sleep Res., 4, 188.

21. Overton, D.A. (1979) Drug discrimination training with progressively lowered doses. Science, 205, 720-721.

22. Lehmann, D. and Koukkou, M. (1974) Computer analysis of EEG wakefulness-sleep patterns during learning of novel and familiar sentences. Electroenceph.clin.Neurophysiol., 37, 73-84.

23. Koukkou, M. and Lehmann, D. (1968) EEG and memory storage in sleep experiments with humans. Electroenceph.clin.Neurophysiol., 25, 455-468.

AUTHOR INDEX

SUBJECT INDEX

A

B

C

D

E

F

G

H

I

K

L

M

N

O

P

Q

R

S

T

U

V

W

Z